Lecture Notes in Computer Science   9845

Commenced Publication in 1973
Founding and Former Series Editors:
Gerhard Goos, Juris Hartmanis, and Jan van Leeuwen

More information about this series at http://www.springer.com/series/7408

Sabine Wittevrongel · Tuan Phung-Duc (Eds.)

Analytical and Stochastic Modelling Techniques and Applications

23rd International Conference, ASMTA 2016
Cardiff, UK, August 24–26, 2016
Proceedings

 Springer

Editors
Sabine Wittevrongel
Ghent University
Gent
Belgium

Tuan Phung-Duc
University of Tsukuba
Tsukuba
Japan

ISSN 0302-9743　　　　　ISSN 1611-3349　(electronic)
Lecture Notes in Computer Science
ISBN 978-3-319-43903-7　　　ISBN 978-3-319-43904-4　(eBook)
DOI 10.1007/978-3-319-43904-4

Library of Congress Control Number: 2016946630

LNCS Sublibrary: SL2 – Programming and Software Engineering

Printed on acid-free paper

This Springer imprint is published by Springer Nature
The registered company is Springer International Publishing AG Switzerland

Preface

It is our privilege to present the proceedings of the 23rd International Conference on Analytical and Stochastic Modelling Techniques and Applications (ASMTA 2016), held in the city of Cardiff, UK, during August 24–26, 2016. The ASMTA conference is a main forum for bringing together researchers from academia and industry to discuss the latest developments in analytical, numerical, and simulation techniques for stochastic systems, including Markov processes, queueing networks, stochastic Petri nets, process algebras, game theoretical models, mean field approximations, etc.

We are proud of the high scientific level of this year's program. We had submissions from many European countries including Belgium, France, Germany, Greece, Hungary, Italy, Lithuania, Portugal, Spain, The Netherlands, and the UK, but also received contributions from Algeria, Brazil, Canada, Colombia, China, India, Japan, Russia, and the USA. The international Program Committee reviewed these submissions in detail and assisted the program chairs in making the final decision to accept 21 high-quality papers. The selection procedure was based on at least three and on average 3.7 reviews per submission. These reviews also provided useful feedback to the authors and contributed to an even further increase of the quality of the final versions of the accepted papers.

We would like to thank all the authors who submitted their work to the conference. We also would like to express our sincere gratitude to all the members of the Program Committee for their excellent work and for the time and effort devoted to this conference. We wish to thank Khalid Al-Begain and Dieter Fiems for their support during the organization process. Finally, we would like to thank the EasyChair team and Springer for the editorial support of this conference series. Thank you all for your contribution to ASMTA 2016.

June 2016

Sabine Wittevrongel
Tuan Phung-Duc

Organization

Program Committee

Sergey Andreev	Tampere University of Technology, Finland
Jonatha Anselmi	Inria, France
Konstantin Avrachenkov	Inria, France
Christel Baier	Technical University of Dresden, Germany
Simonetta Balsamo	Università Ca' Foscari di Venezia, Italy
Koen De Turck	CentraleSupélec, France
Ioannis Dimitriou	University of Patras, Greece
Antonis Economou	University of Athens, Greece
Dieter Fiems	Ghent University, Belgium
Jean-Michel Fourneau	Université de Versailles St Quentin, France
Marco Gribaudo	Politecnico di Milano, Italy
Yezekael Hayel	University of Avignon, France
András Horváth	University of Turin, Italy
Gábor Horváth	Budapest University of Technology and Economics, Hungary
Stella Kapodistria	Eindhoven University of Technology, The Netherlands
Helen Karatza	Aristotle University of Thessaloniki, Greece
William Knottenbelt	Imperial College London, UK
Lasse Leskelä	Aalto University, Finland
Daniele Manini	Università di Torino, Italy
Andrea Marin	University of Venice, Italy
Yoni Nazarathy	University of Queensland, Australia
José Niño-Mora	Carlos III University of Madrid, Spain
António Pacheco	Instituto Superior Tecnico, Portugal
Tuan Phung-Duc	University of Tsukuba, Japan
Balakrishna J. Prabhu	LAAS-CNRS, France
Juan F. Pérez	University of Melbourne, Australia
Marie-Ange Remiche	University of Namur, Belgium
Anne Remke	WWU Münster, Germany
Jacques Resing	Eindhoven University of Technology, The Netherlands
Marco Scarpa	University of Messina, Italy
Bruno Sericola	Inria, France
Ali Devin Sezer	Middle East Technical University, Turkey
János Sztrik	University of Debrecen, Hungary
Miklós Telek	Budapest University of Technology and Economics, Hungary
Nigel Thomas	Newcastle University, UK

Dietmar Tutsch	University of Wuppertal, Germany
Jean-Marc Vincent	Inria, France
Sabine Wittevrongel	Ghent University, Belgium
Verena Wolf	Saarland University, Germany
Katinka Wolter	Freie Universität Berlin, Germany
Alexander Zeifman	Vologda State University, Russia

Steering Committee

Khalid Al-Begain (chair)	University of South Wales, UK
Dieter Fiems (secretary)	Ghent University, Belgium
Simonetta Balsamo	Università Ca' Foscari di Venezia, Italy
Herwig Bruneel	Ghent University, Belgium
Alexander Dudin	Belarusian State University, Belarus
Jean-Michel Fourneau	Université de Versailles St Quentin, France
Peter Harrison	Imperial College London, UK
Miklós Telek	Budapest University of Technology and Economics, Hungary
Jean-Marc Vincent	Inria, France

Contents

Stochastic Bounds and Histograms for Active Queues Management and Networks Analysis

Farah Aït-Salaht[1]([⊠]), Hind Castel-Taleb[2], Jean-Michel Fourneau[3], and Nihal Pekergin[4]

[1] LIP6, Pierre et Marie Curie University, UMR7606, Paris, France
farah.ait-salaht@lip6.fr
[2] SAMOVAR, UMR 5157, Télécom Sud Paris, Evry, France
hind.castel@telecom-sudparis.eu
[3] DAVID, Versailles St-Quentin University, Versailles, France
jean-michel.fourneau@uvsq.fr
[4] LACL, Paris Est-Créteil University, Créteil, France
nihal.pekergin@u-pec.fr

Abstract. We present an extension of a methodology based on monotonicity of various networking elements and measurements performed on real networks. Assuming the stationarity of flows, we obtain histograms (distributions) for the arrivals. Unfortunately, these distributions have a large number of values and the numerical analysis is extremely time-consuming. Using the stochastic bounds and the monotonicity of the networking elements, we show how we can obtain, in a very efficient manner, guarantees on performance measures. Here, we present two extensions: the merge element which combine several flows into one, and some Active Queue Management (AQM) mechanisms. This extension allows to study networks with a feed-forward topology.

Keywords: Performance evaluation · Histograms · Stochastic bounds · Queue management

1 Introduction

Measurements are now quite common in networks. But they are relatively difficult to use for performance modeling in an efficient manner. Indeed, the measurements for traffics are extremely huge and this precludes to use them directly in a model. Of course it is still possible to use traces in a simulation, but this is not really an abstract model and we want to be very fast when we solve models and this is not possible with simulations.

One possible solution consists in fitting a complex stochastic process (such as a PH process or a Cox process [8]) from the experimental data and use this parametrized process in a queueing theory model. Here we advocate another solution: the histogram based models. We propose to combine this type of models with stochastic ordering theory to obtain performance guarantees in an efficient manner. Such an approach provides a trade-off between the accuracy of the results and the time complexity of the computations. In the last nine years,

© Springer International Publishing Switzerland 2016
S. Wittevrongel and T. Phung-Duc (Eds.): ASMTA 2016, LNCS 9845, pp. 1–16, 2016.
DOI: 10.1007/978-3-319-43904-4_1

Hernández et al. [5–7] have proposed a new performance analysis to obtain buffer occupancy histograms. This new stochastic process called HBSP (Histogram Based Stochastic Process) works directly with small histograms using a set of specific operators on discrete time. The time interval is denoted as a slot. The input traffic is obtained by a heuristic from real traces and it is modeled by a discrete distribution. The arrivals during one time slot are supposed to be identically independently distributed (i.i.d.). The service is supposed to be deterministic, corresponding to the traffic capacity of the link. The buffer is supposed to be finite. Thus, the theoretical model is a $Batch/D/1/K$ queue. In their papers, Hernández et al. do not use the Markovian framework associated with the queue and they develop a numerical algorithm based on the convolution of the distributions. As they named their approach "Histograms", we use the same terminology here. We sometimes write "discrete distributions", which is a more proper term. In this paper, these terms and probability mass function (pmf) are used interchangeably. The analysis proposed by Hernández et al. is only applied to one node because they do not derive properties for the output process of the node. Another problem is that the convergence of their numerical algorithm is not proved. Finally, they use an heuristic to construct reduced histograms from the traces. This is extremely important because their method is fast, but it does not give any guarantees on the results. More precisely, they proceed as follows: they assume the stationarity of the arrivals. Thus, they obtain from the trace, a histogram for the distribution of the number of arrivals during one time slot. But the size of the histogram is too large for a numerical algorithm based on convolution operations. Therefore, they simplify the histogram dividing the space into n sub-intervals (n is a small number) to obtain only n bins (states) in the histograms. And they obtain approximate solutions which can be computed efficiently, if n is small. But there is no guarantee on the quality or the accuracy of the approximations.

To illustrate the approach, we present now a trace used by Hernández et al. and in this work. Figure 1 shows a plot of MAWI traffic trace [11] corresponding to a 1-hour trace of IP traffic of a 150 Mb/s transpacific line (samplepoint-F) for the 9th of January 2007 between 12:00 and 13:00. This traffic trace has an average rate of 109 Mb/s. Using a sampling period of T = 40 ms (25 samples

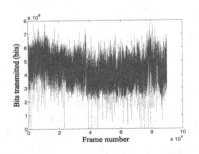

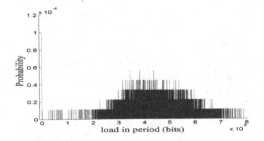

Fig. 1. MAWI traffic trace   **Fig. 2.** MAWI arrival load histogram

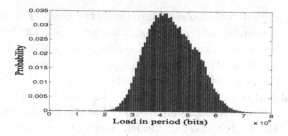

Fig. 3. HBSP approximation of MAWI arrival load histogram with bins = 100.

per second), the resulting traffic trace has 90,000 frames (periods) and an average rate of 4.37 Mb per frame (the corresponding histogram is given in Fig. 2). The number of bins in this histograms is 80511. Finally, the HBSP approximation with 100 bins is given in Fig. 3. The key idea here is the reduction of the number of bins from 80511 bins in the trace to only 100 bins to have the fast numerical analysis.

For our approach, we propose to apply the stochastic bounding method to the histogram based models [2,3]. The goal is to generate bounding histograms with smaller sizes which can be used to analyze queueing elements with some guarantees on the results. We use the strong stochastic ordering (denoted by $\leq_{st}$) [9]. We have proposed to use the algorithm developed in [4] to obtain optimal lower and upper stochastic bounds of the input histogram. This algorithm allows to control the size of the model and it computes the most accurate bound with respect to a given non decreasing, positive reward function. The bounding histograms are then used in the state evolution equations to derive bounds for performance measures for a single queue.

An extension of our approach to a queueing network was also investigated. A queueing network is a set of interconnected queues where the departures from one (or more) queue enter one (or more) other queue, according to a specified routing, or leave the system. Here, we focus on queueing networks with finite capacity. We have decomposed the network nodes into: Traffic sources (input flows), Finite capacity queues, Merge elements and Splitters. Monotonicity of networking elements is the key property for our methodology (the formal definition will be given in the paper). In [2] we have proved that some splitters which divide a flow into several sub-flows routing to distinct nodes are also monotone. Thus, we have generalized the method to networks with a tree topology.

In this paper, we further generalize our methodology in two directions. First, we prove that the merge elements which combine several flows into a global one is also monotone. This first result allows to consider feed-forward networks (i.e. the graph of the networking elements and the links is a Directed Acyclic Graph (DAG)). We use a decomposition approach based on the network topology and the monotonicity allows to obtain approximate results faster than the traditional approach. We remind that the decomposition approach let us to decompose the network and to study the networking elements in a sequential and greedy manner following the topological ordering associated with the DAG. This approach gives approximations on performance measures. The use of our methodology in this case aims to accelerate the computational times of this approach with a similar accuracy. Secondly, we study some Active Queue Management mechanisms to extend the modeling applicability of our method.

The technical part of the paper is organized as follows: in Sect. 2, we describe our methodology: the stochastic comparison of histograms, the reduction of the histogram sizes, the basic queueing model, and the monotonicity. In Sect. 3, we introduce the routing elements: splitter and merge and we prove that they are monotone. Section 4 is devoted to the AQM mechanisms. Finally in Sect. 5, we give numerical results for a single node analysis (to compare with HBSP algorithm), and a feed forward network.

2   Methodology for Bounds and Performances

We briefly introduce a well known ordering, called "strong stochastic ordering" for comparing distributions on $\mathbb{R}$. One may note that this comparison is called "first order stochastic dominance" in the economics literature. We show how one can compute the optimal lower bound and upper bound of a given size. The optimality criterion is the expectation of an arbitrary positive and increasing reward chosen by the modeler. We first define the stochastic comparison.

2.1   Stochastic Bounds

We refer to Stoyan's book [9] for theoretical issues of the stochastic comparison method. We consider state space $\mathcal{G} = \{1, 2, \ldots, n\}$ endowed with a total order denoted as $\leq$. Let X and Y be two discrete random variables taking values on $\mathcal{G}$, with probability mass functions (pmf in the following) $d2$ and $d1$.

Definition 1. *We can define the strong stochastic ordering by non decreasing functions or by some inequalities involving pmf.*

- **generic definition:** $X \leq_{st} Y \iff \mathbb{E}f(X) \leq \mathbb{E}f(Y)$,
  for all non decreasing functions $f : \mathcal{G} \to \mathbb{R}$ whenever expectations exist.
- **probability mass functions**

$$X \leq_{st} Y \Leftrightarrow \forall i, 1 \leq i \leq n, \ \sum_{k=i}^{n} d2(k) \leq \sum_{k=i}^{n} d1(k) \qquad (1)$$

Note that we use interchangeably $X \leq_{st} Y$ and $d2 \leq_{st} d1$.

In order to reduce the computation complexity for computing the steady-state distribution, we propose to decrease the number of bins in the histogram. We apply a bounding approach rather than an approximation. Unlike approximation, the bounds allow us to check if QoS requirements are satisfied or not.

More formally, for a given distribution d, defined as a histogram with N bins, we build two bounding distributions $d1$ and $d2$ defined on $n < N$ bins such that $d2 \leq_{st} d \leq_{st} d1$. Moreover, $d1$ and $d2$ are constructed to be the closest distributions with n bins with respect to a given non decreasing, positive reward function chosen by the modeler. Note that this optimality is not necessary in our approach, but it helps to obtain tight bounds. In [4], three algorithms to construct reduced size bounding distributions have been presented: an optimal algorithm based on dynamic programming with complexity $O(N^2 n)$, a greedy algorithm [4] with complexity $O(NlogN)$ and a linear complexity algorithm. There is no optimality for the last two ones but they are faster. The modeler can use any of them, thus he has the ability to choose between the accuracy and the computation times. In the numerical experiments, we give only results for the optimal one. We emphasize that the important property we need is the construction of a stochastic bound of the experimental distribution extracted from the trace.

We present an example to illustrate our stochastic bounding approach. We consider the histogram associated to the MAWI traffic trace (see Fig. 2) which is defined on 80511 states (bins) and we propose to derive bounding distributions $d1$ (stochastic upper bound distribution) and $d2$ (stochastic lower bound distribution) having reduced-size number of states i.e. $n = 10$ states. By taking the identity function as rewards, and using the optimal algorithm present in [4], we illustrate in Fig. 4 the cumulative distribution functions (cdf). The curve *Exact* is the original histogram on 80511 bins, where curves "Lower bound", "Upper bound" are computed on 10 bins. We can clearly observe that we derive bounds: "Lower bound" (resp. "Upper bound") is always over or equal (resp. below or equal) of "Exact".

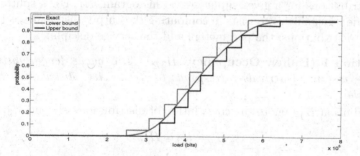

Fig. 4. Cdfs of the histograms for the MAWI traffic trace, and of the reduced-size bounds.

The traces are measured in bits. To keep the model size reasonable, we convert the values in data units. A data unit is D bits. Typically for the numerical analysis we present here, $D = 1000\,bits$. Hence, in the histograms representing the amount of data, the bins are integer multiples of D.

2.2   Stochastic Monotonicity of Networking Elements

The basic networking element is a finite queue associated with one server, a scheduling discipline and an access control. Let B be the buffer size. We assume that the queueing discipline is FCFS and work-conserving. The system evolves in discrete time. The service capacity (the number of data units that can be served during a slot) is constant and denoted by C. During each slot, the events occur in this order: arrivals and then service. The buffer length (buffer occupancy) evolution in the queue is given by a time-homogenous Discrete Time Markov Chain (DTMC) $\{X_n, n \geq 0\}$ taking values in a totally ordered state space, $\{0, 1, 2, \ldots, B\}$. The number of data units received during a time slot is independently, identically distributed (i.i.d.) random variable A specified by distribution H_1. Therefore, the *evolution equation* of the networking element with finite queue operating with Tail Drop policy [8] is:

$$X_{n+1} = \min(B, (X_n + A_n - C)^+), \tag{2}$$

where operator $(X)^+ = \max(X, 0)$.

The output of the analysis will be the buffer occupancy denoted by H_3 which is defined on state space $\{0 \cdots B\}$ and the departure process given by histogram H_5 defined on state space $\{0 \cdots C\}$. For a histogram H, we denote by E^H the set of states. For simplicity, H will be considered as the probability vector corresponding to the probabilities of the ordered elements of E^H.

We now give the main results of [2] about the stochastic monotonicity of the elements. All the proofs are omitted here. At each queuing element, the analysis consists in computing the distributions of H_3 and H_5 or bounds of these distributions for a given input arrival histogram, H_1. For a splitter and a merge node, the analysis consists in computing the output distributions knowing the input distributions, the parameters and the service discipline.

Proposition 1 (Buffer Occupancy, H_3).   *The queue length distribution before the instant of arrivals corresponds to steady state distribution π of the Markov chain.*

Let distribution H_q denote the convolution of distributions H_1 and H_3:

$$H_q = H_3 \otimes H_1.$$

Proposition 2 (Batch Departure, H_5). *The departure histogram H_5 is computed as follows:*

$$\begin{cases} H_5(w) = H_q(w), & \text{if } w < C; \\ H_5(C) = \sum_{w \geq C} H_q(w), & \text{otherwise.} \end{cases}$$

Proposition 3 (Losses, H_L). *The histogram of losses under the Tail Drop policy is:*

$$\begin{cases} H_L(k - B) = H_q(k), & \text{if } k > B + C; \\ H_L(0) = \sum_{k \leq B + C} H_q(k), & \text{otherwise.} \end{cases}$$

Then, the loss probability P_L can be defined as follow: $P_L = \frac{\mathbb{E}[H_L]}{\mathbb{E}[H_1]}$.

Definition 2. *A finite capacity queue is H-monotone, if the following holds:*

$$\text{if } H_1^a \leq_{st} H_1^b, \text{ then } H_3^a \leq_{st} H_3^b, \ H_5^a \leq_{st} H_5^b, \ and \ H_L^a \leq_{st} H_L^b.$$

Theorem 1. *A finite capacity queue which is operating with work-conserving FCFS service policy and Tail Drop policy is H-monotone.*

3 Analysis of a Network with a DAG Topology

In this section, we study network operations involving multiple streams as in [10]. First, we consider the split operation which has already been partially presented in [2]. Then, we introduce the merge operations. We note that the splitters, and merge elements do not have either processing element or queue to store data units. They execute routing decisions instantaneously.

3.1 Splitter

When the input flow modeled by a distribution H_S crosses a splitter, it is divided into m flows: $H_{S,1}, \ldots, H_{S,m}$. We assume that the batches observed after the splitter are still i.i.d. for each flow. This precludes the representation of Round Robin mechanism which may introduce the non stationarity in the flows. We define the H-monotonicity of the split element as follows:

Definition 3. *A splitter is said to be H-monotone, iff*

$$H_S^a \leq_{st} H_S^b \Rightarrow \forall i, \ H_{S,i}^a \leq_{st} H_{S,i}^b.$$

We study two cases of splitter:

- each batch arriving at the splitter is sent completely to one of the output flows. The output is randomly chosen according to a routing probability. This was previously presented in [2].
- the batch is divided into all the outputs according to a distribution for the repartition of the data units. This part is studied in this current paper.

Complete Batch Routing with Probabilities. We study a split element where all the data units of a batch arriving as the input flow are routed to an output flow with a routing probability. Let $p_i, 1 \leq i \leq m$ (such that $\sum_{i=1}^{m} p_i = 1$), be the routing probability of the batch to the output flow i of the split. If the set of states of H_S does not include 0, it will be added with probability 0, and the set of states for output flows will be the same as E^{H_S}.

$$E^{H_{S,i}} = \{0\} \cup E^{H_S}, \ 1 \leq i \leq m.$$

The probability distribution of any output flow i can be computed as follows:

$$1 \leq \forall i \leq m, H_{S,i}(k) = p_i \, H_S(k), \ k > 0; \ \text{and} \ H_{S,i}(0) = 1 - \sum_{k \neq 0} H_{S,i}(k).$$

Example 1. Let us consider histogram H with set of states, $E^H = \{0, 3, 4, 7, 10\}$ and the corresponding probability vector $H = [0.1, 0.2, 0.4, 0.1, 0.2]$. Assume that the batch is routed on two directions with equal probabilities. Each of the routed batch by this splitter has the following histogram: the set of states: $E^{H_i} = \{0, 3, 4, 7, 10\}$ and the probabilities: $H_i = [0.55, 0.1, 0.2, 0.05, 0.1]$, where $1 \leq i \leq 2$.

For an efficient implementation of histograms, the set of states are constituted of the elements with non null probabilities. However, in the sequel, for the proofs, we assume that the histograms are defined on set of states $E^H = \{0, \cdots n\}$ thus, the probability vectors may contain null probabilities.

Theorem 2. *If the batch is routed completely to a flow according to routing probabilities, then the split is H-monotone.*

Proof: Since $H_S^a \leq_{st} H_S^b$, we have $\sum_{k=l}^{n} H_S^a(k) \leq \sum_{k=l}^{n} H_S^b(k)$, $1 \leq l \leq n$. From the splitting property, for each flow $i, 1 \leq i \leq m$:

$$1 \leq \forall l \leq n, \quad \sum_{k=l}^{n} H_{S,i}^a(k) = \sum_{k=l}^{n} p_i \, H_S^a(k) \leq \sum_{k=l}^{n} H_{S,i}^b(k) = \sum_{k=l}^{n} p_i \, H_S^b(k).$$

Thus $H_{S,i}^a \leq_{st} H_{S,i}^b$.

Batch Division and Dispatching Among the Links. We now assume that the data units are dispatched among the m flows. The proportion of data received by each flow is given by the probability p_i which must be understood now as a ratio. Due to this multiplication by p_i, this amount of data can be a non integer amount of data units. Then, we assume that the data units are added with null bits and we obtain an integer number of data units.

$$E^{H_{S,i}} = \{ \lceil p_i * q \rceil, \, q \in E^{H_S} \}.$$

The histogram of output flow i, $1 \leq \forall i \leq m$, can be computed as

$$H_{S,i}(k) = \sum_{q \in E^{H_S}, q \neq 0} H_S(q) \mathbb{1}_{\lceil p_i * q \rceil = k}, \quad \forall k > 0, \quad \text{and} \quad H_{S,i}(0) = 1 - \sum_{k \neq 0} H_{S,i}(k).$$

Example 2. Consider the same example, but assume now that the data units are distributed among the flows. We also assume an equal repartition, thus the output flows have the same distribution with $E^{H_i} = \{0, 2, 4, 5\}$ and the probabilities are $H_i = [0.1, 0.6, 0.1, 0.2]$. Notice that the probability that the batch size is 2 is the sum of the probabilities that the input batch size (before division) is 3 or 4.

Theorem 3. *If the batch is splitted into batches according to dispatching probabilities, then the split is H-monotone.*

Proof: For each flow $i, 1 \leq i \leq m$, we can write

$$1 \leq \forall l \leq n, \qquad \sum_{k=l}^{n} H_{S,i}^{a}(k) = \sum_{k=l}^{n} \sum_{q=0}^{n} H_{S}^{a}(q) \mathbb{1}_{\lceil p_i * q \rceil = k}.$$

After exchanging the summations: $\sum_{k=l}^{n} H_{S,i}^{a}(k) = \sum_{q=0}^{n} H_{S}^{a}(q) \sum_{k=l}^{n} \mathbb{1}_{\lceil p_i * q \rceil = k}$. We can write $\sum_{k=l}^{n} \mathbb{1}_{\lceil p_i * q \rceil = k} = \mathbb{1}_{q \geq Q}$, for some constant Q. Thus,

$\sum_{k=l}^{n} H_{S,i}^{a}(k) = \sum_{q=0}^{n} H_{S}^{a}(q) \mathbb{1}_{q \geq Q} = \sum_{q \geq Q} H_{S}^{a}(q)$.

Since $H_{S}^{a} \leq_{st} H_{S}^{b}$, due to the st-ordering: $\sum_{q \geq Q} H_{S}^{a}(q) \leq \sum_{q \geq Q} H_{S}^{b}(q)$.

Therefore, $\sum_{k=l}^{n} H_{S,i}^{a}(k) \leq \sum_{k=l}^{n} H_{S,i}^{b}(k)$. Thus for all i, $H_{S,i}^{a} \leq_{st} H_{S,i}^{b}$.

3.2   Merge

In a merge element, a set of independent flows with distributions $H_{M,i}, 1 \leq i \leq m$ are aggregated to a flow with distribution H_M. We suppose that the links have a finite capacity, where C_i is the capacity of link i. In this subsection, we present the monotonicity properties for the merge elements by means of random variables corresponding to these histograms. Thus, X_i is the random variable with pmf $H_{M,i}$ representing the number of data units of input flow i of the merge element.

Definition 4. *A merge is a function* $\mathfrak{m} : \times_{i=1}^{m} \{0, \ldots, C_i\} \rightarrow \{0, \ldots, C\}$ *(i.e. the full convolution of m distributions).* $\mathfrak{m}(X_1, \ldots, X_m)$ *represents the state of the output flow of the merge element under independent input flows X_i. In fact it is a random variable with pmf H_M representing the number of data units leaving the merge element and taking values in* $\{0, 1, \cdots, C\}$ *where* $C \leq \sum_{i=1}^{m} C_i$.

Obviously, for the merge operation, the number of departed data units must be lower or equal to the number of arrived data units.

Definition 5. *The merge is causal, if* $\mathfrak{m}(X_1, \ldots, X_m) \leq \sum_{i=1}^{m} X_i$.

We can also define the traffic monotonicity for a merge element as follows:

Definition 6. *A merge element is traffic monotone iff for all couple* $(X_1, \ldots, X_m)$ *and* $(Y_1, \ldots, Y_m)$*, if* $X_k \leq Y_k, \forall k$*, then* $\mathfrak{m}(X_1, \ldots, X_m) \leq \mathfrak{m}(Y_1, \ldots, Y_m)$.

In the sequel, we consider causal merge elements. The merge operation may have the Tail Drop property which is defined as follows:

Definition 7. *A merge element is said to be Tail Drop, iff*

$$\mathfrak{m}(X_1, \ldots, X_m) = \min(C, \sum_{i=1}^{m} X_i).$$

We study now the monotonicity property of the merge elements.

Definition 8. *A merge element is said to be H-monotone, iff*

$$\forall i, \ H_{M,i}^{a} \leq_{st} H_{M,i}^{b} \Rightarrow H_{M}^{a} \leq_{st} H_{M}^{b}.$$

Theorem 4. *If the merge element is traffic monotone then it is H-monotone.*

Proof: We suppose that $\forall i$, $H_{M,i}^a \leq_{st} H_{M,i}^b$, thus the corresponding random variables are comparable: $\forall i$, $X_i^a \leq_{st} X_i^b$. The traffic monotonicity of the merge element means indeed that the function $\mathfrak{m}$ is an increasing function. Since the output flows H_M^a and H_M^b are defined as increasing functions of comparable independent random variables, they are also comparable (see page 7 of [9]).

Corollary 1. *A merge element operating with Tail Drop (i.e. $\mathfrak{m}(X_1,\ldots,X_m) = \min(C, \sum_{i=1}^m X_i)$) is causal and traffic monotone. Therefore, it is H-monotone.*

We now consider loss processes in merge elements. A merge element may delete some data units due to a bandwidth limitation or an access control. First we define the number of data units lost by loss function l which depends on the merge function $\mathfrak{m}$.

Definition 9. *The number of data units lost in a merge element can be defined by a function $l : \times_{i=1}^m \{0,\ldots,C_i\} \rightarrow \{0,\ldots \sum_{i=1}^m C_i\}$:*

$$l(X_1,\ldots,X_m) = \sum_{i=1}^m X_i - \mathfrak{m}(X_1,\ldots,X_m).$$

Indeed, the number of losses is the difference between the number of data units arrived on the m links (i.e. $\sum_{i=1}^m X_i$) and the number of units accepted by the merge element (i.e. $\mathfrak{m}(X_1,\ldots,X_m)$). The loss distribution can be given as follows, since the arrivals are independent. Let us remark that small letters denote the realizations of the corresponding random variables X_i.

Proposition 4 (Loss Distribution for a merge, H_L).

$$H_L(k) = \sum_{(x_1,\ldots,x_m)} \mathbb{1}_{\sum_{j=1}^m x_j - \mathfrak{m}(x_1,\ldots,x_m)=k} \prod_{i=1}^m H_{M,i}(x_i)$$

Property 1. *If $C = \sum_i C_i$ and the merge element is Tail Drop then there is no loss.*

Proof: The element is Tail Drop then, $\mathfrak{m}(X_1,\ldots,X_m) = \min(C, \sum_{i=1}^m X_i)$. But by construction $x_i \leq C_i$. Therefore $\sum_{i=1}^m X_i \leq \sum_{i=1}^m C_i = C$. Thus, there is no loss at the merge element.

Theorem 5. *If the loss function l of the merge element is non decreasing, then the histogram of losses, H_L of the merge element is monotone, which means that*

$$\text{if } \forall i, H_{M,i}^a \leq_{st} H_{M,i}^b, \text{ then } H_L^a \leq_{st} H_L^b.$$

Proof: The proof is similar to that of Theorem 4, and follows from the non decreasing property of the loss function, l.

Property 2. *For a Tail Drop, merge element with output capacity C, if $C < \sum_i C_i$, the distribution of losses is monotone.*

Proof: The number of data units lost is $l(X_1, \cdots X_m) = \max(0, \sum_{i=1}^m X_i - C)$. Thus l is non decreasing and H_L is monotone.

4   Analysis of Some AQM Mechanisms

The queue presented in Sect. 2 is operated under Tail Drop policy, which is a particular case of AQM (Active Queue Management). Indeed, the data units are accepted in the queue until the queue is full. In this section, we also present some conditions for AQM to be H-monotone in order to derive performance measure bounds. We illustrate this approach with a Random Early Detection mechanism (RED in the sequel).

We restrict ourselves to some AQMs where the probabilities of rejection depend on the size of the queue just before the insertion.

Definition 10. *The AQM is immediate if it operates independently and sequentially for each data unit in the batch and if the probabilities of rejection take into account the state of the queue just before the insertion.*

Note that this is a restricted version of AQM. We do not represent some mechanisms like explicit congestion notification. And, in mechanisms like RED, one does not use the instantaneous queue size to compute the acceptation probability, but a moving average of the queue size. However our definition can be used as an approximation.

More formally, we define an AQM acceptation by a function $q(X)$ which equals to 1, if the data unit is accepted and 0 if the data unit is rejected when the buffer size is X.

Definition 11. *The AQM is decreasing if function $q(X)$ is not increasing.*

Example 3. The Tail Drop policy is described by the acceptation function: $q(X) = \mathbb{1}_{\{X < B\}}$.

Thus, Tail Drop at the packet level is clearly immediate and decreasing.

Definition 12 (IRED). *The Immediate Random Early Detection policy is an example of AQM. We assume that it operates at data unit level. Contrary to Tail Drop, the acceptation for RED is given with probabilities. Many RED implementations are based on cubic functions or on the following piece-wise linear function to compute the acceptation probabilities:*

- *if $X \le \frac{B+C}{2}$: $Prob(q(X) = 1) = 1$;*
- *if $\frac{B+C}{2} \le X < B + C$: $Prob(q(X) = 1) = \frac{2(B+C)-2X}{B+C}$;*
- *if $X \ge (B + C)$: $Prob(q(X) = 1) = 0$;*

Thus, the probability that $q(X) = 1$ decreases with the queue length, X.

We extend the definition for H-monotonicity to network elements with an AQM.

Definition 13. *The AQM is H-monotone, iff*

$$H_1^a \le_{st} H_1^b \Rightarrow H_3^a \le_{st} H_3^b \text{ and } H_L^a \le_{st} H_L^b$$

We suppose that the queue works with an immediate AQM specified with a decreasing admission function $q(X)$. We denote by X_n the length of the queue at slot n and by $Y_{n,j}$ the length of the queue at slot n after the admission of the jth data unit. We take the same assumptions for the parameters as in the analysis of a queue (Sect. 2.2), and the maximum arrival batch size is denoted by K. The evolution equation of the queue length can be given as follows in the case when arrivals are taken into account before the services.

$$\begin{cases} Y_{n+1,0} &= X_n; \\ Y_{n+1,j+1} &= Y_{n+1,j} + \mathbb{1}_{\{A_n > j \text{ and } q(Y_{n+1,j})=1\}}; \\ X_{n+1} &= (Y_{n+1,K} - C)^+. \end{cases} \qquad (3)$$

Theorem 6. *If the AQM is immediate and the acceptation function is decreasing, then the AQM is H-monotone.*

Proof: The proof is based on the sample-path property of the strong stochastic ordering [9]. We prove by induction on the number of slot (n) that the realizations of the random variables for the evolution of queue lengths (see Eq. 2) satisfy:

$$x_n^a \le x_n^b, \quad \forall n.$$

We assume that queue lengths are the same for slot 0. Suppose that $x_n^a \le x_n^b$. To prove that $x_{n+1}^a \le x_{n+1}^b$, we proceed by induction on j indicating the data unit accepted during slot $n + 1$ ($y_{n+1,j}$). It follows from the definition that $y_{n+1,0}^a \le y_{n+1,0}^b$. Suppose that $y_{n+1,j}^a \le y_{n+1,j}^b$, and prove that $y_{n+1,j+1}^a \le y_{n+1,j+1}^b$. There are two cases:

1. $y_{n+1,j}^a < y_{n+1,j}^b$: since the data units are accepted one by one, we have $y_{n+1,j+1}^a \le y_{n+1,j+1}^b$.
2. $y_{n+1,j}^a = y_{n+1,j}^b$: acceptation functions $q(y_{n+1,j}^a) = q(y_{n+1,j}^b)$. By hypothesis, $H_1^a \le_{st} H_1^b$, since the arrivals are iid for each slot, we have the inequalities for the number of data units arrived during slot n: $A_n^a \le_{st} A_n^b$. Due to the $\le_{st}$ ordering, $\forall j : \mathbb{1}_{A_n^a > j} \le \mathbb{1}_{A_n^b > j}$. It follows from Eq. 3 that $y_{n+1,j+1}^a \le y_{n+1,j+1}^b$.

So, we deduce that: $x_{n+1}^a = y_{n+1,K}^a \le y_{n+1,K}^b = x_{n+1}^b$. Therefore, we have the stochastic comparison of the queue length evolutions: $X_n^a \le_{st} X_n^b$, $\forall n$. At the limiting case, the stationary processes are also comparable: $H_3^a \le_{st} H_3^b$.
The number of data units lost during slot $n + 1$ can be given as:

$$\sum_{j=1}^{K} \mathbb{1}_{\{A_n > j \text{ and } q(Y_{n+1,j})=0\}}.$$

It follows from the above proof that $Y_{n,j}^a \le_{st} Y_{n,j}^b$. Since the acceptation functions $q()$ are decreasing functions, and $H_1^a \le_{st} H_1^b$, if the above indicator function is 1 under arrival H_1^a then it is also 1 under arrival H_1^b. Thus, the number of data units lost in each slot and in the limit will be comparable: $H_L^a \le_{st} H_L^b$.

5   Examples

We consider respectively a node with an IRED mechanism and a' network of nodes. For all the experiments, we suppose that the monotonicity property is used for the convergence proof of our method [2] for $\epsilon = 10^{-6}$. The reward function used here is defined by $r(i) = i, \forall i \in E^H$. We note that the implementation is performed on Matlab and the experiences were computed on a laptop computer Intel Core I7, 2.53 GHz.

5.1   A RED Node

We give a simple example to illustrate the impact of our method on single node with IRED mechanism. We consider input histogram $H_1 = [0.10, 0.05, 0.10, 0.10, 0.15, 0.15, 0.10, 0.10, 0.05, 0.10]$ defined on state space $E^{H_1} = \{1, \ldots, 10\}$ and deterministic service $C = 2$. The performance measures (blocking probabilities, average queue length and execution time) are calculated by varying the buffer size from 4 to 30 data units. In Figs. 5, 6 and 7, we present the performance measures by using the exact computation (with out size reduction) and optimal lower bound for the number of bins equals to 3 and 5. In this example we illustrate the lower bounds but the upper bounds can also be calculated.

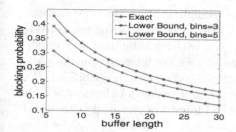

Fig. 5. Results on blocking probabilities.

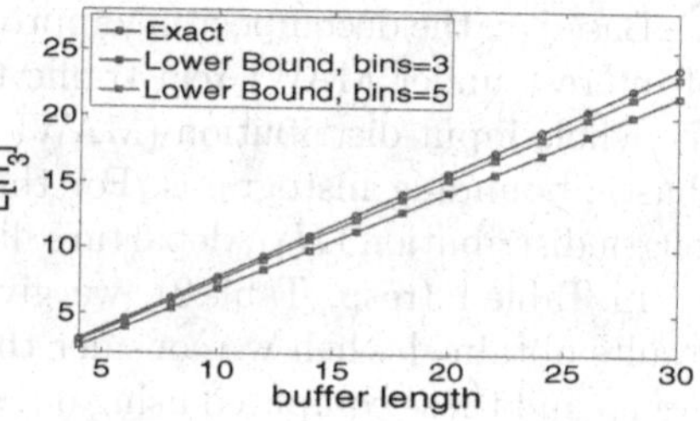

Fig. 6. Results on mean buffer length.

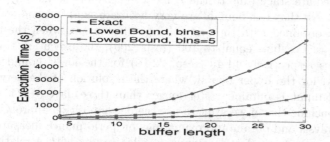

Fig. 7. Execution time (s).

Through these figures, we see that the use of bounding method allows us to obtain accurate results within reduced execution time. We remark that when the number of bins increases the accuracy of the bound is improved.

5.2   A Feed-Forward Network

Unlike HBSP method, our approach can be extended to the study of feed-forward networks as shown in the following example. We consider a feed-forward network model depicted in Fig. 8 with 6 nodes. Each node is a split (resp. merge) element or a finite capacity queue ($B_i = 10$ Mb, $i = 1, 3, 4, 6$). The service for each queue is taken respectively equal to $110\,Mb/s$, $67.5\,Mb/s$, $90\,Mb/s$ and $117.5\,Mb/s$.

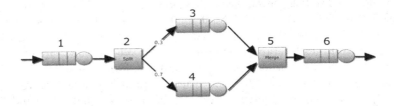

Fig. 8. An example of Feed Forward Network.

Based on the decomposition approach, we compute the performance measures of interest under MAWI real traffic traces (Fig. 1) by considering respectively: the whole input distribution (MAWI histogram without reduction) and our stochastic bounding histograms. For this example, we are interested in the queue length distribution (H_3), departure distribution (H_5) and loss probabilities (P_L).

In Table 1 (resp. Table 2), we give for the four queues of the network, the results obtained when we consider the original input histogram (denoted by O. input) and those computed using our stochastic bounds (denoted by $L.b$ for lower bound and $U.b$ for upper bound) for the number of bins equals to 100 (resp. 200).

From these tables, we remark that the bounds on the results are provided for each intermediate stage (due to the H-monotonicity of the network elements). We can also see that our bounds are very accurate, and become very close to the solution obtained with the original input histogram, when the number of bins increases. For bins equal to 100 (resp. 200), the execution times of the bounds takes respectively 14.4 s (resp. 22.1 s) for the lower bound and 15.9 s (resp. 25.9 s) for the upper bound, where the resolution of the network using the original input is obtained after longer than three days 314248 s. We can therefore conclude that if we want to use the decomposition approach for DAG network analysis and obtain approximations on performance measures, we can use the proposed method and compute similar results with a relatively small computation complexity.

Table 1. Results for bins = 100.

		E[H_1]	E[H_3]	E[H_5]	P_L
	O. input	4375620	4332130	4352390	0.00530
Queue 1	L. b	4356310	3850300	4339640	0.00382
	U. b	4397080	4890670	4364590	0.00739
	O. input	1305720	875834	1305030	0.00052
Queue 2	L. b	1300650	863705	1300010	0.00049
	U. b	1309460	884739	1308730	0.00055
	O. input	3046670	2256310	3037710	0.00293
Queue 3	L. b	3034840	2190650	3026910	0.00260
	U. b	3055400	2304630	3045640	0.00318
	O. input	4342730	2519100	4327660	0.00346
Queue 4	L. b	4313200	2340670	4301450	0.00272
	U. b	4357650	2593470	4341260	0.00375

Table 2. Results for bins = 200.

		E[H_1]	E[H_3]	E[H_5]	P_L
	O. input	4375620	4332130	4352390	0.00530
Queue 1	L. b	4366450	4099970	4346550	0.00455
	U. b	4386860	4622920	4359020	0.00633
	O. input	1305720	875834	1305030	0.00052
Queue 2	L. b	1302780	868692	1302120	0.00050
	U. b	1307780	880729	1307060	0.00054
	O. input	3046670	2256310	3037710	0.00293
Queue 3	L. b	3039820	2217560	3031480	0.00273
	U. b	3051480	2282840	3042080	0.00306
	O. input	4342730	2519100	4327660	0.00346
Queue 4	L. b	4322810	2398430	4310020	0.00295
	U. b	4350010	2555410	4334300	0.00360

6 Conclusions

The results developed in this paper are very promising: they allow to mix in an efficient and accurate manner measurements and stochastic modeling to analyze some networks (simple queue, AQM and DAG networks via decomposition approach). As future works, we want to extend our methodology and state some stochastic comparison results in feed-forward networks [1] (and also general topology networks). Note that the approach is not limited to performance evaluation of networks, it can be applied to any problem (reliability, statistical model checking) where we have large measurements and where the model is monotone in some sense.

Acknowledgement. This work was partially supported by grant ANR MARMOTE (ANR-12-MONU-0019) and DIGITEO.

References

1. Aït-Salaht, F., Castel Taleb, H., Fourneau, J.M., Mautor, T., Pekergin, N.: Smoothing the input process in a batch queue. In: Abdelrahman, O.H., Gelenbe, E., Gorbil, G., Lent, R. (eds.) ISCIS 2015. Lecture Notes in Electrical Engineering, vol. 363, pp. 223–232. Springer, Heidelberg (2015)
2. Aït-Salaht, F., Castel Taleb, H., Fourneau, J.M., Pekergin, N.: A bounding histogram approach for network performance analysis. In: HPCC, China (2013)
3. Aït-Salaht, F., Castel-Taleb, H., Fourneau, J.-M., Pekergin, N.: Stochastic bounds and histograms for network performance analysis. In: Balsamo, M.S., Knottenbelt, W.J., Marin, A. (eds.) EPEW 2013. LNCS, vol. 8168, pp. 13–27. Springer, Heidelberg (2013)

4. Aït-Salaht, F., Cohen, J., Castel-Taleb, H., Fourneau, J.M., Pekergin, N.: Accuracy vs. complexity: the stochastic bound approach. In: 11th International Workshop on Disrete Event Systems, pp. 343–348 (2012)

5. Hernández-Orallo, E., Vila-Carbó, J.: Network performance analysis based on histogram workload models. In: MASCOTS, pp. 209–216, 2007

6. Hernández-Orallo, E., Vila-Carbó, J.: Web server performance analysis using histogram workload models. Comput. Netw. **53**(15), 2727–2739 (2009)

7. Hernández-Orallo, E., Vila-Carbó, J.: Network queue and loss analysis using histogram-based traffic models. Comput. Commun. **33**(2), 190–201 (2010)

8. Kleinrock, L.: Queueing Systems, Volume I: Theory. Wiley, Hoboken (1975)

9. Muller, A., Stoyan, D.: Comparison Methods for Stochastic Models and Risks. Wiley, New York (2002)

10. Schleyer, M.: Discrete time analysis of batch processes in material flow systems. Wissenschaftliche Berichte des Institutes für Fördertechnik und Logistiksysteme des Karlsruher Instituts für Technologie. Univ.-Verlag Karlsruhe (2007)

11. Cho Sony, K., Cho, K.: Traffic data repository at the wide project. In: Proceedings of USENIX 2000 Annual Technical Conference on FREENIX Track, pp. 263–270 (2000)

Subsampling for Chain-Referral Methods

Konstantin Avrachenkov, Giovanni Neglia, and Alina Tuholukova[✉]

Inria Sophia Antipolis, 2004 Route des Lucioles, Sophia Antipolis, France
{k.avrachenkov,giovanni.neglia,alina.tuholukova}@inria.fr

Abstract. We study chain-referral methods for sampling in social networks. These methods rely on subjects of the study recruiting other participants among their set of connections. This approach gives us the possibility to perform sampling when the other methods, that imply the knowledge of the whole network or its global characteristics, fail. Chain-referral methods can be implemented with random walks or crawling in the case of online social networks. However, the estimations made on the collected samples can have high variance, especially with small sample size. The other drawback is the potential bias due to the way the samples are collected. We suggest and analyze a subsampling technique, where some users are requested only to recruit other users but do not participate to the study. Assuming that the referral has lower cost than actual participation, this technique takes advantage of exploring a larger variety of population, thus decreasing significantly the variance of the estimator. We test the method on real social networks and on synthetic ones. As by-product, we propose a Gibbs like method for generating synthetic networks with desired properties.

1 Introduction

Online social networks (OSNs) are thriving nowadays. The most popular ones are: Google+ (about 1.6 billion users), Facebook (about 1.28 billion users), Twitter (about 645 million users), Instagram (about 300 million users), LinkedIn (about 200 million users). These networks gather a lot of valuable information like users' interests, users' characteristics, etc. Great part of it is free to access. This information can facilitate the work of sociologists and give them modern instrument for their research. Of course, real social networks continue to be of great interest to sociologists as well as online social networks. For example, the Add Health study [2] has built the networks of the students at selected schools in the United States, which served as the basis of much further research [10].

The network, besides being itself an object of study, is also an instrument for collecting data. Starting just from one individual that we observe we can reach other representatives of this network. The sampling methods that use the contacts of known individuals of a population to find other members are called *chain-referral methods*. Crawling of online social networks can be viewed as automatisation of chain-referral methods. Moreover, it is one of the few methods to collect information about *hidden populations*, whose members are,

S. Wittevrongel and T. Phung-Duc (Eds.): ASMTA 2016, LNCS 9845, pp. 17–31, 2016.
DOI: 10.1007/978-3-319-43904-4_2

by definition, hard to reach. A lot of research has targeted the study of HIV prevalence in hidden populations like drug users, female sex workers [11], gay men [12]. Another study [9] considered the population of jazz musicians. Even if jazz musicians have no reasons to hide them, it is still hard to access them with the standard sampling methods.

The problem of the chain-referral methods is that they do not achieve independent sampling from the population. It is frequently observed that friends tend to have similar interests. It can be the influence of your friend that leads you to listening the rock music or the opposite: you became friends because you were both fond of it. One way or another, social contacts influence each other in different ways. The fact that people in contact share common characteristics is usually observed in real networks and is called *homophily*. For instance, the study [6] evaluated the influence of social connections (friends, relatives, siblings) on obesity of people. Interestingly, if a person has a friend who became obese during some fixed interval of time, the chances that this person becomes obese are increased by 57 %.

The population sample obtained through chain-referral methods is different from the ideal uniform independent sample and, because of homophily, leads to increased variance of the estimators as we are going to show. The main contribution of this paper is the proposed chain-referral method that allows to decrease the dependency of the collected values by subsampling. Subsampling is done via asking/inferring only contact details of some users without taking any further information.

As by-product of our numerical studies, we develop a Gibbs-like method for generating synthetic attributes' distribution over networks with desired properties. This approach can be used for extensive testing of methods in social network analysis and hence can be of independent interest.

The paper is organized as follows. In Sect. 2 we discuss different estimators of the population mean and the problem of correlated samples. Section 3 presents the subsampling method, that can help to reduce the correlation. In Sect. 4 we evaluate the subsampling method formally, starting from the simple, but intuitive example of a homogeneous correlation (Sect. 4.1), and then moving to the general case (Sect. 4.2). The theoretical results are then validated by the experiments in Sect. 5. Section 5 presents also the method for generating synthetic networks that we used for the experiments together with the real data.

2   Chain-Referral Methods and Estimators

Chain-referral methods take advantage of the individuals connections to explore the network: each study participant provides the contacts of other participants. The sampling continues in this way until the needed size of participants is reached.

In order to study formally chain-referral methods we will model the social network as a graph, where the individuals are represented by nodes and a contact between two individuals is represented by an edge between the corresponding nodes. We will make the following assumptions:

1. One individual can refer exactly another individual, selected uniformly at random from his contacts;
2. The same individual can be recruited multiple times;
3. If individual A knows individual B then individual B knows A as well (the network can be represented as an undirected graph);
4. Individuals know and report precisely their number of connections (i.e. their degree);
5. Each individual is reachable from any other individual (the network is connected).

Under these assumptions the referral process can be regarded as a *random walk* on the graph. For the real social networks some of these assumptions are arguable. There can be inaccuracy in the reported degree, and the choice of the contact to refer can be different from uniform. The sensitivity to violation of some assumptions is studied in [7]. However, it is simpler to design chain-referral methods for online social networks, that satisfy all these assumptions. For example, the individual may be asked to disclose his whole list of contacts (if not already public) and the next participant can then be selected uniformly at random from it.

The random walk is represented by the transition matrix P with elements:

$$p_{ij} = \begin{cases} \frac{1}{d_i} & \text{if } i \text{ and } j \text{ are neighbors,} \\ 0 & \text{if } i \text{ and } j \text{ are not neighbors,} \\ 0 & \text{if } i = j, \end{cases}$$

where d_i is the degree of the node i.

We denote as g_j the value of interest at node j. We are interested to estimate the population average $\mu = \frac{\sum_{i=1}^{m} g_i}{m}$, where m is the population size.

Moreover, let us denote the value that is observed at step i of the random walk as y_i. Some estimators were developed in order to draw conclusions about the population average μ from the collected sample $y_1, y_2, ...y_n$. The simplest estimator of the population mean is the **Sample Average** (SA) estimator:

$$\widehat{\mu}_{SA} = \frac{y_1 + y_2 + ... + y_n}{n}.$$

This estimator is biased towards the nodes with large degrees. Indeed the individuals with more contacts are more likely to be sampled by the random walk. In particular, the probability at a given step to encounter node i is proportional to its degree d_i. To correct this bias the **Volz-Heckathon** (VH) estimator, which was introduced in [13], weights the responses from individuals according to their number of contacts:

$$\widehat{\mu}_{VE} = \frac{1}{\sum_{i=1}^{n} 1/d_i} \sum_{i=1}^{n} \frac{y_i}{d_i}.$$

Problem of Samples Correlation. Due to the way the sample was collected the variance of both estimators will be increased in comparison to the case of independent sampling. Our theoretical analysis will focus on the SA estimator, as for the VH estimator it becomes too complicated and we leave its analysis for future research. However, we consider the VH estimator in the simulations.

The variance of the estimator in the case of independent sampling with replacement is approximated by σ^2/n for large population size, where σ^2 is the population variance. If samples are not independently selected, then a correlation factor $f(n, \mathcal{S})$ should be considered as follows:

$$\sigma^2_{\hat{\mu}_S} = \frac{\sigma^2}{n} f(n, \mathcal{S}). \tag{1}$$

This correlation factor $f(n, \mathcal{S})$ depends on the sampling method $\mathcal{S}$ as well as on the size of the sample. We observe that $f(n, \mathcal{S})$ is an increasing function of n bounded by 1 and n. The less the samples obtained through the sampling method $\mathcal{S}$ are correlated, the smaller we expect $f(n, \mathcal{S})$ to be.

In what follows we consider chain-referral methods when only one individual out of k is asked for his value. Among these methods the correlation factor $f(n, \mathcal{S})$ will be a function of the number of values collected, n, and of k, so we can write $f(n, k)$. We expect $f(n, k)$ to be decreasing in k.

3   Subsampling Technique

In order to reduce correlation between sampled values we will try to decrease the dependency of the samples. Our idea is to thin out the sample. Indeed, the farther are the individuals in the chain from each other, the smaller is the dependency between them. Imagine to have contacted an even number h individuals, but to ask the value of interest only to every second of them. We can use then the $n = h/2$ values. It should be observed that, while we reduce in this way the correlation factor (because $f(h/2, 2) < f(h, 1)$), we also reduce by 2 the number of samples used in the estimation. Then while $f(n, k)$ becomes smaller in Eq. (1) because of the reduction of the correlation, it is not clear if $\frac{f(n,k)}{n}$ becomes smaller.

Another potential advantage originates from the fact that the cost of the referring is less than the cost of the actual sampling. For example, the information about the friends in Facebook is generally available, thus you can serf through the Facebook graph by writing a simple crawler. On the contrary retrieving the information of interest can be more costly and one may need to provide some form of incentives to participants to encourage them to answer some questionnaires. In other context, one may need to pay the users also to reveal the identity of one of his contacts.

Among the individuals in the collected chain some of them will be asked both: to participate in the tests and provide the reference, let us call them *participants*. Some of them will be asked only to recruit other participants, let us call them *referees*. We will look at the strategy when only each k-th individual in the chain is a participant. Thus between 2 participants there are always $k-1$ referees. We will call this approach *subsampling with step k*. Let C_1 be the payment for providing the reference and C_2 the payment for the participation in the test. In this way, every referee receives C_1 units of money and every participant receives $C_1 + C_2$ units of money (C_1 for the reference and C_2 for the test). In this way, for a fixed budget B, if $C_2 > 0$, the subsampling decreases less in the number of samples.

It is evident that the bigger is k, the lower is the correlation between the selected samples. However the choice of the k is not evident: if we take it too small the dependency can be still high; if we take it too big the sample size will be inadequate to make conclusions. It also depends on the level of homophily in the network: with the low level of homophily the best choice would be to take k equal to 1, what means no referees only participants. In the following section we formalize the qualitative results derived here and we determine the value k, such that the profit from the subsampling is maximal.

4 Analysis

In this section we study formally the effect of subsampling. We start with a case when the collected samples are correlated in a known and homogeneous way. While being a too simplified model for the chain-referral methods, it illustrates the main idea of subsampling. We proceed then with the general case, when the samples are collected through the random walk on a general graph.

4.1 Simple Example: Variance with Geometric Correlation

First we will quantify the variance of the estimator for a simple case with defined correlation between the samples in the chain. We will assume that collected samples $Y_1, Y_2, ..., Y_n$ are correlated in the following way:

$$\text{corr}(Y_i, Y_{i+l}) = \rho^l.$$

In this way the nodes that are at the distance 1 in the chain have correlation ρ, at distance 2 have correlation ρ^2 an so on[1]. We will refer to this model as the *geometric model*[2]. If the population variance is σ^2, then we can obtain the variance of the SA estimator in the following way:

$$\sigma^2_{\hat{\mu}_{SA}} = \text{Var}\left[\bar{Y}\right] = \text{Var}\left[\frac{Y_1 + Y_2 + ... + Y_n}{n}\right] = \frac{1}{n^2}\sum_{i=1}^{n}\sum_{j=1}^{n}\text{Cov}(Y_i, Y_j)$$

$$= \frac{\sigma^2}{n^2}\left(n + 2\sum_{i=1}^{n-1}(n-i)\rho^i\right) = \frac{\sigma^2}{n^2}\left(n + 2n\sum_{i=1}^{n-1}\rho^i - 2\sum_{i=1}^{n-1}i\rho^i\right)$$

---

[1] We are ignoring here the effect of resampling.
[2] It could be adopted to model the case where nodes are on a line and social influences are homogeneous.

$$= \frac{\sigma^2}{n} \left(n + 2n \frac{\rho - \rho^n}{1 - \rho} - 2\rho \left(\frac{\rho - \rho^n}{1 - \rho} \right)' \right) = \frac{\sigma^2}{n^2} \frac{n - n\rho^2 - 2\rho + 2\rho^{n+1}}{(1 - \rho)^2}.$$

From here we can get that correlation factor as:

$$f(n, 1) = \frac{1 - \rho^2 - 2\rho/n + 2\rho^{n+1}/n}{(1 - \rho)^2}.$$

It can be shown that this factor $f(n)$ is an increasing function of $n \in \mathbb{N}$ and it achieves its minimum value 1 when $n = 1$. It is clear, when there is only one individual there is no correlation, because we consider single random variable Y_1. When new participants are invited, the correlation increases due to homophily as we explained earlier.

Let us consider what happens to the correlation factor when n goes to infinity:

$$f(n, 1) = \frac{1 - \rho^2 - 2\rho/n + 2\rho^{n+1}/n}{(1 - \rho)^2} \xrightarrow{n \to \infty} \frac{1 - \rho^2}{(1 - \rho)^2} = \frac{1 + \rho}{1 - \rho},$$

and then $f(n, 1) \le \frac{1+\rho}{1-\rho}$ $\forall n$. Using this upper bound the expression for the SA estimator variance can be bounded as $\sigma^2_{\hat{\mu}_{SA}} \le \frac{\sigma^2}{n} \frac{1+\rho}{1-\rho}$.

This bound is very tight when n is large enough, so that it can be used as a good approximation:

$$\sigma^2_{\hat{\mu}_{SA}} \simeq \frac{\sigma^2}{n} \frac{1 + \rho}{1 - \rho}.$$

Figure 1 compares the approximated expression with original one, when the parameter ρ is 0.6. As it is reasonable to suppose that the sample size is bigger than 50, we can consider this approximation good enough in this case. The reason to use this approximation is that the expression becomes much simpler to illustrate the main idea of the method.

Variance for Subsampling. Here we will quantify the variance of the SA estimator on the subsample. For simplicity let us take $h = nk$, where the collected

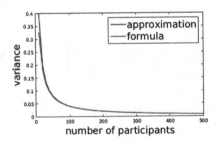

Fig. 1. $\rho = 0.6$

samples $Y_1, Y_2, Y_3, ..., Y_{nk}$ have again geometric correlation. We will take each k sample and look at the variance of the following random variable:

$$\bar{Y}^k = \frac{Y_k + Y_{2k} + Y_{3k} + ... + Y_{nk}}{n}.$$

Let us note that the correlation between the variables Y_{ik} and $Y_{(i+l)k}$ is:

$$\text{corr}(Y_{ik}, Y_{(i+l)k}) = \rho^{kl}.$$

Using the result of Sect. 4.1, we obtain:

$$\text{Var}\left[\bar{Y}^k\right] = \frac{\sigma^2}{n} \frac{1 - \rho^{2k} - 2\rho^k/n + 2\rho^{k(n+1)}/n}{(1 - \rho^k)^2}.$$

or the approximate form:

$$\text{Var}\left[\bar{Y}^k\right] \simeq \frac{\sigma^2}{n} \frac{1 + \rho^k}{1 - \rho^k}. \tag{2}$$

Limited Budget. Equation (2) gives the expression for the variance of the subsample, where the number of actual participants is n and two consecutive participants in the chain are separated by $k - 1$ referees. It is evident that in order to decrease the variance, one needs to take as many participants as possible separated by as many referees as possible. However both of them have their cost. If limited budget B is available, then a chain of length $h = nk$ with n participants is restricted by the following equality:

$$B \geq hC_1 + nC_2,$$

where each reference costs C_1 units of money and each test costs C_2 units of money. We can express the maximum length of the chain as: $h = \frac{kB}{kC_1 + C_2}$, where the number of actual participants is $n = \frac{h}{k} = \frac{B}{kC_1 + C_2}$.

The approximate variance of SA estimator becomes as follows:

$$\sigma^2_{\hat{\mu}_{SA}}(k) = \frac{\sigma^2}{\frac{B}{kC_1 + C_2}} \frac{1 + \rho^k}{1 - \rho^k}. \tag{3}$$

Let us observe what happens to the factors of the variance when we increase k. The first factor in (3) increases in k: the variance increases due to smaller sample size. The second factor decreases in k: the observations become less correlated. Finally, the behavior of the variance depends on which factor is "stronger".

We can observe the trade-off in Fig. 2: initially increasing the subsampling step k can help to reduce the estimator variance. However, after some threshold the further increase of k will only add to the estimator variance. Moreover, this threshold depends on the level of correlation, that is expressed here by the parameter ρ. We observe from the figure that the higher is ρ the higher is the desired k. This coincides with our intuition: the higher is the dependency, the more values we need to skip. Finally we see, that in case of no correlation ($\rho = 0$) skipping nodes is useless.

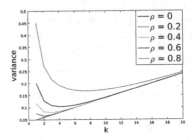

Fig. 2. Variance with Eq. 3 when $B = 100, C_1 = 1, C_2 = 4$

4.2   General Case

Even if the geometric model is not realistic, it allowed us to better understand the potential improvement from subsampling. This section will generalize this idea to the case where the samples are collected through a random walk on a graph with m nodes. We consider first the case without subsampling ($k = 1$).

Let $g = (g_1, g_2, ..., g_m)$ be the values of the attribute on the nodes $1, 2, ..., m$. Let P be the transition matrix of the random walk.

The stationary distribution of the random walk is:

$$\pi = \left(\frac{d_1}{\sum_{i=1}^{n} d_i}, \frac{d_2}{\sum_{i=1}^{n} d_i}, ..., \frac{d_n}{\sum_{i=1}^{n} d_i} \right),$$

where d_i is the degree of the node i.

Let Π be the matrix that consists of m rows, where each row is the vector π. If the first node is chosen according to the distribution π, then variance for any sample Y_i[3] is the following:

$$\mathrm{Var}(Y_i) = <g, g>_\pi - <g, \Pi g>_\pi, \text{where} <a, b>_\pi = \sum_{i=1}^{m} a_i b_i \pi_i.$$

and covariance between the samples Y_i and Y_{i+l} is the following [5, chapter 6]:

$$\mathrm{Cov}(Y_i, Y_{i+l}) = <g, (P^l - \Pi)g>_\pi,$$

Using these formulas we can write the formula for the variance of the estimator as:

$$\mathrm{Var}\left[\bar{Y}\right] = \frac{1}{n^2} \left(n\mathrm{Var}(Y_i) + 2 \sum_{i=1}^{n} \sum_{j|i<j}^{n} \mathrm{Cov}(Y_i, Y_j) \right)$$

$$= \frac{1}{n^2} \left(n(<g, g>_\pi - <g, \Pi g>_\pi) + 2 \sum_{i=1}^{n} \sum_{j|i<j}^{n} <g, (P^{j-i} - \Pi)g>_\pi \right) \qquad (4)$$

---

[3] Note that $Y_i = g_j$ if the random walk is on node j at the i-th step.

Equation (4) is quite cumbersome: computing large powers of the m by m matrix P can be unfeasible. Using the spectral theorem for diagonalizable matrices:

$$\text{Var}\left[\bar{Y}\right] = \frac{1}{n}\sum_{i=2}^{m}\frac{1 - \lambda_i^2 - 2\frac{\lambda_i}{n} + 2\frac{\lambda_i^{n+1}}{n}}{(1-\lambda_i)^2} <g, v_i>_\pi^2, \tag{5}$$

where $\lambda_i, v_i, u_i (i = 1..m)$ are respectively eigenvalues, right eigenvectors and left eigenvectors of the auxiliary matrix P^{*4}, defined as $P^* \triangleq D^{\frac{1}{2}}PD^{-\frac{1}{2}}$, where D is the $m \times m$ diagonal matrix with $d_{ii} = \pi_i$.

In the case of subsampling similar calculation can be carried on leading to:

$$\text{Var}\left[\bar{Y}^k\right] = \frac{1}{\frac{B}{kC_1+C_2}}\sum_{i=2}^{m}\frac{1 - \lambda_i^{2k} - 2\frac{\lambda_i^k}{\frac{B}{kC_1+C_2}} + 2\frac{\lambda_i^{k\left(\frac{B}{kC_1+C_2}+1\right)}}{\frac{B}{kC_1+C_2}}}{(1-\lambda_i)^{2k}} <g, v_i>_\pi^2 . \tag{6}$$

As in the geometric model Eq. (6) can be approximated as follows:

$$\sigma_{\hat{\mu}_{SA}}^2 = \text{Var}\left[\bar{Y}^k\right] = \frac{1}{\frac{B}{kC_1+C_2}}\sum_{i=2}^{m}\frac{1+\lambda_i^k}{1-\lambda_i^k} <g, v_i>_\pi^2 .$$

Interestingly, the expression for the variance in the general case has the same structure as for the geometric model. Therefore, the interpretation of the formula is the same. There are two factors, that "compete" with each other. If we try to decrease the first factor, we will increase the second one and the opposite. In order to find the desired parameter k we need to find the minimum of the estimator function for variance. Even if it is difficult to obtain the explicit formula for k, the fact that k is integer allows us to find it through binary search.

The quality of an estimator does not depend only on its variance, but also on its bias:

$$\text{Bias}(\hat{\mu}_{SA}) = E[\hat{\mu}_{SA}] - \mu \ = <g, \pi> -\mu. \tag{7}$$

Then the mean squared error of the estimator, $MSE(\hat{\mu}_{SA})$, is:

$$MSE(\hat{\mu}_{SA}) = \text{Bias}(\hat{\mu}_{SA})^2 + \text{Var}(\hat{\mu}_{SA}). \tag{8}$$

This bias can be non-null if the quantity we want to estimate is correlated with the degree. In fact, we observe that the random walk visits the nodes with more connections more frequently. Subsampling has no effect on such bias, hence minimizing the variance leads to minimizing the mean squared error.

---

[4] Matrix P^* is always diagonalizable for RW on undirected graph.

5   Numerical Evaluation

To validate our theoretical results we performed numerous simulations. We considered both real datasets from the Project 90 [3] and Add health [2], as well as synthetic datasets, obtained through the Gibbs sampler. Both the Project 90 and the Add health datasets contain the graph describing the social contacts as well as information about the users.

Data from the Project 90. Project 90 [3] studied how the network structure influences the HIV prevalence. Besides the data about social connections the study collected some data about drug users, such as race, gender, whether he/she is a sex worker, pimp, sex work client, drug dealer, drug cook, thief, retired, housewife, disabled, unemployed, homeless. For our experiments we took the largest connected component from the available data, which consists of 4430 nodes and 18407 edges.

Data from the Add Health Project. The National Longitudinal Study of Adolescent to Adult Health (Add Health) is a huge study that began surveying students from the 7–12 grades in the United States during the 1994–1995 school year. In general 90,118 students representing 84 communities took part in this study. The study kept on surveying students as they were growing up. The data include, for example, information about social, economic, psychological and physical status of the students.

The network of students' connections was built based on the reported friends by each participant. Each of the students was asked to provide the names of up to 5 male friends and up to 5 female ones. Then the network structure was built to analyze if some characteristics of the students indeed are influenced by their friends.

Though these data are very valuable, they are not freely available. However a subset of the data can be accessed through the link [1] but only with few attributes of the students, such as: sex, race, grade in school and, whether they attended middle or high school. There are several networks available for different communities. We took the graph with 1996 nodes and 8522 edges.

Synthetic Datasets. To perform extensive simulations we needed more graph structures with node attributes.

There is no lack of available real network topologies. For example, the Stanford Large Network Dataset Collection [4] provides data of Online-Social Networks (we will use part of Facebook graph), collaboration networks, web graphs, Internet peer-to-peer network and a lot of others. Unfortunately, in most of the cases, nodes do not have any attribute.

At the same time random graphs can be generated with almost arbitrary characteristics (e.g. number of nodes, links, degree distribution, clustering coefficient). Popular graph models are Erdős-Rényi graph, random geometric graph, preferential attachments graph. Still, there is no standard way to generate synthetic attributes for the nodes and in particular providing some level of homophily (or correlation).

In the same way we can generate numerous random graphs with desired characteristics, we wanted to have mechanism to generate the values on the nodes of the given graph which will represent needed attribute, which will satisfy the following properties:

1. Nodes attributes should have the property of homophily
2. We should have the mechanism to control the level of homophily

These properties are required to evaluate the performance of the subsampling methods. In what follows we derive a novel (to the best of our knowledge) procedure for synthetic attributes generation.

First we will provide some definitions. Let us imagine that we already have a graph with m nodes. It may be the graph of a real network or a synthetic one. Our technique is agnostic to this aspect. To each node i, we would like to assign a random value G_i from the set of attributes $V, V = \{1, 2, 3, ..., L\}$. Instead of looking at distributions of the values on nodes independently, we will look at the joint distribution of values on all the nodes.

Let us denote $(G_1, G_2, ..., G_m)$ as $\dot{G}$. We call $\dot{G}$ a *random field on graph*. When random variables $G_1, G_2, ..., G_m$ take respectively values $g_1, g_2, ..., g_m$, we call $(g_1, g_2, ..., g_m)$ a *configuration* of the random field and we denote it as $\dot{g}$. We will consider random fields with a Gibbs distribution [5].

We can define the *global energy* for a random field $\dot{G}$ in the following way:

$$\varepsilon(\dot{G}) \triangleq \sum_{i \sim j, i \leq j} (G_i - G_j)^2,$$

where $i \sim j$ means that the nodes i and j are neighbors in the graph.

The *local energy* of node i is defined as:

$$\varepsilon_i(G_i) \triangleq \sum_{j | i \sim j} (G_i - G_j)^2.$$

According to the Gibbs distribution, the probability that the random field $\dot{G}$ takes the configuration $\dot{g}$ is:

$$p(\dot{G} = \dot{g}) = \frac{e^{-\frac{\varepsilon(\dot{g})}{T}}}{\sum_{\dot{g}' \in |V|^m} e^{-\frac{\varepsilon(\dot{g}')}{T}}}, \tag{9}$$

where $T > 0$ is a parameter called the temperature of the Gibbs field.

The reason why it is interesting to look at this distribution follows from [5, Theorem 2.1]: *when a random field has distribution (9) then the probability that the node has particular value depends only on the values of its neighboring nodes and does not depend on the values of all other nodes.*

Let N_i be the set of neighbors of node i. Given a subset L of nodes, we let $\dot{G}_L$ denote the set of random variables of the nodes in L. Then the theorem can be formulated in the following way:

$$p(G_i = g_i | \dot{G}_{N_i} = \dot{g}_{N_i}) = p(G_i = g_i | \dot{G}_{\{1,2,...,m\}\backslash i} = \dot{g}_{\{1,2,...,m\}\backslash i}).$$

This property is called *Markov property* and it will capture the homophily effect: the value of a node is dependent on the values of the neighboring nodes. Moreover, for each node i, given the values of its neighbors, the probability distribution of its value is:

$$p(G_i = g_i) = \frac{e^{-\frac{\varepsilon_i(g_i)}{T}}}{\displaystyle\sum_{g' \in V} e^{-\frac{\varepsilon_i(g')}{T}}}.$$

The temperature parameter T plays a very important role to tune the homophily level (or the correlation level) in the network. Low temperature gives us network with highly correlated values. Increasing temperature we can add more and more "randomness" to the attributes.

In Fig. 3 we present the same random geometric graph with 200 nodes and radius 0.13, $RGG(200, 0.13)$ where the values $V = \{1, 2, ..., 5\}$ are chosen according to the Gibbs distribution and depicted with different colors. From the figure

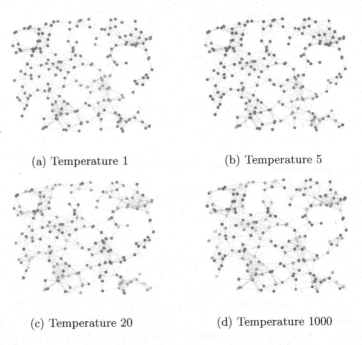

(a) Temperature 1   (b) Temperature 5

(c) Temperature 20   (d) Temperature 1000

Fig. 3. RGG(200, 0.13) with generated values for different temperature (Color figure online)

we can observe that the level of correlation between values of the node changes with different temperature. When temperature is 1 we can distinguish distinct clusters. When the temperature increases ($T = 5$ and $T = 20$), the values of neighbors are still similar but with more and more variability. When the temperature is very high then the values seem to be assigned independently.

5.1   Experimental Results

We performed simulations for two reasons: first, to verify the theoretical results; second, to see if subsampling gives improvement on the real datasets and on the synthetic ones.

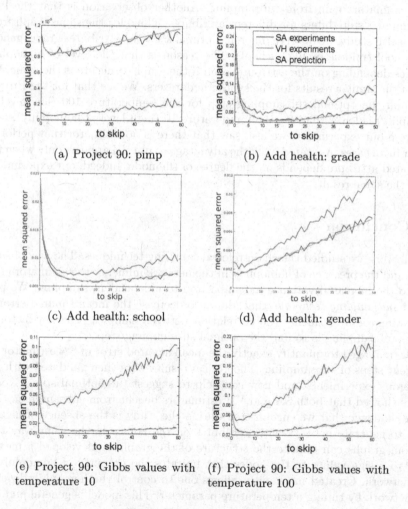

(a) Project 90: pimp

(b) Add health: grade

(c) Add health: school

(d) Add health: gender

(e) Project 90: Gibbs values with temperature 10

(f) Project 90: Gibbs values with temperature 100

Fig. 4. Experimental results

The simulations for a given dataset are performed in the following way. For the fixed budget B, rewards C_1 and C_2, we first collect the samples through the random walk on the graph for the subsampling step 1. We estimate the population average with the SA and VH estimators. Then we repeat this operation in order to have multiple estimates for the subsampling step 1, that we can count the mean squared error of the estimator. The same process is performed for different subsampling steps. In this way we can compare the mean squared error for different subsampling steps and choose the optimal one.

Figure 4 presents the experimental mean squared error of the SA and VH estimators and also the mean squared error of the SA obtained through Eqs. (6), (7) and (8) for different subsampling steps. From the figure we can observe that the experimental results are very close to the theoretical ones. We can notice that both estimators gain from subsampling. Another observation is that the best subsampling step differs for different attributes. Thus, for the same graph from Add health study, we observe different optimal k for the attributes grade, gender and school (middle or high school). The reason is that the level of homophily changes depending on the attribute, even if the graph structure is the same. We obtain the similar results for the synthetic datasets. We see that for the Project 90 graph the optimal subsampling step for the temperature 100 (low level of homoplily) is lower than for the temperature 10 (high level of homophily).

From our experiments we also saw that there is no estimator that performs better in all cases. As stated in [8] the advantage to use VH appears only when the estimated attribute depends on the degree of the node. Indeed, our experiments show the same result.

6   Conclusion

In this work we studied the chain-referral sampling techniques. The way of sampling and the presence of homophily in the network influence the estimator error due to the increased variance in comparison to independent sampling. We proposed *subsampling technique* that allows to decrease the mean squared error of the estimator by reducing the correlation between samples. The key-factor of successful sampling is to find the optimal subsampling step.

We managed to quantify exactly the mean squared error of SA estimator for different steps of subsampling. Theoretical results were then validated with the numerous experiments, and now can help to suggest the optimal step. Experiments showed that both SA and VH estimators benefit from subsampling.

A challenge that we encountered during the study is the absence of mechanism to generate network with attributes on the nodes. In the same way that random graphs can imitate the structure of the graph we developed a mechanism to assign values to the nodes that imitates the property of homophily in the network. Created mechanism allows one to control the homophily level in the network by tuning a temperature parameter. This model is general and can also be applied in other tests.

Acknowledgements. This work was supported by CEFIPRA grant no. 5100-IT1 "Monte Carlo and Learning Schemes for Network Analytics," Inria Nokia Bell Labs ADR "Network Science," and Inria Brazilian-French research team Thanes.

References

1. Freeman, L.C.: Research Professor, Department of Sociology and Institute for Mathematical Behavioral Sciences School of Social Sciences, University of California, Irvine. http://moreno.ss.uci.edu/data.html. Accessed 01 July 2015
2. The National Longitudinal Study of Adolescent to Adult Health. http://www.cpc.unc.edu/projects/addhealth. Accessed 01 July 2015
3. The Office of Population Research at Princeton University. https://opr.princeton.edu/archive/p90/. Accessed 01 July 2015
4. Stanford Large Network Dataset Collection. https://snap.stanford.edu/data/ Accessed 01 July 2015
5. Brémaud, P.: Markov Chains: Gibbs Fields, Monte Carlo Simulation, and Queues, vol. 31. Springer Science & Business Media, Berlin (2013)
6. Nicholas, A.: Christakis and James H Fowler.: The spread of obesity in a large social network over 32 years. New Engl. J. Med. **357**(4), 370–379 (2007)
7. Gile, K.J., Handcock, M.S.: Respondent-driven sampling: an assessment of current methodology. Sociol. Methodol. **40**(1), 285–327 (2010)
8. Goel, S., Salganik, M.J.: Assessing respondent-driven sampling. Proc. Natl. Acad. Sci. **107**(15), 6743–6747 (2010)
9. Heckathorn, D.D., Jeffri, J.: Jazz networks: using respondent-driven sampling to study stratification in two jazz musician communities. In: Unpublished Paper Presented at American Sociological Association Annual Meeting (2003)
10. Jeon, K.C., Goodson, P.: US adolescents' friendship networks and health risk behaviors: a systematic review of studies using social network analysis and Add Health data. PeerJ **3**, e1052 (2015)
11. Musyoki, H., Kellogg, T.A., Geibel, S., Muraguri, N., Okal, J., Tun, W., Raymond, H.F., Dadabhai, S., Sheehy, M., Kim, A.A.: Prevalence of HIV, sexually transmitted infections, and risk behaviours among female sex workers in Nairobi, Kenya: results of a respondent driven sampling study. AIDS Behav. **19**(1), 46–58 (2015)
12. Ramirez-Valles, J., Heckathorn, D.D., Vázquez, R., Diaz, R.M., Campbell, R.T.: From networks to populations: the development and application of respondent-driven sampling among IDUs and Latino gay men. AIDS Behav. **9**(4), 387–402 (2005)
13. Volz, E., Heckathorn, D.D.: Probability based estimation theory for respondent driven sampling. J. Off. Stat. **24**(1), 79 (2008)

System Occupancy of a Two-Class Batch-Service Queue with Class-Dependent Variable Server Capacity

Jens Baetens[1]([⊠]), Bart Steyaert[1], Dieter Claeys[1,2], and Herwig Bruneel[1]

[1] SMACS Research Group,
Department of Telecommunications and Information Processing,
Ghent University, Ghent, Belgium
jens.baetens@telin.ugent.be
[2] Department of Industrial Systems Engineering and Product Design,
Ghent University, Zwijnaarde, Belgium

Abstract. Due to their wide area of applications, queueing models with batch service, where the server can process several customers simultaneously, have been studied frequently. An important characteristic of such batch-service systems is the size of a batch, that is the number of customers that are processed simultaneously. In this paper, we analyse a two-class batch-service queueing model with variable server capacity, where all customers are accommodated in a common first-come-first served single-server queue. The server can only process customers that belong to the same class, so that the size of a batch is determined by the number of consecutive same-class customers. After establishing the system equations that govern the system behaviour, we deduce an expression for the steady-state probability generating function of the system occupancy at random slot boundaries. Also, some numerical examples are given that provide further insight in the impact of the different parameters on the system performance.

Keywords: Discrete time · Batch service · Two classes · Variable server capacity · Queueing

1 Introduction

In telecommunication applications, a single server can often process multiple customers (i.e. data packets) simultaneously in a single batch. An important characteristic of such batch-service systems is the maximum size of a batch, that is the maximum number of customers processed simultaneously. In many batch-service systems this number is assumed to be a constant [1–5]. However, in practice, the maximum batch size or capacity of the server can be variable and stochastic, a feature that has been incorporated in only a few papers. Chaudhry and Chang analysed the system content at various epochs in the $Geo/G^Y/1/N + B$ model in discrete time, where Y denotes the stochastic capacity of the server,

© Springer International Publishing Switzerland 2016
S. Wittevrongel and T. Phung-Duc (Eds.): ASMTA 2016, LNCS 9845, pp. 32–44, 2016.
DOI: 10.1007/978-3-319-43904-4_3

which is upper-bounded by B, and N is the maximum queue capacity [6]. Furthermore, Pradhan et al. obtained closed-form expressions for the queue length distribution at departure epochs for the discrete-time $M/G_r^Y/1$ queue where the service process depends on the batch size [7]. A similar feature in the models from Chaudhry and Chang, and Pradhan et al. is that the capacity of a batch is independent of the queue length and of the capacities of the previous batches. On the contrary, Germs and van Foreest have recently developed an algorithmic method for the performance evaluation of the continuous-time $M(n)^{X(n)}/G(n)^{Y(n)}/1/K + B$ queue [8]. In that model, both the arrival rate and service process (the service times as well as the capacities) depend on the queue size.

Another feature of the above models is that customers are indistinguishable, i.e., they all are of the same type. Although in many types of queueing systems several customer classes are included to account for customer differentiation, only a few papers on batch service consider multiple customer classes. Reddy et al. study a multi-class batch-service queueing system with Poisson arrivals and a priority scheduling discipline, in the context of an industrial repair shop where the most critical machines are repaired first [9]. Boxma et al. study a polling system with Poisson arrivals and batch service [10]. In this case, each customer class has a dedicated queue and the server visits the different queues in a cyclic manner. Boxma et al. focus on the influence of a number of different gating policies on the performance. Dorsman et al. study a polling system with a renewal arrival process and batch service, where the batches are created by accumulation stations before they are added to a queue [11]. Such a system can be used when a single server processes multiple product types with batching constraints. Dorsman et al. focus on optimizing the batch sizes of each class.

In this paper, we analyse a two-class discrete-time batch-service queueing model, with a variable service capacity that depends on the queue size and on the specific classes of the successive customers. To the best of our knowledge, the combination of batch service with variable capacity and multiple customer classes has not appeared in the literature before. Whereas in the mentioned papers about priority queueing and polling systems the customers of different classes are accomodated in different queues, the customers of all classes are accommodated in a common queue here. When the server becomes available, it will simultaneously process the customer at the head of the queue, and all successive customers that are of the same class as the head customer. This, for instance, means that if the first customer is of class A, all of the following class A customers are also grouped in the batch that will be taken into service, until the next customer is of class B. Applications of this server can be found in manufacturing environments or telecommunication systems, where customers with the same system parameter, such as the required temperature or the destination of the customer, can be processed simultaneously on a FCFS-basis.

The paper is structured as follows. In Sect. 2 we describe the discrete-time two-class queueing model with batch service in detail. This system consists of a single First-Come-First-Served (FCFS) queue of infinite size, and a single batch server with a variable capacity. In Sect. 3 we establish the system equations, from which we deduce the stability condition, and derive a closed-form expression for the steady-state probability generating function (pgf) of the system occupancy at random slot boundaries. Next, using the expressions obtained in Sect. 3, we evaluate the behaviour of the system through some numerical examples in Sect. 4. Our conclusions are presented in Sect. 5.

2 Model Description

Let us consider a discrete-time two-class queueing system with infinite queue size, and a batch server whose capacity is stochastic. The classes of the customers are denominated as A and B. Arriving customers are inserted at the tail of the queue. When the server is or becomes available and finds a non-empty queue, a new service is initiated. The size of the batch is then determined by the number of consecutive customers at the front of the system that are of the same class. More specifically, the server starts serving a batch of n customers if and only if one of the following two cases occurs:

- Exactly n customers are present and they are all of the same class.
- More than n customers are present, the n customers at the front of the queue are of the same class and the $(n + 1)$-th customer is of the other class.

We define the class of a batch as the class of the customers within it.

The aggregated numbers of customer arrivals in consecutive slots are modelled as a sequence of independent and identically distributed (i.i.d.) random variables, with common probability mass function (pmf) $e(n)$ and pgf $E(z)$. The mean aggregated number of customer arrivals per slot is denoted as λ. A random customer is of class A with probability (w.p.) σ and of class B w. p. $1 - \sigma$ regardless of the classes of other customers. The service time of a batch is always a single slot, independently of both the class of the batch and its size.

3 Analysis

In this section, we first determine the system equations that capture the system behaviour. Then we analyse the conditions for stability, and we establish the steady-state pgf of the system occupancy, that is the number of customers in the system at the beginning of a slot, including those in the batch that will be served during this slot (if any).

3.1 System Equations

In this subsection, we give the system equations that capture the behaviour of the system at successive slot boundaries. The number of customers in the system or the system occupancy at random slot boundaries is denoted by u_k. We also define the random variables $u_{I,k}$, $u_{A,k}$ and $u_{B,k}$ as the system occupancy at the boundary of slot k when the server respectively is idle or processes a class A or B batch.

If the server is idle during slot k, then the next slot is also an idle slot if there are no new arrivals. On the other hand, when there is at least one arrival, then the server will process a class A or B batch based on the class of the first arrival. This leads to the system equations if $u_{I,k} = 0$:

$$u_{I,k+1} = 0, \text{ if } e_k = 0$$
$$u_{A,k+1} = e_k, \text{ if } e_k > 0 \ \& \ \text{first arrival of class A, w.p. } \sigma$$
$$u_{B,k+1} = e_k, \text{ if } e_k > 0 \ \& \ \text{first arrival of class B, w.p. } 1 - \sigma, \qquad (1)$$

where e_k is the number of customers that arrive during slot k, with pmf $e(n)$ and pgf $E(z)$.

On the other hand, if a class A batch is processed during a random slot k and there are $u_{A,k}$ customers in the system, then the system equations also depend on the size of the processed batch. In the first case, all waiting customers are served meaning that the batch size c_k is equal to $u_{A,k}$. This leads to a similar behaviour as for an idle slot. If the size of the batch is less than $u_{A,k}$ meaning that not all customers are processed simultaneously, then the customer at the head of the queue must be of the opposite class, implying that a class B batch is always processed during slot $k + 1$. Summarized, we have

$$u_{I,k+1} = 0, \text{ if } e_k = 0 \ \& \ c_k = u_{A,k}$$
$$u_{A,k+1} = e_k, \text{ if } e_k > 0 \ \& \ c_k = u_{A,k} \ \& \ \text{first arrival of class A (w.p. } \sigma)$$
$$u_{B,k+1} = e_k, \text{ if } e_k > 0 \ \& \ c_k = u_{A,k} \ \& \ \text{first arrival of class B (w.p. } 1 - \sigma)$$
$$u_{B,k+1} = u_{A,k} - c_k + e_k, \text{ if } c_k < u_{A,k}, \qquad (2)$$

where $c_k > 0$ is the size of the batch being processed during slot k.

The case that a class B batch was processed during a random slot k leads to the counterpart system equations

$$u_{I,k+1} = 0, \text{ if } e_k = 0 \ \& \ c_k = u_{B,k}$$
$$u_{A,k+1} = e_k, \text{ if } e_k > 0 \ \& \ c_k = u_{B,k} \ \& \ \text{first arrival of class A (w.p. } \sigma)$$
$$u_{B,k+1} = e_k, \text{ if } e_k > 0 \ \& \ c_k = u_{B,k} \ \& \ \text{first arrival of class B (w.p. } 1 - \sigma)$$
$$u_{A,k+1} = u_{B,k} - c_k + e_k, \text{ if } c_k < u_{B,k}. \qquad (3)$$

3.2 Stability Condition

In order to find the stability condition, we analyse the system under the condition that the queue is saturated. In such a system, the batch server is never idle and the size of the processed batches is not limited by a lack of customers and therefore geometrically distributed. Because the server processes all same-class customers at the head of the queue, the server will alternate between processing class A and B batches, which means we can limit ourselves to considering 2 consecutive slots. The system is stable when the mean number of customer arrivals during two consecutive slots, which is equal to 2λ, is less than the mean number of customers processed during the same slots. The mean number of processed customers during two consecutive slots is the sum of the mean batch size of a class A and B batch. The batch size follows a geometric distribution with parameter σ (class A) or $1 - \sigma$ (class B) respectively. The stability condition is then given by

$$2\lambda < \frac{1}{1 - \sigma} + \frac{1}{\sigma}.$$

If σ is either 0 or 1, then the stability condition is reduced to $\lambda < \infty$, i.e., the system is always stable. This is as expected, since in this case all customers are of the same class, which means that no matter how many customers arrive, the server will always aggregate all waiting customers in a single batch. Also, if σ is equal to 0.5 then the maximum tolerable arrival rate reaches a minimum value.

We can also define the load ρ of the system as the fraction of the average number of arrivals versus the maximum allowed arrival rate, which leads to

$$\rho := \frac{2\lambda}{\frac{1}{1-\sigma} + \frac{1}{\sigma}} = 2\lambda\sigma(1 - \sigma) < 1. \tag{4}$$

3.3   System Occupancy

Assuming the stability condition is met, we can define the pmf of u_k, the system occupancy at random slot boundaries, as

$$u(i) := \lim_{k\to\infty} \Pr[u_k = i],$$

with corresponding pgf

$$U(z) := \sum_{i=0}^{\infty} u(i)z^i.$$

We can split the generating function of the system occupancy $U(z)$ in three parts based on the state of the server (idle, processing a class A batch or class B batch). This leads to

$$U(z) = u_I + U_A(z) + U_B(z), \tag{5}$$

where we introduced the following definitions

$$u_I := \lim_{k\to\infty} \Pr[u_{I,k} = 0],$$

$$U_A(z) := \sum_{i=1}^{\infty} \lim_{k\to\infty} \Pr[u_{A,k} = i]z^i,$$

$$U_B(z) := \sum_{i=1}^{\infty} \lim_{k\to\infty} \Pr[u_{B,k} = i]z^i.$$

The first term in the right-hand side of Eq. 5 corresponds to the probability that the server is idle during a random slot. This probability u_I is found by invoking the system equations in Sect. 3.1.

$$u_I = u_I E(0) + \lim_{k \to \infty} \sum_{i=1}^{\infty} \Pr[u_{A,k} = i] E(0) \sigma^{i-1}$$

$$+ \lim_{k \to \infty} \sum_{i=1}^{\infty} \Pr[u_{B,k} = i] E(0)(1 - \sigma)^{i-1}$$

$$= u_I E(0) + E(0) \frac{U_A(\sigma)}{\sigma} + E(0) \frac{U_B(1 - \sigma)}{1 - \sigma},$$

leading to

$$u_I = \frac{E(0)}{(1 - E(0))} \left(\frac{U_A(\sigma)}{\sigma} + \frac{U_B(1 - \sigma)}{1 - \sigma} \right). \tag{6}$$

Based on the state of the server during the previous slot, we can split the second term of Eq. 5 as

$$U_A(z) = E[z^{u_{A,k+1}}] = E[z^{u_{A,k+1}} I_{\{u_{I,k}=0\}}] + E[z^{u_{A,k+1}} I_{\{u_{A,k}>0\}}]$$

$$+ E[z^{u_{A,k+1}} I_{\{u_{B,k}>0\}}], \tag{7}$$

where $I_{\{C\}}$ are indicator functions which are equal to 1 if event C occurs and zero otherwise. The first part of this equation gives the partial generating function in case of the server being idle in the previous slot. Using Eq. 1 we can write this function as

$$E[z^{u_{A,k+1}} I_{\{u_{I,k}=0\}}] = \sigma E[z^{e_k} I_{\{u_{I,k}=0, e_k>0\}}] = \sigma u_I (E(z) - E(0)). \tag{8}$$

Analogously we can write the second part, invoking the system equations in Eq. 2, as

$$E[z^{u_{A,k+1}} I_{\{u_{A,k}>0\}}] = \sigma E[z^{e_k} I_{\{u_{A,k}>0, c_k=u_{A,k}, e_k>0\}}]$$

$$= (E(z) - E(0)) \lim_{k \to \infty} \sigma \sum_{i=1}^{\infty} \sigma^{i-1} \Pr[u_{A,k} = i]$$

$$= (E(z) - E(0)) U_A(\sigma). \tag{9}$$

The last part of $U_A(z)$ corresponds to the case that a class B batch is processed during the previous slot. Using Eq. 3, we obtain the following equation

$$E[z^{u_{A,k+1}} I_{\{u_{B,k}>0\}}]$$

$$= \sigma E[z^{e_k} I_{\{u_{B,k}>0, c_k=u_{B,k}, e_k>0\}}] + E[z^{u_{B,k}-c_k+e_k} I_{\{u_{B,k}>1, c_k<u_{B,k}\}}].$$

If the number of customers in the system is equal to i, then the probability that the size of the batch is equal to i is given by $(1 - \sigma)^{i-1}$ since we know the class of the first customer and the next $i - 1$ customers must be of class B. On the other hand, the probability that the size of the served batch is equal to $j < i$ is given by $\sigma(1 - \sigma)^{j-1}$ because the $(j + 1)$-th customer must be of class A. This leads to

$$E[z^{u_{A,k+1}} I_{\{u_{B,k}>0\}}]$$

$$= \sigma(E(z) - E(0)) \sum_{i=1}^{\infty} (1 - \sigma)^{i-1} \lim_{k \to \infty} \Pr[u_{B,k} = i]$$

$$+ \sigma E(z) \sum_{i=2}^{\infty} \sum_{j=1}^{i-1} z^{i-j} (1 - \sigma)^{j-1} \lim_{k \to \infty} \Pr[u_{B,k} = i]$$

$$= \sigma(E(z) - E(0)) \frac{U_B(1 - \sigma)}{1 - \sigma}$$

$$+ \sigma E(z) \sum_{i=2}^{\infty} \lim_{k \to \infty} \Pr[u_{B,k} = i] \frac{(1 - \sigma)z^i - z(1 - \sigma)^i}{(1 - \sigma)(z - (1 - \sigma))}$$

$$= \sigma(E(z) - E(0)) \frac{U_B(1 - \sigma)}{1 - \sigma}$$

$$+ \frac{\sigma E(z)}{(1 - \sigma)(z - (1 - \sigma))} \left((1 - \sigma)U_B(z) - z U_B(1 - \sigma) \right). \tag{10}$$

By combining Eqs. 8, 9 and 10, we obtain the partial pgf of the system occupancy in a slot where a class A batch is processed.

$$U_A(z) = \sigma(E(z) - E(0)) \left(u_I + \frac{U_A(\sigma)}{\sigma} + \frac{U_B(1 - \sigma)}{1 - \sigma} \right)$$

$$+ \frac{\sigma E(z)}{(1 - \sigma)(z - (1 - \sigma))} \left((1 - \sigma)U_B(z) - z U_B(1 - \sigma) \right). \tag{11}$$

An analogous analysis leads to the partial generating function $U_B(z)$.

$$U_B(z) = (1 - \sigma)(E(z) - E(0)) \left((u_I + \frac{U_A(\sigma)}{\sigma} + \frac{U_B(1 - \sigma)}{1 - \sigma} \right)$$

$$+ \frac{(1 - \sigma)E(z)}{\sigma(z - \sigma)} \left(\sigma U_A(z) - z U_A(\sigma) \right). \tag{12}$$

By substituting Eq. 12 in Eq. 11 we obtain for $U_A(z)$

$$U_A(z) = \sigma(E(z) - E(0)) \left(u_I + \frac{U_A(\sigma)}{\sigma} + \frac{U_B(1 - \sigma)}{1 - \sigma} \right) - \frac{\sigma z E(z) U_B(1 - \sigma)}{(1 - \sigma)(z - (1 - \sigma))}$$

$$+ \frac{\sigma E(z)}{z - (1 - \sigma)} \left((1 - \sigma)(E(z) - E(0)) \left(u_I + \frac{U_A(\sigma)}{\sigma} + \frac{U_B(1 - \sigma)}{1 - \sigma} \right) \right.$$

$$\left. + \frac{(1 - \sigma)E(z)}{z - \sigma} U_A(z) - \frac{(1 - \sigma)z E(z) U_A(\sigma)}{\sigma(z - \sigma)} \right).$$

We now multiply by $(z - \sigma)(z - 1 + \sigma)$ and put all terms that contain $U_A(z)$ in the left-hand side of the equation. Also using Eq. 6 to substitute u_I leads to

$$U_A(z)\Big((z - \sigma)(z - (1 - \sigma)) - \sigma(1 - \sigma)E(z)^2\Big)$$

$$= \sigma(z - \sigma)\Big(z - 1 + \sigma + (1 - \sigma)E(z)\Big)\frac{E(z) - E(0)}{1 - E(0)}\left(\frac{U_A(\sigma)}{\sigma} + \frac{U_B(1 - \sigma)}{1 - \sigma}\right)$$

$$- \frac{\sigma(z - \sigma)zE(z)U_B(1 - \sigma)}{1 - \sigma} - (1 - \sigma)zE(z)^2 U_A(\sigma). \tag{13}$$

The analogous expression for class B then satisfies

$$U_B(z)\Big((z - \sigma)(z - (1 - \sigma)) - \sigma(1 - \sigma)E(z)^2\Big)$$

$$= (1 - \sigma)(z - 1 + \sigma)\Big(z - \sigma + \sigma E(z)\Big)\frac{E(z) - E(0)}{1 - E(0)}\left(\frac{U_A(\sigma)}{\sigma} + \frac{U_B(1 - \sigma)}{1 - \sigma}\right)$$

$$- \frac{(1 - \sigma)(z - 1 + \sigma)zE(z)U_A(\sigma)}{\sigma} - \sigma z E(z)^2 U_B(1 - \sigma). \tag{14}$$

The sum of Eqs. 6, 13 and 14 lead to the pgf of the system occupancy at random slot boundaries. This generating function is equal to

$$U(z)\Big((z - \sigma)(z - (1 - \sigma)) - \sigma(1 - \sigma)E(z)^2\Big)$$

$$= u_I\Big((z - \sigma)(z - (1 - \sigma)) - \sigma(1 - \sigma)E(z)^2\Big) + \frac{E(z) - E(0)}{1 - E(0)}$$

$$\cdot \Big((z - \sigma)(z - 1 + \sigma) + \sigma(1 - \sigma)(2z - 1)E(z)\Big)\left(\frac{U_A(\sigma)}{\sigma} + \frac{U_B(1 - \sigma)}{1 - \sigma}\right)$$

$$- (1 - \sigma)zE(z)U_A(\sigma)\Big(E(z) + \frac{z - 1 + \sigma}{\sigma}\Big)$$

$$- \sigma z E(z)U_B(1 - \sigma)\Big(E(z) + \frac{z - \sigma}{1 - \sigma}\Big). \tag{15}$$

The two remaining unknowns $U_A(\sigma)$ and $U_B(1 - \sigma)$ in the pgf $U(z)$ are yet to be determined. With the theorem of Rouché, we can easily prove that the denominator of $U(z)$ has two zeros inside or on the unit circle. Each zero of the denominator must also be a zero of the numerator since generating functions are analytical functions inside the complex unit disk and bounded for $|z| = 1$. In Eq. 15 we can easily see that $z = 1$ is a zero of the denominator. The other zero can be calculated numerically. The equations provided by the zeros constitute a set of two linear equations for two unknowns. We also note that $z = 1$ leads to the normalisation condition. By evaluating Eq. 15 at $z = 1$ and applying l'Hôpital's rule we obtain

$$1 = u_I + U_A(\sigma)\frac{1 + \frac{2(1-\sigma)\lambda E(0)}{1 - E(0)}}{1 - 2\sigma(1 - \sigma)\lambda} + U_B(1 - \sigma)\frac{1 + \frac{2\sigma\lambda E(0)}{1 - E(0)}}{1 - 2\sigma(1 - \sigma)\lambda}.$$

4   Numerical Results

In this section, we illustrate the results obtained in the previous section through numerical examples. In Figs. 1, 2 and 3, we consider a geometric arrival process with mean arrival rate λ. The pgf $E(z)$ is equal to

$$E(z) = \frac{1}{1 + \lambda(1 - z)}.$$

The influence of the parameter σ on the average system occupancy $E[U]$ is shown in Fig. 1 as function of both the mean arrival rate and the load of the system. In Fig. 1a we note that smaller values of σ lead to a significant improvement of the performance of the system. This is caused by the inverse proportionality of the parameter σ to the size of the batches being processed. On the other hand, we observe in Fig. 1b that the mean system occupancy is larger for smaller values of σ when the server is operating under the same load, as defined in Eq. 4. This is the result of two conflicting effects. A smaller value of σ leads to a higher average number of customers that the server can process due to larger sequences of same-class customers, but also to a higher arrival rate to obtain the same load in the system. In Fig. 1b it is clear that the influence of the increased arrival rate is most significant, partially because the number of customers in the server are also part of the system occupancy and partially because there must be more customers waiting in the queue to create the larger batches that can be processed.

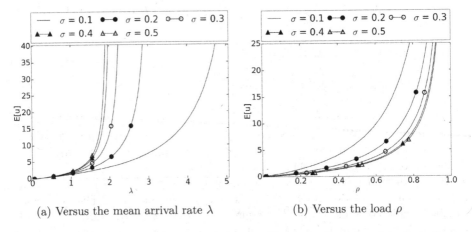

(a) Versus the mean arrival rate λ (b) Versus the load ρ

Fig. 1. Influence of σ on the average system occupancy as a function of the arrival rate λ (a) and the load ρ (b).

A more detailed analysis of the influence of σ on the mean system occupancy when the server is operating under a certain load is shown in Fig. 2. We first observe that σ is symmetric around 0.5, which means that a probability of a class A customer being equal to σ or $1 - \sigma$ will lead to the same value for the mean system occupancy. If we look at $0.2 < \sigma < 0.8$ we see that the influence of σ is only significant for larger loads. This is because the performance of the server is limited by a lack of customers at lower loads for these values of σ. For σ closer to 0 or 1 we see that the average system occupancy is increases drastically, even for small loads. This is because the server can process larger batches and the arrival rate must increase to obtain the same load.

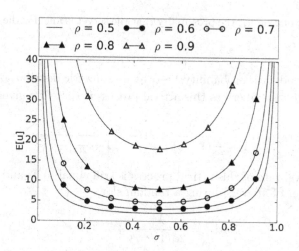

Fig. 2. Influence of σ on the mean system occupancy for a number of different loads

Another important characteristic of the system is the probability that a server is idle during a random slot. This probability is calculated according to Eq. 6 and depends on the probability that there are no arrivals during a slot and the probability that the server processes all customers during the same slot. In Fig. 3 we show this characteristic in terms of σ for a number of different loads. We observe that when σ approximates 0 or 1, that the server is almost never idle regardless of the load of the system. This occurs because the maximum allowed arrival rate is very large so that even small loads lead to a large mean arrival rate. A large mean arrival rate means that the probability that there are no arrivals is very small so that the server will almost always be able to start a service. We also observe that for σ closer to 0.5, the probability u_I is not very sensitive for variations of σ. This is caused by a conflict between the probability that there are no arrivals and the probability that all customers are processed. Values of σ closer to 0.5 means that the probability that there are no arrivals increases but the probability that all customers are processed decreases.

In a last example we examine the influence of the variance in the arrival process. Therefore, we consider an arrival process where the number of arrivals in an arbitrary slot is with probability α determined by a geometric distribution

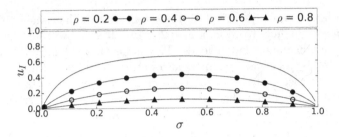

Fig. 3. Influence of σ on u_I, the probability that the server is idle, for the system under different loads

with mean $\frac{\lambda}{2\alpha}$ and with probability $1-\alpha$ by a geometric distribution with mean $\frac{\lambda}{2(1-\alpha)}$. The pgf that describes this arrival process is therefore given by

$$E(z) = \alpha \frac{1}{1 + \frac{\lambda}{2\alpha}(1-z)} + (1-\alpha)\frac{1}{1 + \frac{\lambda}{2(1-\alpha)}(1-z)}.$$

The mean arrival rate of this arrival process is still equal to λ, and the variance of e, the number of arrivals during a random slot,

$$\text{Var}[e] = \frac{\lambda^2}{2\alpha(1-\alpha)} + \lambda - \lambda^2.$$

This equation indicates that the variance is minimal for $\alpha = 0.5$, and approaches infinity for α close to 0 or 1. In Fig. 4, we plot the mean system occupancy as a function of α for values of σ as indicated and with a load of $\rho = 0.9$. We clearly observe the detrimental effect an increasing variance has on the mean system occupancy.

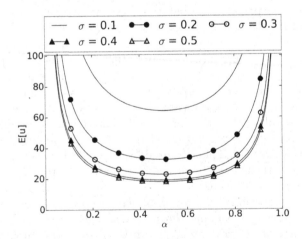

Fig. 4. Influence of α, which determines the variance in the arrival process, on the system occupancy for a number of values of σ and with a load of 0.9

5   Conclusions

In this paper we have analysed a discrete-time two-class single-server queueing system with batch service. The size of the batches that are processed are determined by the number of customers in the queue and their respective classes. We have derived the steady-state pgf of the number of customers in the system at random slot boundaries. Using these results, we have demonstrated the impact of the various parameters of the arrival process on the average system occupancy and the probability that the server is idle.

There are a number of possible extensions that could be considered for this model. A first extension would be to find the probability generating function for the number of customers that are being processed by the batch server. In a second extension we could extend the model to use a class-dependent general service time distribution for class A and B batches. A further extension we mention is that we could introduce bursty behaviour of same class customers by introducing correlation between the class of two consecutive customers. This can for instance be done by using a general 2-state Markov process to assign customer classes. This allows us to tweak the length of class A or B customers that arrive while maintaining a certain ratio of class A and B customers.

Acknowledgment. Dieter Claeys is a Postdoctoral Fellow with the Research Foundation Flanders (FWO-Vlaanderen), Belgium. Part of the research has been funded by the Interuniversity Attraction Poles Programma initiated by the Belgian Science Policy Office.

References

1. Banerjee, A., Gupta, U.C.: Reducing congestion in bulk-service finite-buffer queueing system using batch-size-dependent service. Perform. Eval. **69**(1), 53–70 (2012)
2. Claeys, D., Steyaert, B., Walraevens, J., Laevens, K., Bruneel, H.: Tail distribution of the delay in a general batch-service queueing model. Comput. Oper. Res. **39**, 2733–2741 (2012)
3. Claeys, D., Steyaert, B., Walraevens, J., Laevens, K., Bruneel, H.: Analysis of a versatile batch-service queueing model with correlation in the arrival process. Perform. Eval. **70**(4), 300–316 (2013)
4. Goswami, V., Mohanty, J.R., Samanta, S.K.: Discrete-time bulk-service queues with accessible and non-accessible batches. Appl. Math. Comput. **182**, 898–906 (2006)
5. Weng, W.W., Leachman, R.C.: An improved methodology for real-time production decisions at batch-process work stations. Trans. Semicond. Manufact. **6**(3), 219–225 (1993)
6. Chaudhry, M.L., Chang, S.H.: Analysis of the discrete-time bulk-service queue $Geo/G^Y/1/N+B$. Oper. Res. Lett. **32**(4), 355–363 (2004)
7. Pradhan, S., Gupta, U.C., Samanta, S.K.: Queue-length distribution of a batch service queue with random capacity and batch size dependent service: $M/G_r^Y/1$. OPSEARCH **53**, 329–343 (2016)

8. Germs, R., Van Foreest, N.D.: Analysis of finite-buffer state-dependent bulk queues. OR Spectr. **35**(3), 563–583 (2013)
9. Reddy, G.V.K., Nadarajan, R., Kandasamy, P.R.: A nonpreemptive priority multi-server queueing system with general bulk service and heterogeneous arrivals. Comput. Oper. Res. **20**(4), 447–453 (1993)
10. Boxma, O.J., van der Wal, J., Yechiali, U.: Polling with batch service. Stochast. Models **24**(4), 604–625 (2008)
11. Dorsman, J.L., Van der Mei, R.D., Winands, E.M.M.: Polling with batch service. OR Spectr. **34**, 743–761 (2012)

Applying Reversibility Theory for the Performance Evaluation of Reversible Computations

Simonetta Balsamo, Filippo Cavallin, Andrea Marin$^{(\boxtimes)}$, and Sabina Rossi

DAIS, Università Ca' Foscari Venezia, Via Torino, 155, Venice, Italy
{balsamo,filippo.cavallin,marin,rossisab}@unive.it

Abstract. Reversible computations have been widely studied from the functional point of view and energy consumption. In the literature, several authors have proposed various formalisms (mainly based on process algebras) for assessing the correctness or the equivalence among reversible computations. In this paper we propose the adoption of Markovian stochastic models to assess the quantitative properties of reversible computations. Under some conditions, we show that the notion of time-reversibility for Markov chains can be used to efficiently derive some performance measures of reversible computations. The importance of time-reversibly relies on the fact that, in general, the process's stationary distribution can be derived efficiently by using numerically stable algorithms. This paper reviews the main results about time-reversible Markov processes and discusses how to apply them to tackle the problem of the quantitative evaluation of reversible computations.

1 Introduction

Reversible computations have two execution directions: forward, corresponding to the usual notion of computation, and backward that restores previous states of the execution. Various applications and problems related to reversible computations have been widely studied in different research areas and from different viewpoints, including functional analysis and energy consumption (see, e.g., [17,22] and the references therein). Various formalisms and models have been proposed in the literature to represent and assess qualitative properties of reversible computations such as their correctness or if two reversible processes are equivalent in some terms. Most of the proposed approaches are based on process algebras that do not include any notion of computation time [7,17]. We focus on the quantitative analysis and evaluation of reversible computations based on Markov stochastic processes. The dynamic behaviour of the forward and backward computation may be represented by stochastic models that include the notion of time. Hence, under certain conditions, time-reversibility of stochastic processes can be applied to assess quantitative properties of reversible computations.

Quantitative models based on Markov processes have been widely applied for the analysis and evaluation of complex systems (see e.g., [5,8]). Markov models

© Springer International Publishing Switzerland 2016
S. Wittevrongel and T. Phung-Duc (Eds.): ASMTA 2016, LNCS 9845, pp. 45–59, 2016.
DOI: 10.1007/978-3-319-43904-4_4

and formalisms have the advantage of efficient methods and algorithms for studying their behaviour. In particular, under appropriate stationary conditions one can derive the equilibrium state distribution of a continuous-time Markov chains by applying algorithms with polynomial time complexity in the process state space cardinality [25]. Several higher level formalisms that are widely applied for quantitative analysis are based on Markov processes, including Stochastic Process Algebras (SPA), Stochastic Petri Nets (SPN), Stochastic Automata Networks (SAN) and Queueing Networks (QN). Although the quantitative analysis based on these formalisms can be obtained by the direct solution of the underlying Markov chain, the state space dimension of the process in general grows exponentially with the model dimension. This is known as the state-space explosion problem and becomes intractable from the computational viewpoint as the problem size increases. In order to overcome this problem, various techniques have been proposed in the literature, including the state-space reduction by aggregating (or lumping) methods, approximation techniques, and the identification of product-form solutions for state probabilities of the Markov chain. The product-form theory provides techniques to derive the equilibrium state distribution of a complex model based on the analysis of its components in isolation. Product-form models consist of a set of interacting sub-models whose solutions are obtained by isolating them from the rest of the systems. Then, the stationary state distribution of the entire model is computed as the (normalised) product of the stationary state distributions of the sub-models. Various classes of product-form models have been defined for different formalisms and some of them can be analysed through efficient algorithms with a low polynomial complexity in the model dimension. Product-form has been widely investigated for queueing network models [4, 14]. These product-form models have simple closed-form expressions of the stationary state distributions that lead to efficient solution algorithms. For more general Markov models and by the compositionality property of Stochastic Process Algebra, the Reversed Compound Agent Theorem (RCAT) [2, 11] provides a product-form solution of a stationary CTMC defined as a cooperation between two sub-processes under certain conditions. This result gives a unified view of most of the commonly used product-forms.

The concept of time-reversibility of Markov stochastic processes has been introduced and applied to the analysis of Markov processes and stochastic networks by Kelly [16]. A reversible Markov process has the property that when the process obtained by reversing the direction of time is reversed has the same probabilistic behaviour of the original one. Early applications of these results lead to the characterisation of product-form solutions for some models with underlying time-reversible Markov process, such as closed exponential Queueing Networks [4, 9]. Also the RCAT characterisation of product-form solutions is connected to time-reversibility: the solution is based on the definition of a set of transition rates in the time-reversed process. Further notions of reversibility have been introduced in [16, 26] for dynamically reversible processes where some states of the direct and reversed processes are interchanged, and more recently the ρ-reversibility for reversible processes with arbitrary state renaming [18, 19].

Some results on properties and product-form solutions have been recently derived for this class of time-reversibility [20].

In this paper we survey the main results about time-reversible Markov processes and discuss how to apply them to address the problem of quantitative evaluation of reversible computations. We recall the definition of time reversibility for continuous time Markov processes, the main properties and its application for quantitative analysis. We present an abstract model of continuous time Markov chain for representing and performance evaluating reversible parallel computations. Taking advantage of the process reversibility, the stationary distribution of the model can be efficiently derived by using numerically stable algorithms. In particular we present some product-form results of reversible synchronising automata by applying ρ-reversibility to the underlying Markov process.

The paper is organised as follows. Section 2 introduces the notation for Markov processes and presents the time-reversibility definitions and criteria. The application of ρ-reversible Markov process to model reversible computations is presented in Sect. 3, where we discuss the modelling assumptions and applications of the quantitative analysis. Section 4 presents an abstract model based on continuous time Markov chains and Stochastic Automata for synchronising parallel reversible computations. We discuss the application of ρ-reversibility and the derivation of product-form solution of ρ-reversible synchronised automata that represent reversible computations, and an application example.

2   Theoretical Background

Let $X(t)$ with $t \in \mathbb{R}$ be a Continuous Time Markov Chain (CTMC) with state space $\mathcal{S}$. Then, assuming that the process is irreducible, an *invariant measure* of the CTMC is a collection of positive real numbers $g(s)$ for all $s \in \mathcal{S}$ that satisfies the system of *Global Balance Equations* (GBE):

$$g(s) \sum_{s' \in \mathcal{S}} q(s, s') = \sum_{s' \in \mathcal{S}} g(s')q(s', s), \tag{1}$$

or equivalently $\mathbf{g}\mathbf{Q} = \mathbf{0}$. If the CTMC is ergodic there exists a unique invariant measure $\pi(s)$ which is also a probability distribution over $\mathcal{S}$, i.e., $\sum_{s \in \mathcal{S}} \pi(s) = 1$ and this is the steady-state distribution of the CTMC. The Markov chain $X(t)$ is *stationary* if $P\{X(0) = s\} = \pi(s)$ for all $s \in \mathcal{S}$. In the following two paragraphs we introduce the notion of time-reversibility for stationary Markov chains in the continuous time setting (for the discrete case see [16, 18]).

2.1   Time Reversibility for CTMCs

Given a stationary CTMC, $X(t)$ with $t \in \mathbb{R}$, we call $X(\tau - t)$ its reversed process for all $\tau \in \mathbb{R}$. We denote by $X^R(t)$ the reversed process of $X(t)$. It can be shown that $X^R(t)$ is also a stationary CTMC. Given a state renaming function ρ (a bijection from $\mathcal{S}$ to $\mathcal{S}$), we say that $X(t)$ is *ρ-reversible* if it is stochastically identical to $X^R(t)$ modulo the state renaming ρ [18,19]. Intuitively, an external observer is not able to distinguish $X(t)$ from $X^R(t)$ once the state renaming function ρ is applied to rename the states. Notice that if ρ is the identity then we simply say that $X(t)$ is *reversible*, whereas if ρ is an involution, then we say that $X(t)$ is *dynamically reversible* [16,26].

We can decide if a CTMC is ρ-reversible in two ways: the first involves the steady-state distribution of the CTMC, while the latter is based on an extended formulation of Kolmogorov's criteria [16], i.e., requires the analysis of the cycles in the reachability graph.

Lemma 1. *Given a stationary CTMC $X(t)$ with state space S, if there exists a collection of positive real numbers $\boldsymbol{\pi}$ summing to unity and a bijection ρ from S to S such that:*

$$q_s = q_{\rho(s)} \qquad \text{for all } s \in S \qquad (2)$$
$$\pi(s)q(s,s') = \pi(\rho(s'))q(\rho(s'),\rho(s)) \text{ for all } s, s' \in S, s \neq s' \qquad (3)$$

then $X(t)$ is ρ-reversible and $\pi(s)$ is its steady-state distribution.

Equation (2) states that the residence time of a state and its renaming must be equal. Notice that this condition is trivially satisfied if ρ is the identity, i.e., $X(t)$ is reversible. The set of equations (3) are called *detailed balance equations*. In case the renaming function ρ is known, it is possible to use the detailed balance equations to compute the chain's steady-state distribution instead of the more complex GBE.

Lemma 2. *Given a stationary CTMC $X(t)$ with state space S and let ρ be a renaming on S. $X(t)$ is ρ-reversible with respect to ρ if and only if for every finite sequence $s_1, s_2, \ldots s_n \in S$,*

$$q(s_1, s_2)q(s_2, s_3) \cdots q(s_{n-1}, s_n)q(s_n, s_1) =$$
$$q(\rho(s_1), \rho(s_n))q(\rho(s_n), \rho(s_{n-1})) \cdots q(\rho(s_3), \rho(s_2))q(\rho(s_2), \rho(s_1)) \quad (4)$$

and Eq. (2) holds for all $s \in S$.

Analogously to Kolmogorov's criteria for reversible chains, Lemma 2 requires to check Eq. (4) for all the (minimal) cycles of the CTMC and can be a useful tool for proving the ρ-reversibility of a CTMC. A consequence of ρ-reversibility is that $\pi(s) = \pi(\rho(s))$ for all $s \in S$.

3   Modelling Reversible Computations with ρ-Reversible Markov Processes

Reversible computations are characterised by the fact that they have two execution directions: the forward and the backward that restores past states of the computation. Our idea of the implementation of purely reversible computations[1] is similar to that considered in [22], i.e., the code being executed is naturally reversible. For instance, the programmer may have used Janus [27] which is a programming language for reversible computations or a subset of a standard language equipped with a reversible compiler.

3.1 Modelling Reversible Programming Structures

In this section we describe a modelling methodology for the reversible programming structures such as sequences, branches, cycles and sequences with checkpoints.

Sequential Computations. The simplest reversible computation is the reversible sequential one shown in Fig. 1 where s_i are the states of the computation and the arc labels denote the transition rates, f standing for forward rates and r for the reversed ones. In this model every state can be restored in one step. For each state we define a probabilistic law that decides if the computation will proceed in the forward or backward direction. In practice these probabilities can be derived by the statistical analysis of the software execution or by the knowledge of the intrinsic law that governs the probability of proceeding in one direction or the opposite.

Assume that the residence time in state s_n is exponentially distributed with rate $f_n + r_{n-1}$, then the probability of a forward transition given that $X(t)$ is in state s_n is $f_n/(f_n + r_{n-1})$ and the probability of a backward transition is $r_{n-1}/(f_n + r_{n-1})$. This follows from the properties of the exponential random variable (see, e.g., [24]) and the so called *race policy*.

If the Markov chain depicted in Fig. 1 is ergodic then it is reversible. The ergodicity is trivially satisfied if there exist lower and higher boundary states. The former is a state that does not allow a backward computation while the latter is a state that does not allow a forward computation. According to Lemma 2 we have that the forward cycle $s_n \xrightarrow{f_n} s_{n+1} \xrightarrow{r_n} s_n$ has itself as inverse cycle and therefore the conditions of Lemma 2 are satisfied.

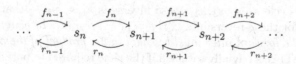

Fig. 1. Model for a reversible sequential computation.

---

[1] By purely reversible computations we mean those computations in which each step can be undone and there are no segments in which the execution direction is forward only.

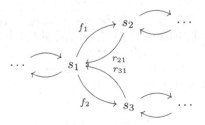

Fig. 2. Model for a reversible branch.

Branches. Branches can be modelled in a similar way to the one used for the sequential computations. Notice that, as commonly done in stochastic modelling, we model the branch by means of the probabilistic behaviour of the executed process. Although a modelling approach taking into account the detailed description of the system state is theoretically possible, in many cases this is not practically feasible due to the high cardinality that would be reached by the state space. Suppose that state s_1 is associated with a branch that proceeds to state s_2 with probability p and to s_3 with probability $1 - p$ (see Fig. 2). In this case, let $1/f$ be the expected residence time in state s_1, then the transition rates are $f_1 = fp$ and $f_2 = f(1 - p)$. Following the reasoning proposed in the previous paragraph on sequential computations, it is easy to see that the conditions of Lemma 2 are satisfied by choosing ρ as the identity.

Cycles. Cycles can be modelled as long as each transition they consist of can be undone. Let us consider the model of Fig. 3 where the computation at state s_1 can proceed by entering the cycle s_1, s_2, s_3, s_4 or by moving to state s_5. The probability of entering the cycle given that the computation will proceed in the forward direction is $f_1/(f_1 + f_1')$ and the number of (forward) iterations are geometrically distributed. Modelling the exact number of iterations of the cycles is possible but, in general, will drastically increase the number of model states. Let us focus on the cycle s_1, s_2, s_3, s_4 and its inverse s_1, s_4, s_3, s_2. If we apply Lemma 2 with ρ being the identity function, we notice that the conditions are satisfied for the cycles consisting of two states (e.g., s_3, s_4, s_3) but we need also to consider the cycle s_1, s_2, s_3, s_4 whose inverse is s_1, s_4, s_3, s_2 that originates a rate-condition for the ρ-reversibility: $f_1 f_2 f_3 f_4 = r_1 r_2 r_3 r_4$. *In general, in cycles, the product of the forward rates must be equal to the product of the corresponding backward rates.* This is trivially satisfied if the time required to perform a forward computation follows the same distribution of that required to undo it.

Sequences with Checkpoints. In the previous paragraphs we have shown how it is possible to model reversible sequential computations, branches and cycles by using a reversible CTMC, i.e., by taking the identify as ρ function. In the context of modelling reversible computations, the notion of ρ-reversibility is important because it allows the specification of atomic sequences that can be only fully reversed. For instance, consider a system atomic transaction whose correctness

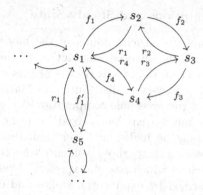

Fig. 3. Model for a reversible cycle.

is tested at a certain checkpoint. If an invalid state is detected, then all the operations performed by the transaction must be undone. An example of such a computation is shown in Fig. 4. In order to prove the ρ-reversibility of the model, we define function ρ as:

$$
\rho(s) = \begin{cases} CK_1 & \text{if } s = CK_1 \\ CK_2 & \text{if } s = CK_2 \\ s_i' & \text{if } s = s_i \quad 1 \le i \le n \\ s_i & \text{if } s = s_i' \quad 1 \le i \le n \end{cases}.
$$

By Lemma 2 we observe that the residence time of s_i must have the same expectation of that of its ρ-renaming, s_i' (and vice versa). Therefore, we have the rate condition for the ρ-reversibility whose interpretation is that the time required to perform an operation in the transaction must follow the same probabilistic law of that required to undo it. For what concerns the cycle analysis, observe that $CK_1, s_1, \ldots, s_n, CK_2, s_n', \ldots, s_1', CK_1$ has itself as inverse and hence the condition (4) is satisfied.

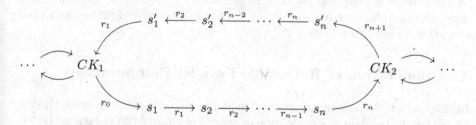

Fig. 4. Model for a reversible computation with checkpoints.

3.2 Modelling Assumptions and Steady-State

In this section we discuss two crucial points of the modelling technique that we propose: How does the exponential assumption of the distribution of the state residence time impact on the expressiveness of this modelling framework? How do we interpret the steady-state distribution of Markov chains in terms of quantitative properties of the reversible computations?

The exponential assumption can be relaxed by using distributions whose coefficient of variation may be higher or lower than that of the exponential. This is achieved by splitting a state whose residence time is not exponential into a set of micro-states each of which has an exponential residence time. Coxian random variables are formed by exponential stages and can approximate any distribution with rational Laplace transform with arbitrary accuracy (see, e.g., [15]). The literature proposing algorithms to fit data statistics to a distribution by means of a combination of the exponential stages is very rich (e.g., [6,21]).

Informally, the steady-state distribution of a CTMC is the probability of observing a given state when the time elapsed since the first observation is very large (the time required to reach the stationary behaviour depends on the magnitude of the second eigenvalue of the infinitesimal generator). For instance, in stationary reversible simulations [22] the state of the process after a period of warm up, is independent of its initial conditions and hence our framework can be applied easily. The assumption that each state transition can be undone includes the transitions that take the model to the state encoding the result of the computation. As a consequence, it is not obvious how the steady-state distribution can give an idea about the time required to obtain the result in a reversible computation. In stochastic analysis this problem is connected to the computation of the (moments of the) distribution of the time to absorption. Basically, we assume that once the chain enters in one of the states encoding the result, then they cannot leave them. Unfortunately, to the best of our knowledge, time-reversibility does not help in the exact computation of the distribution of the time to absorption. Nevertheless, approximating methods which may take advantage from the process' reversibility are available and are quite accurate when the expected computation time is much higher than the expected transition delays of the model (see, e.g., [1,3]). The steady-state distribution may also be interpreted as the fraction of a large number of processes which are in a given state (in the long-run) once they are run in parallel and they restart their computation after terminating it.

4 Cooperation of Reversible Parallel Computations

In this section we present an abstract model based on continuous time Markov chains for the performance evaluation of reversible parallel computations. Differently from those functional models that represent explicitly the parallel composition of reversible computations, we do not consider here any notion of causality. Instead we present a stochastic model for analysing the dynamic behaviour of those computations that can be realized in a reversible fashion, where the underlying conditional probabilities play the role of causality.

4.1   Labelled Stochastic Automata and Synchronisation

We introduce the notion of labelled stochastic automaton as a model for synchronizing computations. In the definition of stochastic automata we distinguish between *active* and *passive* action types, and only active/passive synchronisations are allowed when forming the composition of automata.

Definition 1 (Stochastic Automaton (SA)). *A stochastic automaton P is a tuple $(\mathcal{S}_P, \mathcal{A}_P, \mathcal{P}_P, \leadsto_P, q_P)$ where*

- *$\mathcal{S}_P$ is a denumerable set of states, named* state space *of P*
- *$\mathcal{A}_P$ is a finite set of* active *types*
- *$\mathcal{P}_P$ is a finite set of* passive *types*
- *τ denotes the* unknown *type*
- *$\leadsto_P \subseteq (\mathcal{S}_P \times \mathcal{S}_P \times \mathcal{T}_P)$ is a transition relation where $\mathcal{T}_P = (\mathcal{A}_P \cup \mathcal{P}_P \cup \{\tau\})$ and for all $s \in \mathcal{S}_P$, $(s,s,\tau) \notin \leadsto_P$[2]*
- *q_P is a function from $\leadsto_P$ to $\mathbb{R}^+$ such that $\forall s_1 \in \mathcal{S}_P$ and $\forall a \in \mathcal{P}_P$, $\sum_{s_2:(s_1,s_2,a) \in \leadsto_P} q_P(s_1, s_2, a) \leq 1$.*

Hereafter, we denote by $\to_P$ the relation defined as

$$\to_P = \{(s_1, s_2, a, q) \mid (s_1, s_2, a) \in \leadsto_P \text{ and } q = q_P(s_1, s_2, a)\}.$$

We will use the notation $s_1 \stackrel{a}{\leadsto}_P s_2$ to denote the tuple $(s_1, s_2, a) \in \leadsto_P$; moreover we denote by $s_1 \xrightarrow{(a,r)}_P s_2$ (resp., $s_1 \xrightarrow{(a,p)}_P s_2$) the tuple $(s_1, s_2, a, r) \in \to_P$ (resp., $(s_1, s_2, a, p) \in \to_P$).

For $s, s' \in \mathcal{S}_P$ and for $a \in \mathcal{A}_P \cup \{\tau\}$, $q_P(s, s', a) \in \mathbb{R}^+$ denotes the *rate* of the transition from s to s' with type a. For $s, s' \in \mathcal{S}_P$ and for $a \in \mathcal{P}_P$, $q_P(s, s', a) \in (0, 1]$ denotes the *probability* that the automaton synchronises on type a with a transition from s to s'. In the following, we adopt the convention that $q_P(s, s', a) = 0$ whenever there are no transitions with type a from s to s'. For $s \in \mathcal{S}_P$ and for $a \in \mathcal{T}_P$ we write $q_P(s, a) = \sum_{s' \in \mathcal{S}_P} q_P(s, s', a)$. We say that the automaton P is *closed* if $\mathcal{P}_P = \emptyset$.

Every closed automaton has an underlying continuous time Markov chain as defined below.

Definition 2 (CTMC underlying a closed automaton). *Given a closed automaton P, we denote by $X_P(t)$ the CTMC underlying P, whose state space is $\mathcal{S}_P$ and whose infinitesimal generator matrix $\mathbf{Q}$ is as follows: for all $s_1 \neq s_2 \in \mathcal{S}_P$,*

$$q(s_1, s_2) = \sum_{a,r:s_1 \xrightarrow{(a,r)}_P s_2} r\,.$$

---

[2] We exclude the τ self-loops from the definition of stochastic automaton in order to simplify the semantics of synchronisation. Indeed, the τ self-loops are irrelevant for the equilibrium distribution of the CTMC underlying the automaton.

A closed automaton P is said to be ergodic *(irreducible) if its underlying CTMC is ergodic (irreducible). The equilibrium distribution of the CTMC underlying the automaton P is denoted by* π_P.

Stochastic automata can be composed throughout a synchronisation operator which is defined in the style of the master/slave synchronisation of SANs [23] based on the Kronecker's algebra and the active/passive cooperation operation used in Markovian process algebra such as PEPA [12,13].

Definition 3 (SA synchronisation). *Let P and Q be two stochastic automata and assume that* $\mathcal{A}_P = \mathcal{P}_Q$ *and* $\mathcal{A}_Q = \mathcal{P}_P$. *The parallel composition of P and Q is the automaton* $P \otimes Q$ *defined as follows:*

- $\mathcal{S}_{P \otimes Q} = \mathcal{S}_P \times \mathcal{S}_Q$
- $\mathcal{A}_{P \otimes Q} = \mathcal{A}_P \cup \mathcal{A}_Q = \mathcal{P}_P \cup \mathcal{P}_Q$
- $\mathcal{P}_{P \otimes Q} = \emptyset$
- τ *is the unknown type*
- $\rightsquigarrow_{P \otimes Q}$ *and* $q_{P \otimes Q}$ *are defined according to the rules for* $\rightarrow_{P \otimes Q}$ *depicted in Table 1 where* $\rightarrow_{P \otimes Q}$ *contains the tuples* $((s_{p_1}, s_{q_1}), (s_{p_1}, s_{q_2}), a, q)$ *with* $((s_{p_1}, s_{q_1}), (s_{p_1}, s_{q_2}), a) \in \rightsquigarrow_{P \otimes Q}$ *and* $q = q_{P \otimes Q}((s_{p_1}, s_{q_1}), (s_{p_1}, s_{q_2}), a)$.

Table 1. Operational rules for SA synchronisation

$$\frac{s_{p_1} \xrightarrow{(a,r)}_P s_{p_2} \quad s_{q_1} \xrightarrow{(a,p)}_Q s_{q_2}}{(s_{p_1}, s_{q_1}) \xrightarrow{(a,pr)}_{P \otimes Q} (s_{p_2}, s_{q_2})} \quad (a \in \mathcal{A}_P = \mathcal{P}_Q)$$

$$\frac{s_{p_1} \xrightarrow{(a,p)}_P s_{p_2} \quad s_{q_1} \xrightarrow{(a,r)}_Q s_{q_2}}{(s_{p_1}, s_{q_1}) \xrightarrow{(a,pr)}_{P \otimes Q} (s_{p_2}, s_{q_2})} \quad (a \in \mathcal{P}_P = \mathcal{A}_Q)$$

$$\frac{s_{p_1} \xrightarrow{(\tau,r)}_P s_{p_2}}{(s_{p_1}, s_{q_1}) \xrightarrow{(\tau,r)}_{P \otimes Q} (s_{p_2}, s_{q_1})} \qquad \frac{s_{q_1} \xrightarrow{(\tau,r)}_Q s_{q_2}}{(s_{p_1}, s_{q_1}) \xrightarrow{(\tau,r)}_{P \otimes Q} (s_{p_1}, s_{q_2})}$$

Notice that, according to the above definition, an automaton obtained by a composition does not have passive types. This is reasonable if we consider the fact that in this case the resulting automaton has an underlying CTMC and then we can study its equilibrium distribution. In [20] we show that this semantics for pairwise SA synchronisations can be easily extended in order to include an arbitrary finite number of pairwise cooperating automata.

4.2 Reversible Stochastic Automata

We now introduce the notion of ρ-reversibility for stochastic automata. We present a definition in the style of the Kolmogorov's criteria stated in [16].

We assume the existence of a bijection (renaming) $\bar{\cdot}$ from $\mathcal{T}_P$ to $\mathcal{T}_P$ such that for each forward action type a there is a corresponding backward action type $\bar{a}$ with $\bar{\tau} = \tau$. In most of practical cases, $\bar{\cdot}$ is an involution, i.e., $\bar{\bar{a}} = a$ for all $a \in \mathcal{T}_P$, and hence the semantics becomes similar to the one proposed in [7]. We say that $cev\cdot$ respects the active/passive types of an automaton P if $\bar{\tau} = \tau$ and for all $a \in \mathcal{T}_P \setminus \{\tau\}$ we have that $a \in \mathcal{A}_P \Leftrightarrow \bar{a} \in \mathcal{A}_P$ (or equivalently $a \in \mathcal{P}_P \Leftrightarrow \bar{a} \in \mathcal{P}_P$).

The notion of ρ-reversible automaton is defined as follows.

Definition 4 (ρ-reversible automaton). *Let P be an irreducible stochastic automaton. Assume that*

- *$\rho : \mathcal{S}_P \to \mathcal{S}_P$ is a renaming (permutation) of the states, and*
- *$\bar{\cdot}$ is a bijection from $\mathcal{T}_P$ to $\mathcal{T}_P$ that respects the active/passive typing.*

We say that P is ρ-reversible if

1. *$q(s, a) = q(\rho(s), a)$, for each state $s \in \mathcal{S}_P$;*
2. *for each cycle $\Phi = (s_1 \overset{a_1}{\rightsquigarrow} s_2 \overset{a_2}{\rightsquigarrow} \ldots \overset{a_{n-1}}{\rightsquigarrow} s_n \overset{a_n}{\rightsquigarrow} s_1)$ in P there exists one cycle $\bar{\Phi} = (\rho(s_1) \overset{\bar{a}_n}{\rightsquigarrow} \rho(s_n) \overset{\bar{a}_{n-1}}{\rightsquigarrow} \ldots \overset{\bar{a}_2}{\rightsquigarrow} \rho(s_2) \overset{\bar{a}_1}{\rightsquigarrow} \rho(s_1))$ in P such that:*

$$\prod_{i=1}^{n} q(s_i, s_{i+1}, a_i) = \prod_{i=1}^{n} q(\rho(s_{i+1}), \rho(s_i), \bar{a}_i) \text{ with } s_{n+1} \equiv s_1 .$$

We say that $\bar{\Phi}$ is the inverse of cycle Φ. If ρ is the identity function we simply say that P is reversible.

Notice that the inverse cycle $\bar{\Phi}$ of a cycle Φ is unique. This can be easly derived from the fact that, by Definition 1 of stochastic automaton, there exists at most one transition between any pair of states with a certain type $a \in \mathcal{T}_P$.

The following theorem states that any ρ-reversible automaton satisfies a set of detailed balance equations similar to those presented in Lemma 1.

Theorem 1 (Detailed balance equations). *If P is ergodic and ρ-reversible then for each pair of states $s, s' \in \mathcal{S}_P$, and for each type $a \in \mathcal{T}_P$, we have*

$$\pi_P(s)q(s, s', a) = \pi_P(s')q(\rho(s'), \rho(s), \bar{a}) .$$

The next proposition says that the states of an ergodic ρ-reversible automaton have the same equilibrium probability of the corresponding image under ρ.

Proposition 1 (Equilibrium probability of the renaming of a state). *If P is an ergodic and ρ-reversible automaton then for all $s \in \mathcal{S}_P$,*

$$\pi_P(s) = \pi_P(\rho(s)) .$$

4.3   Product-Form Result

It is well-known that the cardinality of the state space of complex systems can grow exponentially with the structure of the model. Even worse, the numerical algorithms for deriving the equilibrium distribution become numerically unstable and prohibitive in terms of computation time. In this section we present the product-form result for networks of ρ-reversible synchronising automata. First we prove that the parallel composition of ρ-reversible automata is still ρ-reversible. Then, based on this result, we prove that the equilibrium distribution of the composition of two ρ-reversible automata can be derived from the equilibrium distribution of the cooperating automata considered in isolation (i.e., without generating the joint state space and solving the system of global balance equations). The analysis in isolation requires to set a rate for the passive transitions. To this aim, in [20] we prove that, thanks to the *rescaling* property of ρ-reversible automata, we can choose an arbitrary positive constant.

Theorem 2 (Closure under ρ-reversibility). *Let P_1 and P_2 be two ρ_1- and ρ_2-reversible automata with respect to the same function $\stackrel{\cdot}{\cdot}$ on the action types. Then, the composition $P_1 \otimes P_2$ is ρ-reversible with respect to the same $\stackrel{\cdot}{\cdot}$, where, for all $(s_1, s_2) \in \mathcal{S}_{P_1} \times \mathcal{S}_{P_2}$,*

$$\rho(s_1, s_2) = (\rho_1(s_1), \rho_2(s_2)). \tag{5}$$

The next theorem provides the product-form result for networks of ρ-reversible stochastic automata. In order to understand the relevance of this result, consider a set of M cooperating automata and assume that each automaton has a finite state space of cardinality N. The state space of the network may have the size of the Cartesian product of the state space of each single automaton, i.e., in the worst case, its cardinality is N^M. Since the computation of the equilibrium distribution of a CTMC requires the solution of the linear system of global balance equations, its complexity is $\mathcal{O}(N^{3M})$. For ρ-reversible automata, by applying Theorem 1, we can efficiently compute the equilibrium distribution of each automaton in linear time on the cardinality of the state space, and by Theorem 3 the complexity of the computation of the joint equilibrium distribution is $\mathcal{O}(NM)$.

Theorem 3 (Product-form solution). *Let P_1 and P_2 be two ergodic ρ_1- and ρ_2-reversible automata with respect to the same function $\stackrel{\cdot}{\cdot}$ on the action types, and let π_1 and π_2 be the equilibrium distributions of the CTMCs underlying P_1 and P_2, respectively. If $P_1 \otimes P_2$ is ergodic on the state space given by the Cartesian product of the state spaces of P_1 and P_2, then for all $(s_1, s_2) \in \mathcal{S}_{P_1} \times \mathcal{S}_{P_2}$,*

$$\pi(s_1, s_2) = \pi_1(s_1)\pi_2(s_2) \tag{6}$$

where π is the equilibrium distribution of the CTMC underlying $P_1 \otimes P_2$. In this case we say that the composed automaton exhibits a product-form *solution.*

Notice that this analysis, differently from those based on the concepts of quasi-reversibility [10,16] and reversibility, does not require a reparameterisation of the cooperating automata, i.e., the expressions of the equilibrium distributions of the isolated automata are *as if* their behaviours are stochastically independent although they are clearly not.

4.4 Example

In this section we describe a model for the parallel composition of two reversible computations. Consider the stochastic automata P_1 and P_2 depicted in Fig. 5. P_1 and P_2 communicate on the reversible channels a, b and c. Channel a is unreliable, i.e., a packet sent from P_1 to P_2 is recevied by P_2 with probability p and lost with probability $1 - p$. P executes its computations in the forward $(s_0 \rightarrow s_1 \rightarrow s_2 \rightarrow s_3 \rightarrow s_4 \rightarrow s_5$ or $s_0 \rightarrow s_1 \rightarrow s_2 \rightarrow s_3 \rightarrow s_4 \rightarrow s_6)$ or backward $(s_5 \rightarrow s_4 \rightarrow s'_3 \rightarrow s'_2 \rightarrow s_1 \rightarrow s_0$ or $s_6 \rightarrow s_4 \rightarrow s'_3 \rightarrow s'_2 \rightarrow s_1 \rightarrow s_0)$ direction. It has two checkpoints modelled by states s_1 and s_4. P_2 moves from t_0 to t_1 or t_2 with a probabilistic choice upon the synchronisation with type a. P_1 is ρ_1-reversible with $\rho_1(s_i) = s_i$ for $i = 0, 1, 4, 5, 6$ and $\rho_1(s_i) = s'_i$ and $\rho_1(s'_i) = s_i$ for $i = 2, 3$, while P_2 is ρ_2-reversible where ρ_2 is the identity function. Notice that $a, \bar{a}, b, c \in \mathcal{A}_{P_1} = \mathcal{P}_{P_2}$ and $\bar{b}, \bar{c} \in \mathcal{A}_{P_2} = \mathcal{P}_{P_1}$.

We assume that the model encodes the result of the computation in the states $(s_5, t_2), (s_5, t_4), (s_6, t_2), (s_6, t_4)$. We aim to compute the equilibrium probability of these four states that represents the fraction of time that the process spends in the states that encode the desired result.

Now we use Theorem 1 to derive the equilibrium distribution of the isolated automata. Let us consider an arbitrary state in P, say s_0. We can immediately derive $\pi_1(s_1)$ by using the detail balance equation and we obtain:

$$\pi_1(s_0)\lambda(1 - p) = \pi_1(s_1)\mu(1 - p),$$

which gives $\pi_1(s_1) = \pi_1(s_0)\lambda/\mu$. Then, we derive $\pi_1(s_2)$ using the detailed balance equation with s_1 and obtain: $\pi_1(s_2) = \pi_1(s_0)\lambda\gamma_1/(\mu\gamma_2)$. By Proposition 1 we immediately have $\pi_1(s'_2) = \pi_1(s_2)$. Then we derive $\pi_1(s'_3) = \pi_1(s_3) = \pi_1(s_0)\lambda\gamma_1/(\mu\gamma_3)$, $\pi_1(s_4) = \pi_1(s_0)\lambda\gamma_1/(\mu\gamma_4)$, $\pi_1(s_5) = \pi_1(s_0)\lambda\gamma_1\nu q/(\mu\gamma_4)$ and $\pi_1(s_6) = \pi_1(s_0)\lambda\gamma_1\nu(1 - q)/(\mu\gamma_4)$. It remains to derive $\pi_1(s_0)$ that is computed by normalising the probabilities. We can apply the same approach to derive the equilibrium distribution of P_2, obtaining:

$$\pi_2(t_1) = \pi_2(t_3) = \pi_2(t_0)\frac{1}{2}, \quad \pi_2(t_2) = \pi_2(t_0)\frac{1}{2\alpha}, \quad \pi_2(t_4) = \pi_2(t_0)\frac{1}{2\beta}.$$

Again, by normalising the probabilities, we obtain $\pi_2(t_0)$. By applying Theorem 3 we can now easily derive the desired result:

$$\pi(s_5, t_2) + \pi(s_5, t_4) + \pi(s_6, t_2) + \pi(s_6, t_4) =$$
$$\pi_1(s_5)\pi_2(t_2) + \pi_1(s_5)\pi_2(t_4) + \pi_1(s_6)\pi_2(t_2) + \pi_1(s_6)\pi_2(t_4).$$

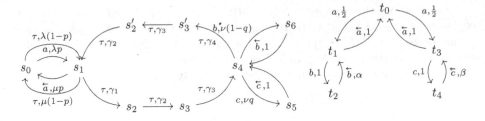

Fig. 5. Models for P_1 and P_2

Notice that we have not build the joint state space and also that the automata P_1 and P_2 are not independent. For example, when P_2 is in state t_1 and P_1 is in checkpoint s_4, P_2 moves to t_2 only if P_1 decides neither to roll back to checkpoint s_1 nor to move to s_5.

5   Conclusion

In this paper we have proposed an abstract modelling framework for the quantitative analysis of reversible computations. The main idea is to exploit the time-reversibility property of Markov processes in order to provide a computationally efficient way of deriving the desired performance indices. We have shown that, under some conditions, the proposed approach is suitable to be applied for a compositional formalism based on labelled stochastic automata. As a consequence the advantages (reduction of time-complexity and improvement of algorithms' numerical stability) of time-reversibility are applicable also for the analysis of the cooperation of automata that are proved to have product-form steady-state distributions.

References

1. Artalejo, J.R., Economou, A., Lopez-Herrero, M.J.: The maximum number of infected individuals in SIS epidemic models: computational techniques and quasi-stationary distributions. J. Comput. Appl. Math. **233**, 2563–2574 (2010)
2. Barbierato, E., Dei Rossi, G.-L., Gribaudo, M., Iacono, M., Marin, A.: Exploiting product forms solution techniques in multiformalism modeling. Electr. Notes Theor. Comput. Sci. **296**, 61–77 (2013)
3. Barbour, A.D.: Quasi-stationary distributions in markov population processes. Adv. Appl. Prob. **8**, 296–314 (1976)
4. Baskett, F., Chandy, K.M., Muntz, R.R., Palacios, F.G.: Open, closed, and mixed networks of queues with different classes of customers. J. ACM **22**(2), 248–260 (1975)
5. Bugliesi, M., Gallina, L., Hamadou, S., Marin, A., Rossi, S.: Behavioural equivalences and interference metrics for mobile ad-hoc networks. Perform. Eval. **73**, 41–72 (2014)

6. Casale, G., Smirni, E.: KPC-toolbox: fitting markovian arrival processes and phase-type distributions with MATLAB. SIGMETRICS Perform. Eval. Rev. **39**(4), 47 (2012)
7. Danos, V., Krivine, J.: Reversible communicating systems. In: Gardner, P., Yoshida, N. (eds.) CONCUR 2004. LNCS, vol. 3170, pp. 292–307. Springer, Heidelberg (2004)
8. Gallina, L., Hamadou, S., Marin, A., Rossi, S.: A probabilistic energy-aware model for mobile ad-hoc networks. In: Al-Begain, K., Balsamo, S., Fiems, D., Marin, A. (eds.) ASMTA 2011. LNCS, vol. 6751, pp. 316–330. Springer, Heidelberg (2011)
9. Gordon, W.J., Newell, G.F.: Cyclic queueing networks with exponential servers. Oper. Res. **15**(2), 254–265 (1967)
10. Harrison, P.G.: Turning back time in Markovian process algebra. Theoret. Comput. Sci. **290**(3), 1947–1986 (2003)
11. Harrison, P.G.: Reversed processes, product forms and a non-product form. Elsevier Linear Algebra Appl. **386**, 359–381 (2004)
12. Hillston, J.: A Compositional Approach to Performance Modelling. Cambridge Press, Cambridge (1996)
13. Hillston, J., Marin, A., Piazza, C., Rossi, S.: Contextual iumpability. In: Proceedings of Valuetools 2013 Conference. ACM Press (2013)
14. Jackson, J.R.: Jobshop-like queueing systems. Manag. Sci. **10**, 131–142 (1963)
15. Kant, K.: Introduction to Computer System Performance Evaluation. McGraw-Hill, New York (1992)
16. Kelly, F.: Reversibility and Stochastic Networks. Wiley, New York (1979)
17. Lanese, I., Mezzina, C.A., Tiezzi, F.: Causal-consistent reversibility. Bull. EATCS **114**, 17 (2014)
18. Marin, A., Rossi, S.: On discrete time reversibility modulo state renaming and its applications. In: Proceedings of Valuetools 2014 Conference (2014)
19. Marin, A., Rossi, S.: On the relations between lumpability and reversibility. In: Proceedings of MASCOTS 2014, pp. 427–432 (2014)
20. Marin, A., Rossi, S.: Quantitative analysis of concurrent reversible computations. In: Sankaranarayanan, S., Vicario, E. (eds.) FORMATS 2015. LNCS, vol. 9268, pp. 206–221. Springer, Heidelberg (2015)
21. Mészáros, A., Papp, J., Telek, M.: Fitting traffic traces with discrete canonical phase type distributions and Markov arrival processes. Appl. Math. Comput. Sci. **24**(3), 453–470 (2014)
22. Perumalla, K.S.: Introduction to Reversible Computing. CRC Press, Boca Raton (2013)
23. Plateau, B.: On the stochastic structure of parallelism and synchronization models for distributed algorithms. SIGMETRICS Perf. Eval. Rev. **13**(2), 147–154 (1985)
24. Ross, S.M.: Stochastic Processes, 2nd edn. Wiley, Hoboken (1996)
25. Stewart, W.J.: Introduction to the Numerical Solution of Markov Chains. Princeton University Press, Princeton (1994)
26. Whittle, P.: Systems in Stochastic Equilibrium. Wiley, Hoboken (1986)
27. Yokoyama, T., Glück, R.: A reversible programming language and its invertible self-interpreter. In: Proceedings of the 2007 ACM SIGPLAN Symposium on Partial Evaluation and Semantics-based Program Manipulation, pp. 144–153, New York, NY, USA. ACM (2007)

Fluid Approximation of Pool Depletion Systems

Enrico Barbierato[1], Marco Gribaudo[1(✉)], and Daniele Manini[2]

[1] Dip. di Elettronica e Informazione, Politecnico di Milano,
via Ponzio 34/5, 20133 Milano, Italy
{enrico.barbierato,marco.gribaudo}@polimi.it
[2] Dip. di Informatica, Università di Torino, Corso Svizzera 185, 10149 Torino, Italy
manini@di.unito.it

Abstract. Today's most of high performance computing applications use parallel programming paradigms to reach the desired efficiency objectives. In particular, they divide the problem into small elements that can be solved in parallel by as many computing devices as available. Some examples are Apache Spark, the evolution of Hadoop and map-reduce, GPGPU (General Purpose Graphical Processing Units) applications, many-core and multi-core embedded systems. In many cases this type of applications can be modeled by pool depletion systems, i.e. queuing models characterized by a set of parallel servers whose goal is to execute a predetermined number of tasks. Although the modeling paradigm is very simple, it suffers from state space explosion, and can be used to model systems with a limited degree of parallelism only. The main contribution provided by this work consists of presenting a fluid approximation approach capturing the main features of the considered pool depletion systems and solving the above mentioned issues.

1 Introduction

High performance computing applications usually rely on parallel execution to divide a large problem in small elements that can be considered concurrently. In this way, the higher is the number of available computing resources, the faster the problem can be solved. In turn this allows to consider very large scale problems, such as searching images in a big database based on visual comparison, detecting anomalies in bank account usage or recommending users products based on the analysis of the choices made by a large set of customers. Big Data and data-analytic applications rely on technologies that are based, for example, on Apache Spark [23], the evolution of Hadoop [1] and map-reduce [12] to parallelize computation and reduce communication overhead among the participating nodes. Spark has been proposed to supports applications reusing a working set of data across multiple parallel operations offering also the scalability and fault tolerance of map-reduce. To achieve these goals, Spark introduces an abstraction called resilient distributed datasets (RDDs), i.e., a read-only collection of objects partitioned across a set of machines that can be rebuilt if a partition is lost. Parallelization of tasks can also be exploited inside a single computer or inside an embedded system. For example, technologies like GPGPU

S. Wittevrongel and T. Phung-Duc (Eds.): ASMTA 2016, LNCS 9845, pp. 60–75, 2016.
DOI: 10.1007/978-3-319-43904-4_5

(General Purpose Graphical Processing Units) [17] allow the use of the computational feature of a graphic adapter to execute parallel tasks and have applications in video editing, image processing, cryptography and scientific calculus. Embedded systems are also based on many-core and multi-core CPUs, which sometimes are implemented in FPGA to push the use of parallelism at extreme levels.

In order to asses the scalability of the considered type of parallel application, modeling approach must be used since the acquisition of very large set of resources has high costs and could require long provisioning times. However, conventional modeling techniques cannot be applied due to size of the problem: standard techniques like Queuing Networks [15], Generalized Stochastic Petri Nets [16] or Performance Evaluation Process Algebras [13] suffer from state space explosion of the Continuous Time Markov Chains that they use to solve the problem. For this reason in the literature several different modeling approaches have been used to address this problem. For example, in [3] advanced simulation techniques have been used to consider large map-reduce applications, and the proposed approach has been extended in [4] to insert it in a multi formalism context in order to address more complex applicative scenarios. Mean-field approach has instead been used in [8] to consider different types of BigData applications. GPGPU has also drawn a lot of attention even if, due to its much simpler architecture, it has been studied with conventional techniques: for example in [9] multi-class fork-join queuing models have been proposed and validated as a tool to capture the main aspects of these systems.

In [19], the authors introduce a framework to address the problem originated by the amount of resources that a user should lease from a service provider (assuring that the data processing is completed under a specific deadline), proposing a new approach to resource sizing and provisioning service in map-reduce environments. The problem of studying the performance prediction of individual jobs is explored in [18] through a framework consisting of a Hadoop job analyzer, while the prediction component exploits locally weighted regression methods. A similar issue is studied in [20] by using instead a hierarchical model including a precedence graph model and a queuing network model to simulate the intra-job synchronization constraints. In [11], the authors study the problem of minimizing the cost involved in the search of the optimal resource provisioning, proposing a cost function modeling acting as a relationship that takes in account (i) the time cost, (ii) the amount of input data, (iii) the available system resources (Map and Reduce slots), and (iv) the complexity of the Reduce function for the target map-reduce job. The usage of a simulator to better understand the performance of map-reduce setups is described in [21] with particular attention to (i) the effect of several component inter-connect topologies, (ii) data locality, and (iii) software and hardware failures. Finally, in [14] the authors discuss a novel approach aiming to improve map-reduce provisioning by using a database of similar resource consumption signatures of different applications.

In this work, we present an approach to study these types of applications using fluid models based on pool depletion systems where a given number of tasks must be completed by a set of computing resources. Indeed, they are queuing models characterized by a set of parallel servers whose goal is to execute a predetermined number of tasks. This paper presents (i) the measurement performed on a real application to motivate the considered class of system, and (ii) a more in-depth analysis on the effect of different task length distributions using a custom built discrete event simulator. In order to overcome the scalability problems caused by the state explosion arising when modeling such systems, a fluid model is proposed to approximate the average task ending time with a fluid variable representing the total number of tasks to be completed by the parallel servers. This model is finally exploited to study in an efficient way the execution times of map-reduce jobs.

The paper is structured as follows: in Sect. 2 a pool depletion systems is presented, and Sect. 3 shows how they are characterized in the considered scenario. The fluid model is described in Sect. 4, and it is exploited in Sect. 5 to analyze a map-reduce job. Section 6 concludes the paper.

2   Pool Depletion Systems

Pool depletion systems are models where a given number of tasks must be completed by a set of computing resources. Figure 1 shows a single class, single resource type pool depletion system. In this case, the system has to execute N tasks, each one requiring an independent, identically distributed service time according to a random variable $\mathbf{S}$. The system is composed by K identical servers: at the beginning, K tasks enter the system and start being served at the same time. The other $N - K$ are forced to wait in the external task pool. As soon as the first task ends, another one is admitted from outside. The process is repeated until no more tasks are waiting in the pool; after that moment, some of the K server starts becoming idle. The job ends when all its task are finished. In this type of models, the study of the total depletion time (i.e. the time required to complete all N tasks and leave all the K servers of the system idle) can be very interesting. In [10], two classes, two resources, single server pool depletion systems where introduced to study energy consumption in large data-centers. In that paper, exponential service with processor sharing nodes were considered, and an analytical solution based on the generation of the state space was proposed.

In this work, which considers a single type of resource represented by K identical parallel servers, the complexity of the analytical solution depends on the chosen service time distribution $\mathbf{S}$. If the service time distribution is exponential, then the system has a relatively simple analytical solution. Let $\mu = 1/E[\mathbf{S}]$

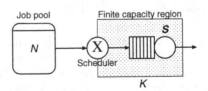

Fig. 1. A single class, single resource type pool depletion system.

denote the rate of the exponential distribution. For the first $N - K$ tasks, the inter-completion time corresponds to the time required by the first server to complete. Since all services are exponential, it corresponds to the minimum of K exponential distributions, which is again exponentially distributed with rate $K \cdot \mu$. When depletion starts, the inter-completion time is still exponentially distributed, but this time with rate $k \cdot \mu$ with $1 \leq k \leq K$. In the end, the total completion $\mathbf{C}$ time is distributed as the sum of K exponential distributions:

$$\mathbf{C} = Erlang(N - K, K \cdot \mu) + \sum_{k=1}^{K} Exp(k \cdot \mu). \tag{1}$$

An analytical expression for Eq. 1 can be found for example in [2]. If $\mathbf{S}$ can be expressed as a Phase Type distribution (PH), an analytical solution is still possible, since the class of PH distributions is closed under both the minimum and the sum. However, the number of states grows linearly with N, and exponentially with both K and the number of phases. In particular, following [7], where multiple servers with (PH) distribution and Markov Arrival Process where considered, let M denote the phases required to describe $\mathbf{S}$, then the total number of phases $\#PH$ required to express the distribution of $\mathbf{C}$ is:

$$\#PH = (N - K) \cdot \binom{K + M - 1}{M - 1} + \sum_{k=1}^{K} \binom{k + M - 1}{M - 1}. \tag{2}$$

Equation 1 shows that phase type distribution is not a feasible solution for realistic systems where the level of parallelism K can be in the $10^1 \sim 10^3$ range, and the number of phases required is also high due to the low coefficient of variation characterizing the task execution time distribution $\mathbf{S}$ in most applications (this claim is not based on existing literature, but it is based on our experience). To solve the scalability issue, we propose a fluid model to capture the behavior of the average completion time in an efficient way.

In the following, the considered class of depletion models (both on measurements performed on a real parallel application run on a multi-core processor, and on discrete event simulation) is analyzed.

3   Scenario Characterization

The pool depletion model presented in this work can be applied to several parallel applications, ranging from Apache Spark [22], the evolution of Hadoop and mapreduce, GPGPU (General Purpose Graphical Processing Units) applications [17], many-core and multi-core embedded systems, and several parallel programming paradigms like the well known consumer-producer model. Firstly, this section presents measurement performed on a real application to motivate the considered class of system, secondly it illustrates an in-depth analysis on the effect of different task length distributions using a custom-built discrete event simulator.

3.1   Real System Scenario

The performance model solution process used to perform what-if analysis on the models presented in [10] is considered as a benchmark for the class of systems described in this paper. The application considers the solution of a large Markov chain used to compute energy-related performance metrics for different parameters configuration. All the models are characterized by the same state-space, but by a different transition matrix. The algorithm generates several scenarios, then it solves them in a first-in-first-out queuing, running as many solution in parallel as available cores. Figure 2 shows the completion time of the considered benchmark on a quad-core, eight-threads MacBook Pro. In particular it considers the case with $K = 8$ simultaneous executions exploiting all the cores and all the hardware threads of the CPU, and the case with $K = 4$ avoiding the use of multithreading limiting the parallelism to the number of available cores. On the y-axis, the number of remaining tasks is presented, while the x-axis shows the ordered times at which tasks complete. The intersection of the curves with the x-axis defines the total job execution time. Even if all runs use the same parameterization, they are characterized by a different running time: this depends on the operating system and on the energy saving configuration of the machine slowing down in traces in non-deterministic ways. However, all cases have a ladder-like shape, with the height of a stair corresponding to the K, the level of parallelism (this is a direct consequence of the fact that all tasks are related to the same model, and are characterized by a very similar running time). The runs with $K = 4$ requires more or less the same time and present the same ladder pattern, even if in this case the steps are smaller and shorter. This is due to the fact that for the considered benchmark the CPU multithreading is not able to provide significant performance increases, determining very similar performances to the $K = 8$ cases.

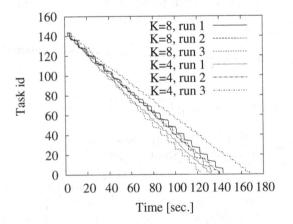

Fig. 2. Execution times of the considered benchmark on a quad-core, eight-threads MacBook Pro for different runs with and without multi-threading.

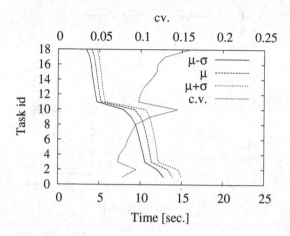

Fig. 3. Average execution time, standard deviation and coefficient of variation (cv) of the benchmark application running $N = 18$ tasks on $K = 8$ simultaneous threads.

Figure 3 shows the average execution time μ with $N = 18$ tasks running on $K = 8$ simultaneous threads over 100 experiments. The standard deviation σ of the execution time is shown added to (curve $\mu + \sigma$) and removed from (curve $\mu - \sigma$) the average. The coefficient of variation (cv) is shown on the secondary axis. As it can be seen, despite the variability outlined in Fig. 2, the cv is very small, and tends to decrease as the number of tasks increases, with jumps only corresponding to the stairs of the ladder-shaped evolution of the average.

The previous results are confirmed in Fig. 4 showing the distribution of the execution time of the benchmark application running $N = 18$ tasks on $K = 8$ simultaneous threads for the first three, the ninth and the last task. The small shift in the distribution of the ending time of the first three tasks emphasizes the little difference in execution time in the considered blocks, while the last tasks show that there is an expected spread in the completion time, even if the increase in the standard deviation is smaller than the one of the mean.

Note that this behaviour is characteristic of most the applications to which pool depletion models applies: for example, map-reduce techniques have the goal to produce small chunks of similar execution times. GPGPU applications instead base their parallelism on SIMD (single-instruction-multiple-data) techniques, which do not allow the different tasks to have significant differences in their execution times.

Figure 5 shows the average ending time of map-reduce tasks taken from the *OpenCloud Hadoop workload*, a public repository of Hadoop traces available on the web[1]. The traces are taken from a Hadoop cluster managed by CMU's Parallel Data Lab, and consider the workload of a cluster used for scientific workloads for a 20-month period. In particular we focused on the trace collected April 2010, and we limited our analysis to jobs composed by 200 map tasks and

---

[1] Available at the time of writing at http://ftp.pdl.cmu.edu/pub/datasets/hla/.

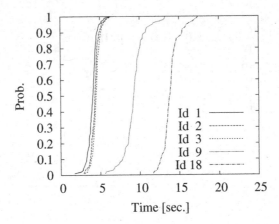

Fig. 4. Distribution of the execution time of the benchmark application running $N = 18$ tasks on $K = 8$ simultaneous threads for the first three, the ninth and the last task.

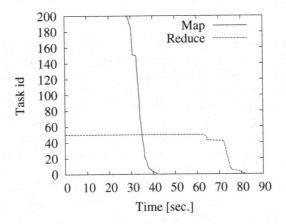

Fig. 5. Average execution time of map-reduce tasks from a public available trace.

50 reduce tasks. As it can be seen, the average decreases in a similar way to the benchmark application we consider. The number of used nodes for each job is not available: however from the peaks on the curves it would seems to be around $K = 50$ nodes.

3.2   Impact of Task-Length Distributions

An analytical scenario is now considered to study the impact of different task duration distributions, all characterized by the same average. The focus concerns different distributions characterized by the same coefficient of variation (where possible). For $cv = 1$ an exponential and a log-normal distribution is taken in account. For $cv = 0.5$ and $cv = 1/6$ two Erlang (respectively with four and thirty-six stages), two uniform and two log-normal distributions are considered, including $cv = 2$ with a Hyper-exponential and a log-normal distribution to have an idea of the impact of $cv > 1$. Larger cv are ignored since they are outside the intended range of applications. As a limiting case, the effect of the deterministic distribution ($cv = 0$) is shown as well. All curves have been computed with discrete event simulation[2]. Although confidence intervals have been generated, they have not been shown for the sake of simplicity. Figure 6 shows the average execution time for the considered distributions of the task duration. In particular, Fig. 6a–c focuses on the case with $K = 20$ and $N = 144$. The whole picture of the tasks ending times is given in Fig. 6a. The distribution with $cv \leq 1$ tends to have a very similar behaviour for the average time at which the single tasks complete. Distributions with $cv > 1$ tends instead to finish the first task earlier, while having a longer job completion time. The most important result that can be appreciated is that the average ending time of each task is mainly influenced by the cv: higher moments of the distribution play an impact only for the very first completion times, and tend to become less and less evident as time passes. Figure 6b zooms on the ending times of the first tasks. As it can be seen, only the deterministic component has a clear ladder behaviour. As the cv increases and as tasks complete, it become less and less evident. Moreover, after the very few tasks, the effect of higher moments also disappears, and curves of distributions with the same cv overlaps almost perfectly. The last tasks ending times are instead considered in Fig. 6c. As it can be seen, as depletion occurs, the higher moments play a significant role again, by creating slightly different tails. To better emphasize the effect of distribution with a $cv > 1$, Fig. 6d considers the case with $K = 80$ and $N = 3600$. When a large number of tasks has to be executed, the task length distribution plays a marginal role, and all scenario evolves at a rate which can be estimated to $E[\mathbf{S}]/K$. The cv performs a shift in the various curves, and determines the speed at which the depletion moves away from the considered average behaviour. This type of evolution motivates the idea of resorting to fluid models to efficiently consider this type of systems.

The effect of the level of parallelism K is studied in Fig. 7 showing the average execution time for the exponential and for a 36 stages Erlang distribution (corresponding to a $cv = 1/6$) required to run 720 tasks for $K \in \{8, 20, 40, 80\}$. The length of the tail of the distribution becomes more important as K increases, and also the effect of the cv becomes more evident. The staircase curve is destroyed very earlier in the execution, and becomes negligible after $4 \cdot K$ tasks have been completed even for a relatively low cv.

Figure 8 shows the distribution of the execution time for several tasks duration distributions. The exponential distribution, the four-stage Erlang distribution ($cv = 1/2$) and the Hyper-Exponential distribution (with $cv = 2$) and three different instances of a log-normal distribution with the same cv as the other are shown. All the curves refer to the completion job time when it is split into

---

[2] In this work we use simulation as a simplified way to define a fluid model. Future work will study how to define the fluid model directly from the parameters of the system, without resorting to simulation.

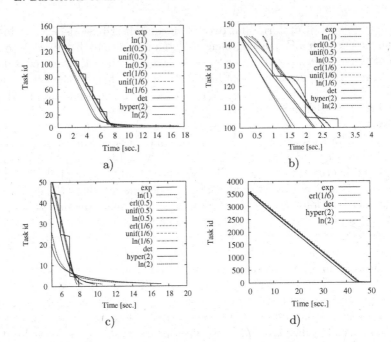

Fig. 6. Average execution time for different task duration distributions: (a), (b) and (c) shows the respectively the complete, the detail of the first tasks and the detail of the last tasks for the case with $K = 20$ and $N = 144$; (d) considers the case with $K = 80$ and $N = 3600$.

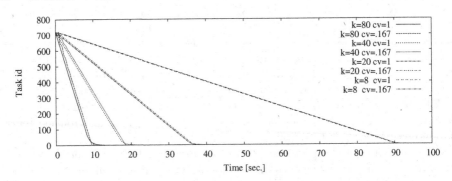

Fig. 7. Distribution of the execution time for the exponential and the 36 stages Erlang distribution $cv = 1/6 \approx 0.167$, for a different level of parallelism $K \in \{8, 20, 40, 80\}$ and $N = 720$ tasks.

$N = 144$ tasks and it is run with parallelism level $K = 8$. Except for the case with $cv = 2$, where the two distributions have very different shapes, the distribution with the same cv presents very similar behaviours, which tend to differ for what concerns the probability of having larger completion times.

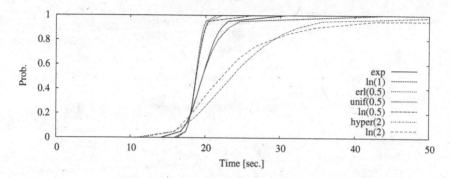

Fig. 8. Distribution of the execution time for several task duration distributions on a system running $N = 144$ tasks with a parallelism level $K = 8$.

4 Fluid Model

Due to the characteristic evolution of the completion time of the considered task, we propose to approximate the average task ending time with a fluid variable. In particular, a fluid variable x starting with the total number of tasks N that needs to be completed by the K parallel servers is added. The continuous variable then decreases at a fluid dependent rate that mimics the behaviour presented in Sect. 3. Following [5], a fluid model is considered where the evolution of the continuous variable is defined by a function $\phi : \mathbb{R}^2 \to \mathbb{R}$, where $\phi(x, t) = x'$ represents a system which has an average of x tasks still remaining to be completed at time 0, will have an average of x' tasks to be completed at time t. The fluid evolution function is defined in such a way that if the value of the continuous variable is $x(t_a)$ at a time instant t_a, then it must be that $x(t_b) = \phi(x(t_a), t_b - t_a)$. This is achieved in the following way: let $\mu(t)$ denote the average number of tasks in the system at time t. For example, $\mu(t)$ could correspond to one of the curves shown in Fig. 6. Then, it is possible to say that:

$$\phi(x, t) = \mu(t + \mu^{-1}(x)) \tag{3}$$

If the fluid variable starts from a point $x(0) \neq \mu(0)$ at time $t = 0$, the definition of $\phi(x, t)$ shifts the fluid evolution curve to allow the continuity of $x(t)$. The average task completion time function $\mu(t)$ can be computed in many ways: for example, it can be determined from measurements taken on a real system. This is shown for example in Fig. 3. In the following, $\mu(t)$ is computed with discrete event simulation, and then by performing linear interpolation among the task completion times.

5 Case Study: Analysis of Map-Reduce Completion Times

This section exploits the fluid model outlined in Sect. 4 to study in an efficient way the execution times of map-reduce jobs. As introduced, *map-reduce* is a

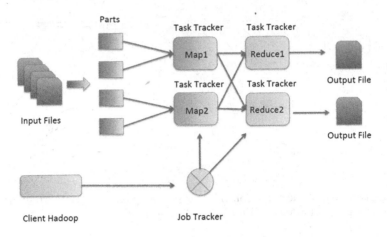

Fig. 9. Logic scheme of map-reduce.

paradigm that allows to create applications able to compute huge amount of data in parallel, and Apache Hadoop is an example of a framework allowing to exploit it. The basis of the framework is a functional programming where data are not shared among threads but are passed as parameters or return values. An application running map-reduce has to specify input and output files, and map and reduce functions. The Hadoop client exploits the service provided by the *JobTracker* that farms out map-reduce tasks to specific nodes in the cluster, ideally the nodes having the data, or at least are in the same rack. A *TaskTracker* is a node in the cluster accepting tasks (i.e. Map, Reduce and Shuffle operations) from a JobTracker. In particular, the Hadoop client provides the jobs and the configurations to the JobTracker distributing them to the nodes for the execution. The JobTracker also defines the number of parts by which the input files must be split into, and activates the TaskTrackers according to their distance to the nodes holding the respective data. TaskTrackers get the data to manage and activate the map function, then they run the reduce phase where data are sorted and aggregated. Then the output is generated and saved in a different file for each tracker. Figure 9 shows a logical scheme of execution of map-reduce jobs.

One of the key-features of map-reduce is that resources can be acquired dynamically during the execution of the tasks: as more computation nodes become available, they start working on the parts until all have been considered. After all the parts of the map phase have finished, the reduce stage can start. From a modeling point of view, a map-reduce job can be considered as the sequence of two pool depletion models representing respectively the map and reduce stages, where the number of available resources K changes with time. Although starting the next job while the previous one is completing can increase the throughput of the system, it complicates a lot the models that should be used to correctly capture the evolution of the system. Thanks to the proposed fluid technique, accurate estimates can be obtained with very simple models.

5.1   A Fluid Petri Net Model of Map-Reduce

Figure 10 shows a Fluid Stochastic Petri Net (FSPN) model describing the considered map-reduce paradigm, where resources are obtained gradually (note that the purpose of this FSPN model is just to describe the underlying fluid model without enumerating its states). The place named Start, initially marked, represents the start of the job. At time $t = 0$, immediate transition Mstarts fires, moving the token to place Map to denote the execution of the map phase. Fluid place Tasks represents the count of tasks that needs to be executed. When the map phase starts with the firing of Mstart, a set arc inserts into place Tasks the number of tasks that will be executed in that stage. Fluid transition Depletion models the execution of tasks. It behaves according to the definitions given in Sect. 4 in a state dependent way as it will be described later. The fluid arc that connects place Tasks to transition Depletion removes tasks as they finish in a continuous way. As soon as the tasks of the map phase end, transition Rstart fires, moving the token into place Reduce to denote the beginning of the reduce stage. Again, the set arc connecting transition Rstart to place Tasks defines the tasks that have to be executed in the reduce phase. When also the reduce phase ends, the model stops with the firing of immediate transitions End. The number of nodes running tasks in parallel is modeled by place Nodes. Acquisition of resources is modeled by transition Nadd, which stop firing thanks to the inhibitor arc coming from place Nodes as soon as the maximum number of available nodes K is reached (note that the addition of resources occurs externally and it is independent from the evolution of the Map-reduce tasks). In the reduce phase, resources can be released due to the firing transition Nrelease. This transition is guarded by a function that depends both on the number of remaining tasks, and on the number of available resources. When the marking of Tasks becomes smaller that the marking of Nodes, a node can be released by allowing transition Nrelease to fire. Resources are also released when the reduce phase ends with the flush-out arc connecting place Nodes to transition End. The state-dependency of transition Depletion is characterized by $2 \cdot K$ fluid evolution functions $\mu_{mr,K}(x,t)$, one for each combination of number of available nodes K, and map or reduce stage mr. The marking of places Nodes, Map and Reduce determines which fluid evolution function should be used in that particular moment of the model evolution.

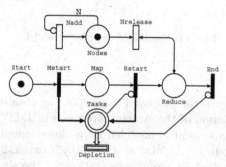

Fig. 10. A Fluid Petri Net model of the map-reduce paradigm.

5.2   Results

Figure 11 shows the results obtained with the solution of the FSPN presented in Fig. 10. In this case we consider that $K = 20$ nodes have to process $N_{Map} = 144$ map tasks, and $N_{Reduce} = 192$ reduce tasks. The duration of map tasks $S_{Map} \sim Erlang_4(1000)$ is assumed to be Erlang distributed with an average of 1000 ms, and a $cv = 0.5$. Reduce tasks $S_{Reduce} \sim Erlang_4(500)$ are instead assumed to be still Erlang distributed with $cv = 0.5$, but this time with an average of 500 ms. Nodes are acquired at regular deterministic time intervals, with a new resource being available every 400 ms. As introduced in Sect. 5.1, the model exploits $2 \cdot K$ fluid evolution functions $\mu_{mr,K}(x,t)$. In this example, such functions are determined by a quick run of discrete event simulation: due to the small cv that characterizes the considered Erlang distributions, acceptable confidence intervals can be achieved in about 100 runs per number of core and map or reduce stage. Figure 11 shows the fluid evolution function for both map and reduce tasks with $K = 20$ (the maximum number of available resources), and $K = 10$ that corresponds respectively to $\mu_{m,10}(0,t), \mu_{r,10}(0,t), \mu_{m,20}(0,t)$ and $\mu_{r,20}(0,t)$. The bold dotted line, referred to the secondary axis, shows the number of available cores as the function of time, while the curve with the largest width represents the fluid evolution of place Tasks. As it can be seen, as soon as number of nodes reaches the maximum available $K = 20$, the evolution corresponds to the corresponding fluid models, translated to make the time evolution continuous. When the reduce phase reaches the end, cores starts be released, and the curve corresponding to K drops very quickly to zero. In the considered scenario, the average time estimated to complete a map-reduce job is $R = 17.219$ s.

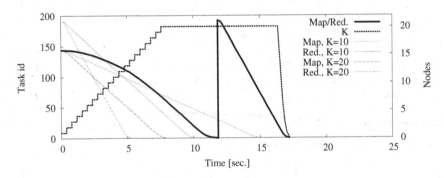

Fig. 11. Results of the FSPN model.

5.3   Validation

To remark the validity of the proposed technique, the obtained results are compared against a simulation of the system performed in JMT [6] using the model shown in Fig. 12. The system is modeled by a closed queuing network with a single job circulating in it. A delay station TinyTerminalDelay with a negligible waiting time is used as a reference station for the single circulating job.

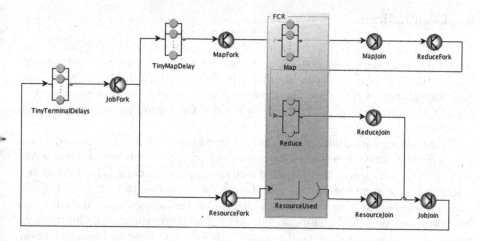

Fig. 12. A Fork/Join with Finite Capacity Regions Queuing Network model of the map-reduce paradigm, used to validate the results obtained with the FSPN model.

As soon as the single job leaves the reference station, it is split in two by the JobFork station: the upper path represents the execution of the map-reduce, while the bottom one models the acquisition of the resources. Again, a negligible delay TinyMapDelay is applied to the upper path, to allow the system to setup the available resources before starting to work on the tasks. The N_{Map} tasks are generated by the MapFork primitive. The execution of the tasks is modeled by the delay station Map, characterized by service time distribution S_{Map}. As soon as all the map tasks are completed, they are united into a single entity by the MapJoin station, and then they are immediately split again into N_{Reduce}. Reduce tasks are executed by the delay station Reduce, whose service time distribution corresponds to S_{Reduce}. The gradual acquisition of resources is modeled by ResourcesFork that inserts in the system $K - 1$ tasks that are immediately routed to the FIFO queue station ResoruceUsed. From this station, service completes every 400 ms. Resources constraints are modeled by the finite capacity region FCR that includes stations Map, Reduce and ResourceUsed, and that is characterized by total capacity K. In this way, at the beginning, only one extra task is allowed to enter the FCS from the MapFork station. As customers leaves the ResourceUsed stations, new computation nodes become available, and new tasks are allowed to start being served. Stations ReduceJoin, ResourceJoin and JobJoin are used to restore the single job and return it to the reference station. In this model, the average response time corresponds to the average time to complete a map-reduce job, with the considered resource acquisition policy, and should correspond to the one computed by the FSPN shown in Fig. 10. Running JMT, we have obtained an average response time $R = 17.528$ s, which differs from the result obtained with the FSPN model for 1.78 %. This shows that the fluid approach can indeed provide good approximation, at a much lower computational effort. Fluid model solution required less than one second, while simulation needed around 6 min and 30 s.

6　Conclusions

This paper has proposed a characterization of single node, single class, multiple server, pool depletion system, and has shown how the average completion time of this type of models can be efficiently described by a fluid approximation. The importance of this type of system has been proven by applying the proposed technique to study incremental resource acquisition in map-reduce task execution.

The main current limitation is the way in which the fluid evolution function is determined, since it still exploits measurements or discrete event simulation. Future work will exploit the proposed characterization to define the fluid model starting from system parameters like the number of nodes, the number of tasks, and the average and coefficient of variation of the service time distribution. Moreover, the proposed technique will be used to study the completion time in Spark jobs characterized by more complex fork-join structures and task execution policies.

Thanks to their efficiency, fluid model will be exploited into optimization algorithm to dynamical configure the system by setting the proper number of nodes, number of tasks and tasks duration to reach proposed KPI (*Key Performance Indicators*) with the lowest possible expense in terms of utilization and energy consumption.

Acknowledgements. The results of this work have been [partially] funded by EUBra-BIGSEA (690116), a Research and Innovation Action (RIA) funded by the European Commission under the Cooperation Programme, Horizon 2020 and the Ministrio de Cincia, Tecnologia e Inovao (MCTI), RNP/Brazil (grant GA-0000000650/04).

References

1. Hadoop website (2013). https://hadoop.apache.org/docs/r1.0.4/
2. Amari, S.V., Misra, R.B.: Closed-form expressions for distribution of sum of exponential random variables. IEEE Trans. Reliab. **46**(4), 519–522 (1997)
3. Barbierato, E., Gribaudo, M., Iacono, M.: A performance modeling language for big data architectures. In: Proceedings of the 27th European Conference on Modelling and Simulation, ECMS 2013, Ålesund, Norway, 27–30 May 2013, pp. 511–517 (2013). http://dx.doi.org/10.7148/2013-0511
4. Barbierato, E., Gribaudo, M., Iacono, M.: Performance evaluation of NoSQL big-data applications using multi-formalism models. Future Gener. Comput. Syst. **37**, 345–353 (2014). http://dx.doi.org/10.1016/j.future.2013.12.036
5. Barbierato, E., Gribaudo, M., Iacono, M.: Modeling hybrid systems in SIMTHESys. In: Proceedings of the 8th International Workshop on Practical Applications of Stochastic Modelling (2016, to appear)
6. Bertoli, M., Casale, G., Serazzi, G.: JMT: performance engineering tools for system modeling. SIGMETRICS Perform. Eval. Rev. **36**(4), 10–15 (2009)
7. Bodrog, L., Gribaudo, M., Horvth, G., Mszros, A., Telek, M.: Control of queues with map servers: experimental results. In: Matrix-Analytic Methods in Stochastic Models: MAM8 - 2016, Kerala, India, 8–10 January 2014, Proceedings, pp. 9–11 (2014)

8. Castiglione, A., Gribaudo, M., Iacono, M., Palmieri, F.: Exploiting mean field analysis to model performances of big data architectures. Future Gener. Comp. Syst. **37**, 203–211 (2014). http://dx.doi.org/10.1016/j.future.2013.07.016

9. Cerotti, D., Gribaudo, M., Iacono, M., Piazzolla, P.: Modeling and analysis of performances for concurrent multithread applications on multicore and graphics processing unit systems. Concurrency Comput.: Pract. Exp. **28**(2), 438–452 (2016). http://dx.doi.org/10.1002/cpe.3504

10. Cerotti, D., Gribaudo, M., Pinciroli, R., Serazzi, G.: Stochastic analysis of energy consumption in pool depletion systems. In: Measurement, Modelling and Evaluation of Dependable Computer and Communication Systems - 18th International GI/ITG Conference, MMB & DFT 2016, Münster, Germany, 4–6 April 2016, Proceedings, pp. 25–39 (2016). http://dx.doi.org/10.1007/978-3-319-31559-1_4

11. Chen, K., Powers, J., Guo, S., Tian, F.: CRESP: towards optimal resource provisioning for mapreduce computing in public clouds. IEEE Trans. Parallel Distrib. Syst. **25**(6), 1403–1412 (2014)

12. Dean, J., Ghemawat, S.: MapReduce: simplified data processing on large clusters. Commun. ACM - 50th Anniversary Issue: 1958–2008 **51**(1), 107–113 (2008)

13. Hillston, J.: A Compositional Approach to Performance Modelling. Cambridge University Press, New York (1996)

14. Kambatla, K., Pathak, A., Pucha, H.: Towards optimizing hadoop provisioning in the cloud. In: Proceedings of the 2009 Conference on Hot Topics in Cloud Computing, HotCloud 2009. USENIX Association, Berkeley (2009). http://dl.acm.org/citation.cfm?id=1855533.1855555

15. Lazowska, E.D., Zahorjan, J., Graham, G.S., Sevcik, K.C.: Quantitative System Performance: Computer System Analysis Using Queueing Network Models. Prentice-Hall Inc., Upper Saddle River (1984)

16. Marsan, M.A., Balbo, G., Conte, G., Donatelli, S., Franceschinis, G.: Modelling with Generalized Stochastic Petri Nets, 1st edn. Wiley, New York (1994)

17. Nickolls, J., Buck, I., Garland, M., Skadron, K.: Scalable parallel programming with CUDA. Queue **6**(2), 40–53 (2008). http://doi.acm.org/10.1145/1365490.1365500

18. Song, G., Meng, Z., Huet, F., Magoules, F., Yu, L., et al.: A hadoop mapreduce performance prediction method. HPCC **2013**, 820–825 (2013)

19. Verma, A., Cherkasova, L., Campbell, R.H.: Resource provisioning framework for mapreduce jobs with performance goals. In: Kon, F., Kermarrec, A.-M. (eds.) Middleware 2011. LNCS, vol. 7049, pp. 165–186. Springer, Heidelberg (2011)

20. Vianna, E., Comarela, G., Pontes, T., Almeida, J., Almeida, V., Wilkinson, K., Kuno, H., Dayal, U.: Analytical performance models for mapreduce workloads. Int. J. Parallel Program. **41**(4), 495–525 (2013). http://dx.doi.org/10.1007/s10766-012-0227-4

21. Wang, G., Butt, A.R., Pandey, P., Gupta, K.: A simulation approach to evaluating design decisions in mapreduce setups. In: MASCOTS, pp. 1–11. IEEE Computer Society (2009). http://dblp.uni-trier.de/db/conf/mascots/mascots2009.html#WangBPG09

22. Zaharia, M., Chowdhury, M., Franklin, M.J., Shenker, S., Stoica, I.: Spark: cluster computing with working sets. In: Proceedings of the 2nd USENIX Conference on Hot Topics in Cloud Computing, HotCloud 2010, p. 10. USENIX Association, Berkeley (2010). http://dl.acm.org/citation.cfm?id=1863103.1863113

23. Zaharia, M., Das, T., Li, H., Hunter, T., Shenker, S., Stoica, I.: Discretized streams: fault-tolerant streaming computation at scale. In: Proceedings of the Twenty-Fourth ACM Symposium on Operating Systems Principles, SOSP 2013, pp. 423–438. ACM (2013)

A Smart Neighbourhood Simulation Tool for Shared Energy Storage and Exchange

Michael Biech, Timo Bigdon, Christian Dielitz, Georg Fromme,
and Anne Remke[(✉)]

Westfälische Wilhelms-Universität, Münster, Germany
{michael.biech,timo.bigdon,christian.dielitz,georg.fromme,
anne.remke}@uni-muenster.de

Abstract. Funding policies and legislation by the European Union for
the installation of photovoltaic (PV) arrays in the residential sector have
led to a steady and successful increase in renewable energy generation by
small-scale private producers. In Germany, even the installation of local
storage systems is being subsidised. Differing feed-in tariffs, as well as
variable production and demand profiles result in very dissimilar amortisation curves for such investments. This paper presents a software tool
which allows for the computation of such curves by means of simulation
in MATLAB/Simulink. It enables the exploration of a wide search space
by manipulating settings on the levels of individual houses, as well as
entire neighbourhoods which might want to share in (the cost of) local
energy storage. Additionally, a case study underlines the tool's potential,
as well as benefits of shared energy storage systems.

1 Introduction

Within the European Union – and Germany in particular – legislation and funding policies have lead to a steady and successful increase in adoption of renewable
energy technology by residential producers. This proves beneficial to consumers
as they need not spend as much money buying electricity from utility companies.
Additionally, in many countries (e.g. Germany, Austria, France to name but a
few) consumers are presented with the opportunity to sell off excess energy to
the grid, receiving reimbursement via feed-in tariffs. At peak production times
however, consumer-produced energy may endanger grid stability. One approach
to mitigate this issue, is to reduce the amount of energy sold to the grid by
installing battery storage, thus greatly increasing the on-premise use of locally
generated energy in a financially viable manner [18].

Another approach to increase local consumption, is through the use of neighbourhood energy exchange, in which one participant's excess production may be
applied towards another member's consumption needs. This trade might take
place by connecting all homes in a given neighbourhood to a *Microgrid Controller* (MGC), which takes care of the electricity routing and load balancing
process. With feed-in tariffs declining, investing into shared energy storage and

© Springer International Publishing Switzerland 2016
S. Wittevrongel and T. Phung-Duc (Eds.): ASMTA 2016, LNCS 9845, pp. 76–91, 2016.
DOI: 10.1007/978-3-319-43904-4_6

a Microgrid Controller may prove increasingly profitable. Setups such as these will be referred to as "smart neighbourhoods" from here on in.

In this paper, we present a software tool which allows for the simulation of such smart neighbourhoods, taking into account the interaction between a large number of factors, e.g. battery storage capacity, photovoltaic array surface area, feed-in tariff compensation and a model for internal energy trade, all in the context of financial viability over extended periods of time (10+ years). The tool helps to answer questions such as "How (much) does internal energy trade impact amortisation?" and "How does battery size impact self-use?". Since the tool permits easy modification of its input parameters, it provides the user with a convenient working environment to conduct case studies. We present sample results generated by our tool for inputs based on German and Dutch markets using realistic neighbourhood scenarios. Specifically, we investigate self-use and the financial consequences of different investment decisions. Furthermore, the presented tool may also be of use to home owners or housing development companies, as it combines complex interactions into an easy-to-use utility, which allows additional research to be conducted by outside parties.

While there is large body of related work available, to the best of our knowledge, such a tool has not yet been made available to the public. Early Simulink models of a stand-alone PV system without battery storage have been presented in [16,27], while one of the first neighbourhood simulation models can be found in [23]. Monetary aspects are considered in [3], however still without local storage facilities. [28] investigates energy trading between houses with local electricity storage. However, economical considerations regarding battery capacity or feed-in tariff compensation are not addressed. More practical approaches include [7], which considers *virtual microgrids*, allowing connected neighbourhoods to sell their generated electricity at more attractive conditions in a larger, competitive marketplace. Exchange of energy in between these virtual microgrids is not addressed. A web application for monitoring and managing Smart Grid neighbourhoods is proposed in [22]. Also, a vast array of research has been conducted regarding analytical and optimisation strategies for smart grids, e.g. [15,20,24,25,31].

The paper is organized as follows and as indicated in Fig. 1: *Input Data* is described in Sect. 2, while *User Scenarios* (i.e. neighbourhoods), and *Simulation Results* are addressed in Sect. 4. Implemented components and *Runnable Model* – which is considered data, since it is generated by our tool – are covered in Sect. 3. Section 5 sums up our findings and briefly outlines potentially interesting future developments.

2 Production and Demand Profiles

To obtain realistic results, it is important to supply detailed and accurate profiles as input data. Sections 2.1 and 2.2 introduce production and demand profiles respectively, while Sect. 2.3 discusses feed-in tariffs.

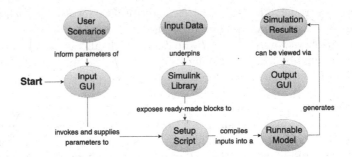

Fig. 1. Overview of the model's workflow. Green items (middle row and bottom centre) represent implemented components, while blue items (top row and bottom right) denote data that goes into or comes out of the model. (Color figure online)

2.1 Production

Solar panel production profiles were obtained from *PVWatts* [2] for *Düsseldorf*, Germany. The data was exported using standard settings, with the exception of *DC System Size (kw)*, which was set to *1* to allow for dynamic scaling of each house's solar panel array within a simulation. The scaling factor x for a $1\,\text{kWp}^1$ array was determined to be $\dot{x} = 0.16 \cdot y$, where y indicates solar panel array surface area. This formula is a simplification of the calculation performed by PVWatts internally [11]. From the exported data set, we extracted the *AC production* column for further use. These values already account for losses incurred through conversion from DC to AC (96 % efficiency) and other factors, such as light-induced degradation, shading and soiling. Note that we always consider PV installation sizes (m²) to mean *effective* array surface areas (i.e. only just the solar cell area). A decrease of efficiency over time, as well as maintenance or repairs that might be necessary during the lifetime of solar panels are not considered. Refer to Fig. 2 for a graphical representation of PV production values over time.

2.2 Demand

Residential demand profiles were obtained from the Dutch organization *NEDU* [17]. Considering that Germany and the Netherlands exhibit very similar electricity consumption per capita [13], we consider the data gathered to be representative for our purposes. From the multicolumn dataset supplied, column *E1A* was chosen, i.e. loads of $< 3 \cdot 25$ amperes, using a single electricity metre. The respective values are given at a resolution of 15 min intervals and represent the percentage of annual energy consumption during this time frame. However, since PVWatts production data only has a resolution of one data point per hour,

---

[1] kWp = Kilowatt-peak, i.e. the power rating of a photovoltaic array in kilowatts (kW) at standard test conditions [2].

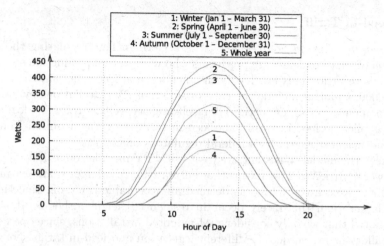

Fig. 2. Energy production in Watts over the course of seasonally typical days for a PV installation producing an average of 11 kWh per day using the data supplied by PVWatts.

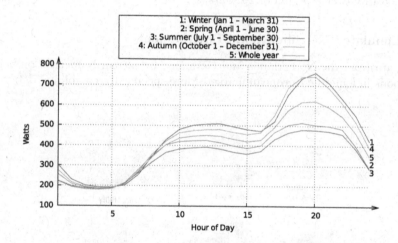

Fig. 3. Consumption in Watts over the course of seasonally average days for an annual residential demand of 3,500 kWh, using the data supplied by NEDU.

four consecutive values of electricity consumption data were added up. Multiplying each resulting value by a given annual consumption yields results in Watts for any given hour of the year. See Fig. 3 for a visual representation of the resulting dataset at 3,500 kWh of consumption per annum.

2.3 Feed-in Tariff

The *Renewable Energy Sources Act* [1] is a German law, regulating the compensation of small-scale energy producers who wish to feed production surpluses into the power grid. This law has been amended multiple times since its inception in 2000. Changes include the categorisation of PV installations by nominal kWp, compensation or deductions for self-use, as well as rates of degression for the applicable feed-in tariff. The feed-in tariff data was acquired from the *Bundesnetzagentur*, Germany's federal regulatory office [6].

To keep the model consistent, we treat every PV installation as if it its power output were $\leq 10\,\mathrm{kWp}$. This means that currently, arrays larger than $62.5\,\mathrm{m}^2$ are not handled accurately. This was deemed an acceptable trade-off, since installations larger than this rarely occur in residential neighbourhoods. It should also be noted that we only consider roof-mounted installations, since open space PV installations are subject to different legislation and feed-in tariffs. For more detailed information regarding the structure and intricacies of feed-in tariffs we refer to [9]. We make no assumptions about the future development of feed-in tariffs. The data we supply ends with December 2015. The feed-in tariff is paid at a fixed rate over a period of 20 years.

2.4 Simulated Neighbourhood Environment

With data from the Netherlands and Germany, we primarily focused on neighbourhoods in a middle European market. More specifically, combining data from

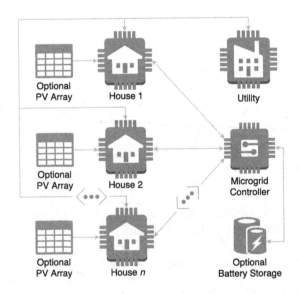

Fig. 4. Schematic representation of the smart neighbourhood model. Arrowheads indicate the flow of electricity. Please note that in one imaginable alternative to this setup, the Microgrid Controller might sit between *Utility* and the houses, thus potentially saving additional wiring and setup cost.

NEDU, the Bundesnetzagentur and PVWatts firmly puts our simulation in a longitude range of 49° and 55° within Germany's latitude. However, please note that the input data described above can easily be exchanged, since it is simply stored using CSV files. This gives rise to much greater flexibility for future experiments.

All houses feature the same orientation (due south), roof tilt (20°), and their respective solar panels do not track the sun. Conditions like these might be found in newly-built or later-equipped uniform housing developments. In practice, a mixture of orientations might provide considerably beneficial [30]. Since we explore the sharing of energy storage as well as trading energy between houses in a neighbourhood, we assume that required communications between houses and the Microgrid Controller are available, safe, secure and stable. The cost for energy storage, given in Euro (€), is split evenly among participants in the neighbourhood. It is decidedly debatable whether certain parties ought to account for more or less of these expenses, depending on what they add to or subtract from the local grid. However, tackling this question was considered to reach beyond the scope of this paper.

Figure 4 depicts a schematic neighbourhood setup as described above.

3   Model Implementation

The model and its components were created using MATLAB version *2015a* and Simulink version *8.5*, the latter of which allows users to build models in a hierarchical manner and to create so-called libraries, which mask the complexity of implemented blocks. This allows for the creation of new, more complex models relatively quickly and easily.

Section 3.1 discusses the *Input GUI*, Sect. 3.2 introduces the newly created Simulink blocks, while Sects. 3.3 and 3.4 present the connecting layer which creates the runnable simulation model and the user interface for outputs respectively.

3.1   Input GUI

We have developed an *Input Graphical User Interface* (Input GUI) through which simulations for given scenarios are configured by means of a large array of settings. Additionally, neighbourhoods – defined by their constituent houses – are defined in this interface as well. The GUI was created using MATLAB's *Graphical User Interface Design Environment* (GUIDE), while its specific functionality – hand-off to the *Setup Script* (cf. Sect. 3.3), adding entries to the "Houses" table etc. – were implemented using standard MATLAB code. In the following, we describe all variables which can be adjusted via the interface.

Houses are defined by three parameters: *Solar panel size*, given in square meters (m^2), *Yearly power demand*, given in kilowatt hours (kWh) and *Date of PV installation*[2], given in the date format MM-YYYY. These values default to $30\,m^2$, 3,500 kWh and 05-2014, respectively. This roughly equals the average consumption of a 2-person-household [12] and a PV output of 4.8 kWp. On average this default installation produces 11 kWh per day, i.e. more than the mean consumption of 9.5 kWh per day.

Battery cost per kWh defines the cost of one kWh of battery capacity. The default value of €500 per kWh was obtained by dividing the advertised cost of a *Tesla Powerwall* [4] by its capacity ($\frac{\$3,000}{7\,kWh} \approx \frac{\$430}{kWh}$) and rounding up to the next hundred, since Tesla's Powerwall is priced very aggressively. Other products might be twice as expensive per kWh of capacity [8].

Battery capacity specifies the effectively available energy storage capacity. This value is then multiplied by *Battery cost per kWh* to obtain the total cost of local storage. The default value of 7 kWh was chosen for two reasons. Firstly, Tesla's Powerwall is primarily advertised at this capacity [4]. More importantly however, this value aligns with results in [29]: choosing a battery which provides enough capacity to provide half of an average day's demand is shown to cover 70 % to almost 80 % of said demand, depending on season. Doubling the battery capacity only increases self-use by another 10 %, while doubling the acquisition cost.

One-time cost subsumes any additional expenses that come with the installation of neighbourhood storage and a Microgrid Controller. For example, this includes costs for the Controller itself, as well as wiring between houses, if necessary. Costs in this field are split evenly among participants in the neighbourhood.

Initial cost per kWh from utility defines the amount charged by the utility company for every kWh that is withdrawn from the external (i.e. non-neighbourhood) grid. €0.2881 was chosen as a default value since it was the average cost per kWh in Germany at the time of implementation (2015) [5].

Annual change of cost per kWh from utility determines the annual rate of change regarding the cost per kWh from the utility. This value is calculated using the compound interest formula $x_1 * (1 + x_2/100)^{x_3}$, where x_1 is the initial cost per kWh (see above), x_2 the percentage of change described here and x_3 the year for which the cost is calculated. The default value of 3.4 % was chosen because it was the average rate of change between the years of 1998 and 2014 in Germany [5]. However, note that for time frames exceeding 15 years, this assumption might no longer hold. In fact, the cost of electricity might or might not stabilise within a given number of years [10].

Date of battery installation provides the date on which the battery becomes operational within the neighbourhood, which must lie *after* every PV installation date in the neighbourhood, since this is the starting date of our simulation. January 2016 was chosen as a default value since it is the first date for which no feed-in tariff data is being provided by us any more.

---

[2] Technically this is inaccurate, since compensation is only paid starting with the first day that electricity is produced and fed into the grid instead of depending on the date of installation [1].

Solar panel efficiency directly influences the amount of energy that is produced by the PV installations. Refer to Sect. 2.1 for the calculations involved. The default value of 16 % for this field reflects the value provided by PVWatts.

Cost per kWh bought from battery and *Compensation per kWh sold to battery* determines the amount of money that is being exchanged when handing off electricity to, or receiving energy from the Microgrid Controller. The values chosen (i.e. 18 cents per kWh) show a slight bias towards buyer's advantage.

Duration of simulation determines the number of years for which the simulation is run. Internally, this number is multiplied by 8,760 (hours in a year), as we simulate at the resolution of hours. The default value of 15 years was chosen because our experiments have shown that this allows for a reasonable assessment of whether, and, if so, when amortisation will occur.

3.2   Simulink Library

This section describes the Simulink blocks which were implemented in more detail. The library provides two blocks: a *House* and a *Microgrid Controller*. The former may exist multiple times in a given scenario, while the latter only exists once within a simulated neighbourhood. As outlined below, the *Setup Script* is used to create a runnable model. All of the blocks described in the following were implemented from scratch, making use only of components available through the basic Simulink block library. In the following, select aspects from each block are highlighted. For detailed usage instructions, we refer the reader to the README file which accompanies the project's files[3].

The *Microgrid Controller* allows for internal energy trade to take place between houses in a neighbourhood and implements different battery management strategies for local storage (if present). A given house may transfer "excess" energy to houses which need additional energy. Such exchange reduces the total amount of energy sold to the grid and increases local self-use. In order to make this a fair process for all parties involved, an internal compensation scheme is necessary. The exact monetary values per kWh are not easily determined, however, appropriate values lie between sale and purchase rates of electricity to and from the grid respectively.

The controller sends a signal to each house with information regarding energy usage at the given time step. This signal consists of three floating point values, which represent the respective percentages of energy that was either traded internally (p2), sold to (p1), or bought from the grid (p3).

We have implemented two battery management strategies, namely *Greedy* and *Smart*. The first makes full use of the battery's capacity at all times, i.e. completely charging and depleting it whenever possible. Using the *Smart* strategy, during normal operation, the battery is only ever drained to the point where 30 % of its charge remains. This is done for two reasons: (1) to prolong the battery's life, as completely depleting and charging it repeatedly has averse

---

[3] Available at http://www.uni-muenster.de/Informatik.AGRemke/forschung/tools/ smart-neighbourhood.html.

effects on its chemistry in the long run, reducing the number of life cycles [14] and (2) increasing survivability for connected houses [14]. For both strategies, the "sold" signal is non-zero only if the battery is fully charged and the "buy" signal is non-zero only if the battery is completely discharged (or retains a charge of 30 % in case of the *Smart* strategy). The decisions regarding what "type" of energy (i.e. external, internal) flows where (i.e. to the grid, to the battery, to the homes) is split between the houses themselves and the Microgrid Controller. See Fig. 5 for the whole logic flowchart.

A *House* has two main functions. Firstly, it adds a normally distributed random value with an average of zero and a variance of 1.301410^{-10} to the demand for more realistic profile behaviour. Secondly, an *Electricity Cost Calculation* is performed, which – based on a wide array of internal parameters – determines cost/compensation, while taking into account the current battery charge. Since connections within the model carry vector information, cost is calculated for all available strategies simultaneously, including a version without Microgrid Controller for later comparison.

3.3   Setup Script and Runnable Model

The *Setup Script* acts as the connecting layer between the library blocks and the input data from the GUI (see above). Written in the MATLAB language, it creates a runnable model, which can then be executed and/or saved for later use by the user. In reality, the *Setup Script* consists of four different scripts: *GridController.m* and *House.m* pull in and set the values of their corresponding blocks from the library. *Simulink.m* merely contains static values which are referenced in other files. Finally, *System.m* connects all the blocks, sets up Workspace Variables and creates the actual runnable simulation model.

3.4   Output GUI

The Output GUI allows users to view and compare simulation results. All generated data can be viewed in the form of four different graphs using a drop-down menu.

Grid Sale Gradient graphs show the amount of energy sold to the utility for each battery strategy. A graph which represents the energy that would be sold to the grid if no Microgrid Controller were present at all is displayed as well.

Split Amortisation graphs show gains and/or losses incurred through the sale and/or purchase of electricity to/from the grid and internal trade. As with the grid sale gradient, a line for each battery strategy is shown, plus an additional one, representing the behaviour if a Microgrid Controller were absent.

Merged Amortisation graphs present money saved and/or directly gained through feed-in tariff compensation and neighbourhood trade. They essentially show the delta between how much would have to be spent if no Microgrid Controller were present and the different battery strategies *with* an MGC.

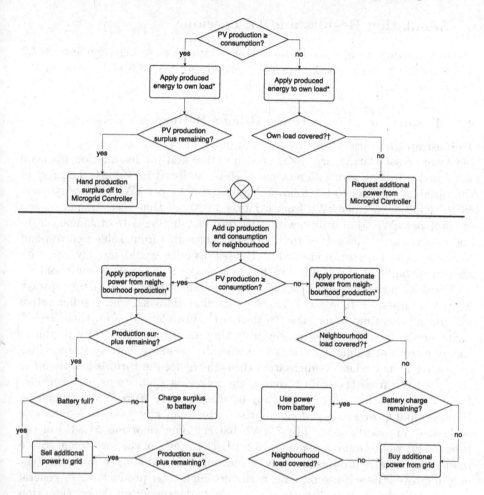

Fig. 5. Logic flowchart for internal trade, storage to and retrieval from battery, as well as sale to and purchase from the utility grid. The horizontal bar delineates the switch between which components are involved in making the decisions. Above the bar, each house applies as much of its production to its own demand as possible. Below the bar, the Microgrid Controller takes over and handles further decisions (see above). *Consumption and/or production can evaluate to 0 after these processes. †Will always evaluate to false, since for these paths PV production < consumption holds.

Neighbourhood Battery Charge Over Time graphs represent the charge stored in the neighbourhood battery over the course of the simulation. For periods of time > 1 year this graph might be difficult to read.

4   Simulation Results and Discussion

Sect. 4.1 presents results for a single house with local storage, while Sect. 4.2 discusses results for a possible neighbourhood and Sect. 4.3 elaborates on the benefits of heterogenous neighbourhoods.

4.1   Results for a Single House Using a Battery

To illustrate the variety that comes with different "types" of houses (i.e. combinations of solar array size, yearly consumption and PV installation size), we simulated 6 houses with different parameters, as listed in Table 1. The rest of the variables were set to the following values: €500,– per kWh battery storage, €0.2881 initial cost per kWh from utility, €1,000 one-time cost, annual change of 3,4 % per kWh from utility, with a battery installation date of January 2016 and a solar panel efficiency of 16 %. Houses 1 through 4 from Table 1 correspond to an average 4 person home, while Houses 5 and 6 would roughly equal an apartment building of 6 families with 2 to 3 people living in each household.

Figure 6 shows the amortisation of the investments for local energy storage for each house over 10 years. It can be seen that Houses 1 and 2 suffer active loss during this time frame. Due to their early installation dates, their feed-in tariffs are so high, they lose money each time they consume locally produced energy instead of selling it to the grid. Generally speaking, battery storage thus only saves money when compensation through the feed-in tariff is less than the cost per kWh from the utility. Hence, the following simulations only consider PV installations that start operation in or after January 2015. When comparing Houses 5 and 6, we see that doubling the battery capacity, leads to a difference of 31.12 % in amortisation. The 7 kWh battery only amortises 51.33 % of the investment, while House 6 manages to get back 37.26 % of the investment. Since most of the locally produced energy is consumed immediately, a larger battery only improves the self-use on days with very high solar production. In general we can state that battery amortisation is highly dependent on the consumption of a given house in relation to its solar panels size and installation date.

Table 1. Settings used to compare the houses in Fig. 6

Identifier	Annual consumption [kWh]	PV surface area [m^2]	PV installation date	Battery capacity [kWh]
House 1	5,000	40	01-2004	7
House 2	5,000	40	01-2010	7
House 3	5,000	40	01-2015	7
House 4	5,000	40	01-2015	14
House 5	27,000	60	01-2015	7
House 6	27,000	60	01-2015	14

4.2 Results for Houses with Different Setups

Trade only occurs when the applied variance has shifted the consumption patterns of houses in such a way where one house produces more energy than it consumes, and at least one other house can not cover its demand by itself. Hence, grouping houses with the same settings into a neighbourhood only results in a minimal amount of internal trading. Therefore, we simulated a more heterogenous neighbourhood consisting of 3 houses, both with and without a shared 14 kWh battery. The exact settings used can be seen in Table 2. To ease comparison, we assume that all houses started producing energy in January 2015.

In Fig. 7 we see that House 2 (with battery) has already amortised its investment within 10 years and has made around €488 of profit on top of that. The other 2 houses (also, with battery) have not amortized yet. House 1 has gained €2,853.– and house 2 has gained €2,413.– during the 10 years of simulation. When comparing the houses that only invested in the Microgrid Controller and no battery, we see that neither house indicates major improvements and that the best of the 3 – House 2 – has only saved €49.94 in total. Note that while in reality consumption patterns will differ even more, [28] suggests that even then, internal neighbourhood trade only slightly improves self-use. The amount of energy sold to the grid was reduced by 30 % to 35 % at the end of the 10-year period (grid-sale gradient graphs are not shown in this paper).

4.3 Adding Houses Without Solar Panels to the Neighbourhood

This section shows how adding houses without solar panels to the smart neighbourhood results in an advantage for all houses in that neighbourhood. Graphs in this section show results for the best and worst houses respectively from Fig. 7. Results for House 5 are not included, since its behaviour is identical to that of House 4, only with a steeper amortisation curve.

Figure 8 shows that Houses 2 and 3 have amortised in both cases. When comparing the results of those 2 houses with the results we saw in Fig. 7, it can be noticed, that a lot more energy was traded between them. House 2 has earned €518.40 employing only trade (i.e. making no use of battery storage). This is more than 10 times the amount it would earn when only houses with solar panels were in the neighbourhood. House 3 has earned €647.44 in 10 years.

Table 2. Settings used for the scenarios depicted in Fig. 8

Identifier	Annual consumption [kWh]	PV surface area [m^2]
House 1	4,200	35
House 2	6,000	45
House 3	3,200	40
House 4	5,000	0
House 5	5,500	0

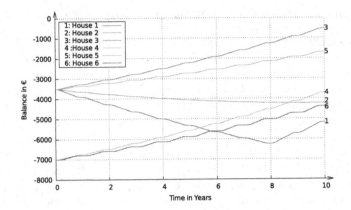

Fig. 6. Amortisation of battery purchases by independent houses

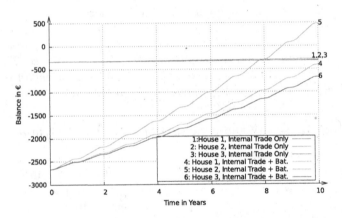

Fig. 7. Amortisation in a heterogeneous neighbourhood with and without shared battery

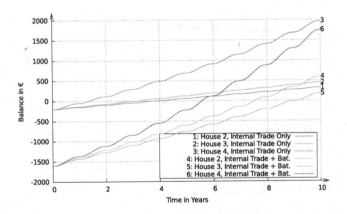

Fig. 8. Amortisation of houses investing in a Microgrid Controller

The immense improvement in profit incurred through trade, as well as the cost of the 14 kWh battery being split evenly between 5 houses, has also led to House 2 amortising its investment in less than 8 years. Even House 3 – which had performed worse in Fig. 7 – has now made money from investing into the Microgrid Controller. Houses without any solar panels are currently saving the most money. This could be somewhat mitigated by agreeing on a higher internal trading price or shifting more of the initial costs for the wiring onto these houses.

Please note that due to the application of a small variance for demand profiles, confidence intervals are insignificant and thus not presented in this paper.

5 Conclusions

In this paper, we have presented: (1) a toolkit for the simulation of smart neighbourhood environments, which allows for energy trade between participants with the possibility of additional battery storage, (2) simulation results of our tool for realistic neighbourhood scenarios based on German and Dutch production and consumption data respectively. The former allows researchers to conduct further experiments on their own, while the latter has shown that (1) houses with recently installed solar panels benefit the most from investing in energy storage, which is caused mainly by feed-in tariff degression, (2) energy trading within a neighbourhood yields the best results when patterns in production and consumption are dissimilar, (3) adding houses that provide no production to a smart neighbourhood, can improve the monetary situation for all houses involved in the neighbourhood, i.e. splitting initial costs while also increasing local energy usage. Ultimately, investing in a Microgrid Controller and central battery storage can indeed be profitable for stakeholders, if certain conditions are met. Substituting input data files, researchers and other interested parties alike could modify the model to fit their needs. This would allow for the analysis of scenarios like the prospective construction of a connected neighbourhood. We have shown that the results our tool generates line up well with the existing literature, suggesting that the described toolkit might well be suited to investigate cases which have not yet been covered by existing research.

Future work might take into account non-linear battery models [21] and additional grid-convenient battery strategies (e.g. *Peak Shaving, Load Balancing* [26]), as well as a dynamic stock exchange model, as proposed in [19].

References

1. Gesetz für den Ausbau erneuerbarer Energien (in German). Bundesgesetzblatt I(33), 1066, July 2014
2. PVWatts. http://pvwatts.nrel.gov/
3. System Advisor Model (SAM). https://sam.nrel.gov/
4. Tesla Powerwall. https://www.teslamotors.com/powerwall
5. BDEW Bundesverband der Energie - und Wasserwirtschaft e.V.: Strompreisanalyse März 2015 (in German), March 2015

6. Bundesnetzagentur für Elektrizität, Gas, Telekommunikation, Post undEisenbahnen: Photovoltaic installations data submissions and EEG-supportedfeed-in tariffs. http://www.bundesnetzagentur.de/cln_1432/EN/Areas/Energy/Companies/RenewableEnergy/PV_data_tariffs/PV_statistic_node.html

7. van der Burgt, J., Sauba, G., Varvarigos, E., Makris, P.: Demonstration of the smart energy neighbourhood management system in the VIMSEN project. In: PowerTech, 2015 IEEE Eindhoven, pp. 1–6, June 2015

8. Bussar, R., Lippert, M., Bonduelle, G., Linke, R., Crugnola, G., Cilia, J., Merz, K.D., Heron, C., Marckx, E.: Battery energy storage for smart grid applications. In: EuroBat, May 2013

9. Couture, T.D., Cory, K., Kreycik, C., Williams, E.: Policymaker's guide to feed-in tariff policy design. Technical report NREL/TP-6A2-44849, National Renewable Energy Laboratory (NREL), Golden, CO, July 2010

10. Darghouth, N.R., Barbose, G., Wiser, R.H.: Customer-economics of residential photovoltaic systems (Part 1): the impact of high renewable energy penetrations on electricity bill savings with net metering. Energy Policy **67**, 290–300 (2014)

11. Dobos, A.P.: PVWatts Version 5 Manual, September 2014. http://www.nrel.gov/docs/fy14osti/62641.pdf

12. EnergieAgentur.NRW GmbH: Erhebung "Wo im Haushalt bleibt der Strom?" (in German), April 2011

13. European Environment Agency: Electricity consumption per capita (in kWh/cap) in 2008. http://www.eea.europa.eu/data-and-maps/figures/electricity-consumption-per-capita-in-1

14. Ghasemieh, H., Haverkort, B., Jongerden, M., Remke, A.: Energy resilience modelling for smart houses. In: 2015 45th Annual IEEE/IFIP International Conference on Dependable Systems and Networks (DSN), pp. 275–286, June 2015

15. Guo, Y., Pan, M., Fang, Y.: Optimal power management of residential customers in the smart grid. IEEE Trans. Parallel Distrib. Syst. **23**(9), 1593–1606 (2012)

16. Hansen, A.D., Sørensen, P.E., Hansen, L.H., Bindner, H.W.: Models for a standalone PV system (2001)

17. Hermans, P.: Role of DATA Handling in the Dutch Energy Market. http://www.smartgrids.eu/documents/eventsandworkshops/2015/Stedin_Hermans_Peter-Role_of_DATA_handling_in_the_Dutch_energy_market.pdf

18. Hoppmann, J., Volland, J., Schmidt, T.S., Hoffmann, V.H.: The economic viability of battery storage for residential solar photovoltaic systems – a review and a simulation model. Renew. Sustain. Energy Rev. **39**, 1101–1118 (2014)

19. Ilic, D., Da Silva, P., Karnouskos, S., Griesemer, M.: An energy market for trading electricity in smart grid neighbourhoods. In: 2012 6th IEEE International Conference on Digital Ecosystems Technologies (DEST), pp. 1–6, June 2012

20. Joe-Wong, C., Sen, S., Ha, S., Chiang, M.: Optimized day-ahead pricing for smart grids with device-specific scheduling flexibility. IEEE J. Sel. Areas Commun. **30**(6), 1075–1085 (2012)

21. Jongerden, M., Haverkort, B.: Which battery model to use? IET Softw. **3**(6), 445–457 (2009)

22. Karnouskos, S., Da Silva, P., Ilic, D.: Developing a web application for monitoring and management of smart grid neighborhoods. In: 2013 11th IEEE International Conference on Industrial Informatics (INDIN), pp. 408–413, July 2013

23. Morvaj, B., Lugaric, L., Krajcar, S.: Demonstrating smart buildings and smart grid features in a smart energy city. In: Proceedings of 2011 3rd International Youth Conference on Energetics (IYCE), pp. 1–8, July 2011

24. Paterakis, N., Pappi, I., Catalao, J., Erdinc, O.: Optimal operational and economical coordination strategy for a smart neighborhood. In: PowerTech, 2015 IEEE Eindhoven, pp. 1–6, June 2015
25. Samadi, P., Mohsenian-Rad, A.H., Schober, R., Wong, V., Jatskevich, J.: Optimal real-time pricing algorithm based on utility maximization for smart grid. In: 2010 First IEEE International Conference on Smart Grid Communications (SmartGridComm), pp. 415–420, October 2010
26. Sterner, M., Eckert, F., Thema, M., Bauer, F.: Der positive Beitragdezentraler Batteriespeicher für eine stabile Stromversorgung (in German), Forschungsstelle Energienetze und Energiespeicher (FENES) OTH Regensburg, Kurzstudie im Auftrag von BEE e.V. und Hannover Messe, Regensburg/Berlin/Hannover (2015)
27. Ursachi, A., Bordeasu, D.: Smart grid simulator. Int. J. Civ. Archit. Struct. Constr. Eng. **8**, 519–522 (2014)
28. Velik, R.: Battery storage versus neighbourhood energy exchange to maximize local photovoltaics energy consumption in grid-connected residential neighbourhoods. Int. J. Adv. Renew. Energy Res. **2**(6) (2013)
29. Velik, R.: The influence of battery storage size on photovoltaics energy self-consumption for grid-connected residential buildings. Int. J. Adv. Renew. Energy Res. **2**(6) (2013)
30. Velik, R.: East-West orientation of PV systems and neighbourhood energy exchange to maximize local photovoltaics energy consumption. Int. J. Renew. Energy Res. (IJRER) **4**(3), 566–570 (2014)
31. Xiao, L., Mandayam, N., Poor, H.: Prospect theoretic analysis of energy exchange among microgrids. IEEE Trans. Smart Grid **6**(1), 63–72 (2015)

Fluid Analysis of Spatio-Temporal Properties of Agents in a Population Model

Luca Bortolussi[1] and Max Tschaikowski[2]([⊠])

[1] University of Trieste and CNR/ISTI, Pisa, Italy
luca@dmi.units.it
[2] IMT School for Advanced Studies Lucca, Lucca, Italy
max.tschaikowski@imtlucca.it

Abstract. We consider large stochastic population models in which heterogeneous agents are interacting locally and moving in space. These models are very common, e.g. in the context of mobile wireless networks, crowd dynamics, traffic management, but they are typically very hard to analyze, even when space is discretized in a grid. Here we consider individual agents and look at their properties, e.g. quality of service metrics in mobile networks. Leveraging recent results on the combination of stochastic approximation with formal verification, and of fluid approximation of spatio-temporal population processes, we devise a novel mean-field based approach to check such behaviors, which requires the solution of a low-dimensional set of Partial Differential Equation, which is shown to be much faster than simulation. We prove the correctness of the method and validate it on a mobile peer-to-peer network example.

1 Introduction

In this paper we focus on stochastic modeling of spatially distributed and large scale systems of interacting agents [16]. Possible examples are cellular networks [2,3], routing protocols [12], ad-hoc networks [10] or ecological and epidemiological models [17]. In particular, we are concerned with checking properties or random individuals, such as the probability of receiving a file within a certain time in a mobile computer network.

A stochastic model of such systems can be formulated as a Markov Population process, discretizing continuous space in a grid and allowing interactions only locally, among agents in the same cell, or among neighboring cells. These models result in Continuous Time Markov Chains with very large state spaces, and with a large number of different populations (one per each cell and agent type), which are very challenging and expensive to analyze, even by stochastic simulation, especially when large populations of agents are involved.

One strategy to deal with large population models, which has gained a lot of momentum in the last years, is to rely on fluid or mean field approximations (see e.g. [4,7,13]). These techniques approximate a stochastic model by a deterministic one, typically given by a low dimensional set of ordinary differential equations (ODEs). Their correctness in rooted in convergence theorems that

S. Wittevrongel and T. Phung-Duc (Eds.): ASMTA 2016, LNCS 9845, pp. 92–106, 2016.
DOI: 10.1007/978-3-319-43904-4_7

guarantee that the approximation is correct in the limit of a divergent population size [7]. When the interest is on the behavior of a random agent in a large population model, one can rely on results known as fast simulation [9], which have been exploited extensively to approximatively verify behavioral properties, in the so called fluid model checking framework [5,6]. When space is also taken into account, there is an increased level of difficulty. In [16], however, the authors propose a limit framework that approximates a large spatio-temporal stochastic model (on a grid), by a low dimensional set of Partial Differential Equations (PDEs), by first taking the fluid limit and then by showing that the resulting ODEs are a finite difference scheme for the PDE model. Solving directly the PDEs rather than the fluid ODEs or the original stochastic model turns out to be much faster, thanks to efficient meshing strategies.

In this paper we combine together the PDE approximation of spatio-temporal population models with the fast simulation and fluid model checking ideas, providing a fast spatial simulation result and using it to investigate complex properties of a random agent moving and interacting in space. We will give a formal proof of convergence and experimental evidence on a model of a mobile peer-to-peer network of the correctness of the method and of the conspicuous speedup achievable.

The paper is organized as follows. Section 2 introduces the modeling framework, the fluid approximation, fast simulation, and the spatio-temporal approximation by PDEs. Section 3, instead, discusses our contribution, introducing the fast spatial simulation framework and proving its correctness. Section 4 shows some experimental evidence to backup the theoretical results. Conclusions are drawn in Sect. 5.

2 Modeling Framework

In this section, we introduce our modeling framework which combines, essentially, the frameworks of fast simulation [5,8,9] and spatial fluid limits [16]. By doing so, we set the scene for the novel result of fast spatial simulation presented in Sect. 3.

2.1 Stochastic Model

We consider Continuous Time Markov Chains (CTMCs) describing a population of interacting agents. Each agent is described as a simple automata, and it can be in one local state among $A_1, \ldots, A_L$, where $L \geq 1$. The dynamics is described at the collective level by a set of J (local) reactions, with each reaction $1 \leq j \leq J$ defined by:

- a function $F_j : \mathbb{R}^{L+P} \to \mathbb{R}$ that describes the rate at which the local interaction occurs. It is dependent on the populations of agents in each of the L local states, and on $P \geq 0$ rate parameters $\beta_1, \ldots, \beta_P$;

- a multiset R_j of atomic transitions of the form $A_l \to A_{l'}$ and $\to A_{l'}$ with $1 \leq l, l' \leq L$, denoting which individual agents are involved and how they change state: $A_l \to A_{l'}$ describes an agent in state A_l interacting and changing state to $A_{l'}$. R_j can alternatively be a singleton $R_j = \{\to A_{l'}\}$, modelling arrivals of new agents into the system.

The collective state of the model is described by an L-vector of variables, counting how many agents are in each local state. By a little but convenient abuse of notation, we will denote by A_l also such variables.

From a multiset R_j, we can extract two integer valued L-vectors d_j and c_j, counting how many agents in each state are transformed during a transition (respectively produced and consumed). Specifically, for each $1 \leq j \leq J$, let $c_{jl}, d_{jl} \in \mathbb{N}_0$ be such that $c_{jl} = \sum_{A_l \to A_{l'} \in R_j} 1$ and $d_{jl'} = \sum_{A_l \to A_{l'} \in R_j} 1$. With these vectors, we can express equivalently each transition in the chemical reaction style as follows:

$$c_{j1}A_1 + \ldots + c_{jL}A_L \xrightarrow{F_j} d_{j1}A_1 + \ldots + d_{jL}A_L,$$

Note that the multiset notation, with the exclusion of arrivals, describes transitions conserving the number of agents. Departure of agents, instead, is modeled by an absorbing state, referred to as coffin state, which we assume to be A_L. In this case, $c_{j,L} = 0$ for all j, and F_j does not depend on A_L. Instead, if the model under study features no departure of agents, A_L is an ordinary state and F_j may depend also on A_L. F_j may also depend on the *scaling parameter* $N \in \mathbb{N}$, which captures a notion of size of the population under consideration. Typically, this is the total number of agents, or the initial population, or a factor modulating rates and proportional to the average population (in case of open systems, like queuing networks), see [7] for a discussion.

Example 1. The model of the peer-to-peer network studied in [14] can be readily identified as a deterministic limit of the following population based CTMC. Let A_1 denote a downloader, that is a node that does not have the full file in question and let A_2 denote a seed, i.e. a node that has the entire file. Denoting the coffin state by A_3, the atomic transitions are given by

$$R_1 = \{\to A_1\}, R_2 = \{A_1 \to A_3\}, R_3 = \{A_2 \to A_3\}, R_4 = \{A_1 \to A_2, A_2 \to A_2\}$$

This and the rate functions $F_1, \ldots, F_4$ then yield

$$1: \quad 0A_1 + 0A_2 + 0A_3 \xrightarrow{F_1} 1A_1 + 0A_2 + 0A_3, \quad F_1 \equiv \lambda N,$$

$$2: \quad 1A_1 + 0A_2 + 0A_3 \xrightarrow{F_2} 0A_1 + 0A_2 + 1A_3, \quad F_2 \equiv \sigma A_1,$$

$$3: \quad 0A_1 + 1A_2 + 0A_3 \xrightarrow{F_3} 0A_1 + 0A_2 + 1A_3, \quad F_3 \equiv \gamma A_2,$$

$$4: \quad 1A_1 + 1A_2 + 0A_3 \xrightarrow{F_4} 0A_1 + 2A_2 + 0A_3, \quad F_4 \equiv \min\{cA_1, \nu(\eta A_1 + A_2)\}.$$

Reaction 1 describes the arrival rate of new downloaders into the network. Here, the arrival rate scales with the network size N, such that the average number of agents in the network is proportional to N. Interactions 2 and 3, instead, describe with which rate downloaders and seeders leave the system. Note that both agents change their state into the absorbing coffin state 3 and that both reaction rates are proportional to the corresponding agent populations. Reaction 4 describes the rate at which a downloader receives the file. The rate is rooted in a bandwidth-sharing argument between the total requested capacity by downloaders, cA_1, with $c > 0$, and the total upload capacity of the system which is given by the total upload seed capacity plus the downloaders who are sharing their copy. In particular, $0 \leq \eta \leq 1$ is the probability that a requested portion of the file is available at a peer.

The local reactions describe the way in which agents *from the same location* communicate with each other. In our spatial model, we make the assumption that agents obey a random walk on a lattice in the unit square with $(K+1)^2$ regions, denoted by $\mathcal{R}_K := \{(i\Delta s, j\Delta s) \mid 0 \leq i, j \leq K\}$, where $\Delta s := 1/K$. The boundary of the lattice is $\Omega_K := \{(x, y) \mid x \in \{0, 1\} \vee y \in \{0, 1\}\}$. An agent in the interior $\mathcal{R}_K \backslash \Omega_K$ can travel to one of its neighboring regions $(x - \Delta s, y)$, $(x + \Delta s, y)$, $(x, y - \Delta s)$ and $(x, y + \Delta s)$. We write $\mathcal{N}(x, y)$ for the neighboring regions of $(x, y) \in \mathcal{R}_K$, e.g. $\mathcal{N}(0, 0) = \{(\Delta s, 0), (0, \Delta s)\}$. In the following, an absorbing boundary condition is assumed, meaning that agents that migrate to the boundary Ω_K disappear.

The overall state descriptor is thus $A := (A_1^{(x,y)}, \ldots, A_L^{(x,y)})_{(x,y) \in \mathcal{R}_K}$. By denoting the transition rate from state A into state A' by $q(A; A')$, the local reactions induce

$$q\left(A; (A_1^{(x,y)} + d_{j1} - c_{j1}, \ldots, A_L^{(x,y)} + d_{jL} - c_{jL})\right)$$
$$= \sum_{j' \in S_j} F_{j'}(A_1^{(x,y)}, \ldots, A_L^{(x,y)}, \beta_1(x, y), \ldots, \beta_P(x, y)),$$

where $\beta_p : [0; 1]^2 \to \mathbb{R}_{\geq 0}$ are parameter functions and S_j is the set of all actions which lead to the same state change as the j-th action, i.e. $S_j = \{1 \leq j' \leq J \mid \forall 1 \leq l \leq L(d_{j'l} - c_{j'l} = d_{jl} - c_{jl})\}$.

Let $\mu_l^K \geq 0$ be the migration rate of an agent in local state l in any region contained in $\mathcal{R}_K$. Then, the transition rates of the migration are given by

$$q\left(A; (\ldots, A_l^{(x,y)} - 1, A_l^{(\tilde{x},\tilde{y})} + \mathbb{1}(\tilde{x}, \tilde{y}))\right)$$
$$= \begin{cases} \mu_l^K A_l^{(x,y)}, & (\tilde{x}, \tilde{y}) \in \mathcal{R}_K \backslash \Omega_K, \\ |\mathcal{N}(x, y) \cap \Omega_K| \, \mu_l^K A_l^{(x,y)}, & \text{otherwise} \end{cases} \tag{1}$$

In order to obtain a fluid limit, we consider a CTMC sequence, denoted by $(A_{N,K}(t))_{t \geq 0}$ and indexed by the scaling parameter N and the grid granularity K such that the initial state of the N-th CTMC is given by $A_l(0) = \lfloor N\alpha_l^0 \rfloor$, where $\alpha_l^0 : [0; 1]^2 \to \mathbb{R}_{\geq 0}$ are zero at the boundary, i.e.

$$\alpha_l^0(x, 1) = 0, \qquad \alpha_l^0(x, 0) = 0, \qquad \alpha_l^0(1, y) = 0, \qquad \alpha_l^0(0, y) = 0 \tag{2}$$

This accounts for absorbing boundary conditions and ensures that no agents are present in Ω_K at $t = 0$.

Example 2. Using the local interactions from Example 1, the above definitions formally induce a sequence of population based CTMCs $(A_{N,K}(t))_{t \geq 0}$ that can be seen as a high-level model of a spatial network in which a file like a song or a smartphone virus is spread across moving agents.

Since the step size $\frac{1}{K}$ is decreasing as $K \to \infty$, it is clear that the impact of migration vanishes for large K if μ_l^K is not increasing as a function of K. In [16] it has been shown that the existence of a spatial limit requires that $\lim_{K \to \infty} \mu_l^K / K^2$ exists and is positive in the presence of migration. Following [16], we thus set $\mu_l^K := \mu_l K^2$, where $\mu_l \geq 0$ is a constant independent of N and K.

2.2 ODE Limit

For any fixed $K \geq 1$, it can be shown [16] that $(\frac{1}{N}A_{N,K}(t))_{t \geq 0}$ converges in probability, as $N \to \infty$, to the solution of a suitable ODE system, provided that the rate functions satisfy common regularity conditions [7]. In particular, we first require that for each $1 \leq j \leq J$ there exist continuous functions $f_j : \mathbb{R}^{L+P} \to \mathbb{R}$ and $g_j : \mathbb{R}^{L+P} \to \mathbb{R}_{\geq 0}$ such that

$$\textbf{(A1):} \quad \frac{1}{N}F_j\big(Na_1, \ldots, Na_L, b_1, \ldots, b_P\big) = f_j\big(a_1, \ldots, a_L, b_1, \ldots, b_P\big)$$
$$+ \mathcal{O}\left(\frac{g_j(a_1, \ldots, a_L, b_1, \ldots, b_P)}{N}\right).$$

for all $N \geq 1$. Second, we require that for all $z_0 \in \mathbb{R}^{L+P}$ there exist an open neighborhood $\mathfrak{O}$ of z_0 and a $C \in \mathbb{R}_{\geq 0}$ such that

$$\textbf{(A2):} \quad \big|f_j\big(z_2\big) - f_j\big(z_1\big)\big| \leq C\|z_2 - z_1\|, \qquad \forall z_1, z_2 \in \mathfrak{O}.$$

In essence, **(A1)** asserts that the CTMC is in the density-dependent form, while **(A2)** requires each f_j to be locally Lipschitz continuous.

Example 3. The rate function $F_4(A_1, A_2, A_3) = \min\{cA_1, \nu(\eta A_1 + A_2)\}$ from Example 1 yields $f_4(a_1, a_2, a_3) = \min\{ca_1, \nu(\eta a_1 + a_2)\}$. Note also that we can incorporate the parameters $\lambda, \sigma, \gamma, \eta, c$ by setting, for instance, $f_4(a_1, a_2, a_3, c, \nu, \eta, \lambda) = \min\{ca_1, \nu(\eta a_1 + a_2)\}$ and $f_1(a_1, a_2, a_3, \nu, \eta, \lambda) = \lambda$.

If $\boldsymbol{a} \in \mathbb{R}^{\mathcal{R}_K \times \{1,\ldots,L\}}$ denotes a vector $(a_1^{(x,y)}, \ldots, a_L^{(x,y)})_{(x,y) \in \mathcal{R}_K}$, the ODE system associated to the normalized CTMC $(\frac{1}{N}A_{N,K}(t))_{t \geq 0}$ is

$$\frac{d}{dt}a_l^{(x,y)}(t) = \mathbb{1}(x, y)\Big(\mathcal{M}_l^{(x,y)}(\boldsymbol{a}(t)) + \mathcal{L}_l^{(x,y)}(\boldsymbol{a}(t))\Big), \tag{3}$$

where $(x, y) \in \mathcal{R}_K$, and, for all $1 \leq l \leq L$,

$$
\mathcal{M}_l^{(x,y)}(\boldsymbol{a}) = \mu_l K^2 \sum_{(\tilde{x},\tilde{y}) \in \mathcal{N}(x,y)} (a_l^{(\tilde{x},\tilde{y})} - a_l^{(x,y)}),
$$

$$
\mathcal{L}_l^{(x,y)}(\boldsymbol{a}) = \sum_{1 \leq j \leq J} (d_{jl} - c_{jl}) f_j\big(a_1^{(x,y)}, \ldots, a_L^{(x,y)}, \beta_1(x,y), \ldots, \beta_P(x,y)\big). \tag{4}
$$

Note that by writing $\beta_p(x, y)$, we implicitly consider *location dependent* parameters. To allow later for the existence of a spatial limit, we additionally require that $\beta_p : [0; 1]^2 \to \mathbb{R}_{\geq 0}$ are continuous functions.

In the above formula, $\mathcal{M}_l^{(x,y)}(\cdot)$ gives the *diffusive* dynamics arising from the random walk of type-l agents in region (x, y), while $\mathcal{L}_l^{(x,y)}(\cdot)$ encodes the *local* dynamics stemming from local reactions.

Example 4. The fluid limit underlying the model from Example 1 is

$$
\frac{d}{dt} a_1^{(x,y)}(t) = \mathbb{1}(x,y)\Big(\mathcal{M}_1^{(x,y)}(\boldsymbol{a}(t)) + \lambda(x,y) - \sigma a_1^{(x,y)}(t) \\
- \min\big(c a_1^{(x,y)}(t), \nu(\eta a_1^{(x,y)}(t) + a_2^{(x,y)}(t))\big)\Big)
$$

$$
\frac{d}{dt} a_2^{(x,y)}(t) = \mathbb{1}(x,y)\Big(\mathcal{M}_2^{(x,y)}(\boldsymbol{a}(t)) - \gamma a_2^{(x,y)}(t) \\
+ \min\big(c a_1^{(x,y)}(t), \nu(\eta a_1^{(x,y)}(t) + a_2^{(x,y)}(t))\big)\Big)
$$

We have skipped the fluid limits of the uninteresting coffin states $a_3^{(x,y)}(t)$. Note also that the arrival rate is assumed to be location dependent, that is $\lambda : [0; 1]^2 \to \mathbb{R}_{\geq 0}$ denotes a continuous function.

The following result can be shown [7, 16].

Theorem 1 (ODE Fluid Limit). *The ODE system* (3) *subjected to the initial condition*

$$
a_l^{(x,y)}(0) = \alpha_l^0(x, y), \qquad (x, y) \in \mathcal{R}_K, \quad 1 \leq l \leq L,
$$

has a unique solution $\boldsymbol{a}$ in $\mathbb{R}^{\mathcal{R}_K \times \{1, \ldots, L\}}$. Moreover, under the assumption that for an arbitrary but fixed $T > 0$ the time domain of $\boldsymbol{a}$ contains $[0; T]$, it holds that

$$
\lim_{N \to \infty} \mathbb{P}\left\{ \sup_{0 \leq t \leq T} \left\| \frac{1}{N} A_N(t) - \boldsymbol{a}(t) \right\|_\infty > \varepsilon \right\} = 0, \qquad \forall \varepsilon > 0,
$$

where $\| \cdot \|_\infty$ on $\mathbb{R}^{\mathcal{R}_K \times \{1, \ldots, L\}}$ is given by $\|\boldsymbol{a}\|_\infty := \max_{(x,y) \in \mathcal{R}_K, 1 \leq l \leq L} |a_l^{(x,y)}|$.

2.3   Fast Simulation

In the present paper we are interested in the probabilistic behavior of a single agent $(Z_{N,K}(t))_{t \geq 0}$ that is part of the population based CTMC $(A_{N,K}(t))_{t \geq 0}$. While being not a CTMC per se, $(Z_{N,K}(t))_{t \geq 0}$ becomes a CTMC if conditioned on $(A_{N,K}(t))_{t\geq 0}$, see [5,6,9] for details. Using this, it is possible to construct a time inhomogeneous CTMC $(z_K(t))_{t\geq 0}$ such that $(Z_{N,K}(t))_{t\geq 0}$ becomes indistinguishable from $(z_K(t))_{t\geq 0}$ if $N \to \infty$. In particular, the following holds [5].

Theorem 2. *Fix any $T > 0$ and assume that (3) has a solution on $[0; T]$ that is positive for all $(x, y) \in \mathcal{R}_K \backslash \Omega_K$. Then, together with $a \mapsto f_j^l(a) := \frac{f_j(a)}{a_l}$, there exists a time inhomogeneous CTMC $(z_K(t))_{t\geq 0}$ with state space $(\mathcal{R}_K \times \{1,\ldots,L\}) \cup \{\square\}$ and time-varying intensities*

$$q^K_{(x,y,l),(\tilde{x},\tilde{y},l)}(t) = \mu_l K^2$$

$$q^K_{(x,y,l),\square}(t) = \mu_l K^2 |\mathcal{N}(x,y) \cap \Omega_K|$$

$$q^K_{(x,y,l),(x,y,l')}(t) = \sum_{1\leq j\leq J \ | \ \{l\to l'\}\subseteq R_j} f_j^l(a(t))$$

where $1 \leq l \leq L$, $(x, y) \in \mathcal{R}_K \backslash \Omega_K$ and $(\tilde{x}, \tilde{y}) \in \mathcal{N}(x,y)\backslash \Omega_K$ which satisfies $\mathbb{P}\{Z_{N,K}(t) \neq z_K(t) \mid$ for some $0 \leq t \leq T\} \to 0$ as $N \to \infty$.

Note that the solution of (3) will be positive whenever inner regions have positive initial conditions and the flux-out from any $(x, y) \in \mathcal{R}_K\backslash\Omega_K$ is proportional to $a^{(x,y)}$. This can be formally established by using differential inequalities.

In the previous theorem, we had to introduce a new special state $\square$, which plays the role of an absorbing *spatial* coffin state to model the situation when an agent leaves the spatial domain after hitting the boundary Ω_K. Note also that the intensities of $(z_K(t))_{t\geq 0}$ depend on $K \geq 1$ in two ways. Apart from having a migration rate that increases with $K \geq 1$, the intensities of local interactions depend also on the solution $a(t)$ of the ODE system (3) of size $L(K+1)^2$.

Remark 1. Note that atomic reactions of the form $\to A_l$ do not show up in the derivation of $(z_K(t))t \geq 0$ because we sum only across atomic jumps of the form $\{A_l \to A_{l'}\}$. This is because, intuitively, no agent present in the system has to change its state in order to create a new agent. The arrival rate to the system itself, however, can depend on the current state of the system because f_j may depend on $a_1,\ldots,a_L$ if $R_j = \{\to A_l\}$.

Example 5. Let us derive the time inhomogeneous forward Kolmogorov equations of $(z_K(t))t \geq 0$. Taking into account Remark 1, the derivation is straightforward and it holds that

$$\frac{d}{dt}\pi(t,x,y,1) = \mathbb{1}(x,y)\left(\mu_1 K^2 \sum_{(\tilde{x},\tilde{y})\in\mathcal{N}(x,y)} (\pi(t,\tilde{x},\tilde{y},1) - \pi(t,x,y,1))\right.$$

$$\left. - \sigma\pi(t,x,y,1) - \pi(t,x,y,1)\min\left(c, \nu\left(\eta + \frac{a_2^{(x,y)}(t)}{a_1^{(x,y)}(t)}\right)\right)\right)$$

The other equations are derived similarly and are omitted due to space reasons.

2.4   PDE Limit

Although solving the ODE system (3) is substantially easier than simulating the CTMC sequence $(\frac{1}{N}A_{N,K}(t))_{t\geq 0}$, the number of ODEs $L(K+1)^2$ is quadratic in K and makes an efficient analysis difficult if K is large. Note that a similar blowup applies also to the system of forward Kolmogorov equations underlying the single agent $(z_K(t))_{t\geq 0}$.

We briefly sketch the approach taken in [16], where the aim was to prove that the sequence of ODE systems (3) converges, as $K \to \infty$, to the solution of a suitable PDE system. To this end, we first observe that the diffusive part of the ODE limit (3) rewrites as

$$\mathcal{M}_l^{(x,y)}(\boldsymbol{a}) = \mu_l \left(\sum_{(\tilde{x},\tilde{y})\in\mathcal{N}(x,y)} \frac{a_l^{(\tilde{x},\tilde{y})} - a_l^{(x,y)}}{(1/K)^2} \right) = \mu_l \,\Delta^d\, a_l^{(x,y)},$$

where Δ^d denotes the *discrete* version of the two dimensional Laplace operator $\Delta = \partial_{xx} + \partial_{yy}$. This allows us to rewrite the system of ODEs (3) into

$$\frac{d}{dt}a_l^{(x,y)}(t) = \mathbb{1}(x,y)\left(\mu_l \,\Delta^d\, a_l^{(x,y)}(t) + \mathcal{L}_l^{(x,y)}(\boldsymbol{a}(t))\right),$$

where $a_l^{(x,y)}(0) = \alpha_l^0(x,y)$ for $(x,y) \in \mathcal{R}_K$ and $1 \leq l \leq L$.

This ODE system can be solved by means of the Euler method. For a fixed time step Δt, the underlying sequence is given by

$$a_l^{(x,y)}(m+1) := a_l^{(x,y)}(m) + \mathbb{1}(x,y)\Delta t\left(\mu_l \,\Delta^d\, a_l^{(x,y)}(m) + \mathcal{L}_l^{(x,y)}(\boldsymbol{a}(m))\right), \quad (5)$$

where $m \geq 0$. Using this key observation from [16], one deduces that the very same sequence corresponds to a finite difference scheme of the *reaction-diffusion* PDE system

$$\partial_t \alpha_l = \mu_l \,\Delta\, \alpha_l + \sum_{1\leq j\leq J} (d_{jl} - c_{jl})f_j(\alpha_1,\ldots,\alpha_L,\beta_1,\ldots,\beta_P), \quad (6)$$

where $(x,y,t) \in [0;1]^2 \times \mathbb{R}_{\geq 0}$ and $1 \leq l \leq L$, subject to the Dirichlet boundary conditions (DBCs)

$$\alpha_l(1,y,t) = 0, \quad \alpha_l(0,y,t) = 0, \quad \alpha_l(x,1,t) = 0, \quad \alpha_l(x,0,t) = 0 \quad (7)$$

for all $1 \leq l \leq L$, $0 \leq t \leq T$, $x \in [0;1]$ and $y \in [0;1]$.

Example 6. For $K \geq 1$ large, (5) can be interpreted as a the Euler sequence of the ODE system from Example 4 and as a finite difference scheme of the PDE system

$$\partial_t \alpha_1 = \mu_1 \,\Delta\, \alpha_1 - \sigma\alpha_1 - \min(c\alpha_1, \nu(\eta\alpha_1 + \alpha_2)) + \lambda$$
$$\partial_t \alpha_2 = \mu_2 \,\Delta\, \alpha_2 - \gamma\alpha_2 + \min(c\alpha_1, \nu(\eta\alpha_1 + \alpha_2)) \quad (8)$$

Remark 2. As has been observed in [16], imposing DBCs on the solution (6) may require some of the parameter functions $\beta_p : [0;1]^2 \to \mathbb{R}_{\geq 0}$ to enjoy the DBCs as well. For instance, in Example 6, the parameter rate functions $\sigma, \gamma, c, \mu, \eta :$ $[0;1]^2 \to \mathbb{R}_{\geq 0}$ can be any continuous functions, while $\lambda : [0;1]^2 \to \mathbb{R}_{\geq 0}$ has to be a continuous function satisfying the DBCs. Otherwise, agents would arrive at the boundary, thus violating the DBCs (7).

Since the ODEs (3) and the PDEs (6) are coupled to each other via the sequence (5), the convergence of the ODE systems (3) to the PDE system can be established by proving that the sequence (5), interpreted as a finite difference scheme for (6), is convergent. This is one of the major results of [16] (see Theorem 4 and proof of Theorem 5) and is stated next.

Theorem 3. *Assume that for a fixed $T > 0$, the family of functions $(\alpha_1, \dots, \alpha_L)$ on $[0;1]^2 \times [0;T]$ describes for all $0 < T' \leq T$ the unique solution of (6) subjected to the DBCs (7) on $[0;1]^2 \times [0;T']$. Further, assume that*

$$\partial_{txx}\alpha_l, \; \partial_{xxxx}\alpha_l, \; \partial_{yxx}\alpha_l, \; \partial_{tyy}\alpha_l, \; \partial_{xyy}\alpha_l, \; \partial_{yyyy}\alpha_l,$$

where $1 \leq l \leq L$, exist and are continuous. Then, the ODE system (3) admits a solution on $[0;T]$ and for any $\varepsilon > 0$, there exists an $K_0 \geq 1$ such that for all $K \geq K_0$ it holds that

$$\sup_{0 \leq t \leq T} \max_{(x,y) \in \mathcal{R}_K, 1 \leq l \leq L} |a_l^{(x,y)}(t) - \alpha_l(x,y,t)| \leq \varepsilon$$

Theorems 1 and 3 can be used to show that the CTMC sequence $(\frac{1}{N}A_{N,K}(t))_{t \geq 0}$ converges in probability to the solution of the PDE system (6) and was the main result of [16]. Instead, we will combine Theorems 2 and 3 to establish a novel result of fast spatial simulation.

We wish to stress that the mathematically simple but inefficient difference scheme (5) should not be used to numerically solve (6), see Sect. 4 and [16].

3 Fast Spatial Simulation

This section establishes the novel result of fast spatial simulation. We begin by introducing the concept of a diffusive switching model that is simplified to our needs, see in [18, Chap. 2] for details.

Let $W(\cdot)$ be the standard Brownian motion on $\mathbb{R}^2$. The stochastic process $(\tau(t), \mathbb{M}(t), \mathbb{L}(t))_{t \geq 0}$ with state space $\mathbb{R} \times \mathbb{R}^2 \times \{1, \dots, L\}$ that satisfies the ordinary differential equation $d\tau(t) = 1$, the stochastic differential equation $d\mathbb{M}(t) = \mu(\mathbb{L}(t))dW(t)$ and

$$\mathbb{P}\{\mathbb{L}(t + \Delta t) = j \mid \mathbb{L}(t) = l, \tau(s), \mathbb{M}(s), \mathbb{L}(s), s \leq t\}$$
$$= q_{(t,\mathbb{M}(t),l),(t,\mathbb{M}(t),l')}\Delta t + o(\Delta t)$$

is called a diffusive switching model. Intuitively, the component $(\mathbb{M}(t))_{t \geq 0}$ models a random walk in $\mathbb{R}^2$ whose diffusive coefficient function $\mu : \{1, \dots, L\} \to \mathbb{R}_{\geq 0}$

may depend on the current mode $\mathbb{L}(t)$. The discrete switching process $(\mathbb{L}(t))_{t\geq0}$, instead, is a time homogeneous CTMC whose intensities may depend on the current state of $\tau(t)$ and $\mathbb{M}(t)$. Note that time is included in the state by requiring that $\tau(0) = 0$. This trick allows us to incorporate, essentially, time inhomogeneous intensities while having formally a time homogeneous Markov process.

It can be proven [15] that $(\tau(t), \mathbb{M}(t), \mathbb{L}(t))_{t\geq0}$ admits a probability density function $(t, x, y, l) \mapsto \rho(t, x, y, l)$ that satisfies the following forward Kolmogorov PDE (also known as Fokker-Planck PDE)

$$\partial_t \rho(t, x, y, l) = \frac{\mu(l)^2}{2}(\triangle\rho)(t, x, y, l) + \sum_{l'=1}^{L} \rho(t, x, y, l')q_{(t,x,y,l'),(t,x,y,l)}, \quad (9)$$

where $\triangle = \partial_{xx} + \partial_{yy}$ denotes the Laplace operator with respect to x and y. At the same time, it is not hard to see that the forward Kolmogorov ODEs of the time inhomogeneous CTMC that describes the single agent $(z_K(t))_{t\geq0}$ at $(x, y) \in \mathcal{R}_K \backslash \Omega_K$ in Theorem 2 are

$$\frac{d}{dt}\pi(t, x, y, l) = \mu_l(\triangle^d\pi)(t, x, y, l) + \sum_{l'=1}^{L} \pi(t, x, y, l')q^K_{(x,y,l'),(x,y,l)}(t) \quad (10)$$

Noting the striking similarity between (9) and (10), we thus aim at proving that the time inhomogeneous CTMC $(z_K(t))_{t\geq0}$ converges in distribution to a certain switching diffusion if $K \to \infty$. Unfortunately, the presence of a *single* spatial coffin state $\square$ and time inhomogeneity make a formal treatment rather involved. We circumvent this by a little trick: we modify the single agent model, allowing it to move in the whole plane, but forcing its state to change very quickly to a local coffin state when outside the unit box. This avoids to deal with boundaries in the convergence theorem, and yet the modified model converges to the original one as the rate of jumping to the local coffin state is sent to infinity, thus guaranteeing the correctness of the approach.

Definition 1. *Let* (3) *have a positive solution on $\mathcal{R}_K \backslash \Omega_K$ for all $t \geq 0$. Then the switching diffusion $(\tilde{z}_K(t))_{t\geq0} := (\tau_K(t), \mathbb{M}_K(t), \mathbb{L}_K(t))_{t\geq0}$ with state space $\mathbb{S}_K := \mathbb{R} \times (\mathbb{Z}/K)^2 \times \{1, \ldots, L, L+1\}$ is given by $d\tau(t) = 1$ and $q^K_{(t,x,y,l),(t,\tilde{x},\tilde{y},l)} = \mu_l K^2$ with $\mu_{L+1} := 1$ and*

$$q^K_{(t,x,y,l),(t,x,y,l')} :=$$

$$\begin{cases} (1 - \phi(x,y))\varsigma & , l \neq L+1, l' = L+1 \\ \phi(x,y)\psi(t) \sum_{\substack{1 \leq j \leq J: \\ \{l \to l'\} \subseteq R_j}} f^l_j(x, y, a^{(x,y)}(t \vee 0)) & , l, l' \neq L+1, (x,y) \in (0;1)^2 \\ 0 & , otherwise \end{cases}$$

for all $(x, y) \in (\mathbb{Z}/K)^2$ and $(\tilde{x}, \tilde{y}) \in \{(x \pm \Delta s, y \pm ds)\}$. Here, $\phi : \mathbb{R}^2 \to \mathbb{R}$ denotes a mollifier with $0 \leq \phi \leq 1$, $\phi_{|(\delta;1-\delta)^2} \equiv 1$ and $\phi_{|\mathbb{R}^2 \backslash (\frac{\delta}{2};1-\frac{\delta}{2})^2} \equiv 0$, while $\psi : \mathbb{R} \to \mathbb{R}$ is a mollifier with $0 \leq \psi \leq 1$, $\psi_{|(-\infty;-\delta)} \equiv 0$ and $\psi_{|\mathbb{R}_{\geq0}} \equiv 1$.

Hence, we will study an agent whose spatial domain is $(\mathbb{Z}/K)^2$ instead of $\mathcal{R}_K$ but that changes with a high intensity ζ into the coffin state $L+1$ as soon as $(x,y) \notin (\delta; 1-\delta)^2$. By choosing $\zeta > 0$ sufficiently large and $\delta > 0$ sufficiently small, one can thus consider $(\tilde{z}_K(t))_{t \geq 0}$ instead of $(z_K(t))_{t \geq 0}$ in practice. An argument similar to the one of [1] may be used to formally state the convergence of $(\tilde{z}_K(t))_{t \geq 0}$ to $(z_K(t))_{t \geq 0}$, in a suitable limit of $\zeta \to \infty$ and $\delta \to 0$. The mollifier functions ϕ and ψ are needed for the existence of a forward Kolmogorov PDE.

Remark 3. Note that it suffices to assume in Definition 1 that (3) has a positive solution on some finite time interval $[0; T]$. Indeed, thanks to (**A1**) and (**A2**), (3) has a unique solution. It may, however, explode in finite time. This issue can be addressed by replacing $a^{(x,y)}(t)$ in the definition of $q^K_{(t,x,y,l),(t,x,y,l')}$ by a bounded locally Lipschitz continuous function $\underline{a}^{(x,y)}(t)$ that coincides with $a^{(x,y)}(t)$ for all $t \in [0; T]$, see proof of Theorem 4 in [16].

The next theorem is the main result of the present work. It shows that $(\tilde{z}_K(t))_{t \geq 0}$ converges in distribution, as $K \to \infty$, to a switching diffusion whose probability density distribution satisfies a forward Kolmogorov PDE. The underlying proof makes use of Theorem 3 which ensures that the family of ODE systems (3) converges, as $K \to \infty$, to the solution of the PDE system (6). We need the following mild assumption.

(**A3**):   Given any $T > 0$ and $\delta > 0$, there is some $\xi > 0$ such that
$$\alpha_l(x,y,t) \geq \xi \text{ for all } (x,y) \in (\delta, 1-\delta)^2, 0 \leq t \leq T \text{ and } 1 \leq l \leq L$$

Assumption (**A3**) can be verified numerically and can be expected to hold true when the initial conditions are positive on $(0; 1)^2$ and the local flux-out from any inner point (x,y) is proportional to $\alpha(x,y,\cdot)$. A sufficient condition for (**A3**) could be derived via the maximum principle for PDEs. Also, it is worth noting that (**A3**) can be dropped if, for all $\{l \to l'\} \subseteq R_j$, $a \mapsto f^l_j(a)$ can be extended to a locally Lipschitz continuous function on $\mathbb{R}^L_{\geq 0}$.

Theorem 4. *Assume that* (6) *has a unique positive solution on* $(0; 1)^2$ *for all* $t \geq 0$. *Then there exists a switching diffusion* $(\tau(t), \mathbb{M}(t), \mathbb{L}(t))_{t \geq 0}$ *on* $\mathbb{S} := \mathbb{R} \times \mathbb{R}^2 \times \{1, \ldots, L+1\}$ *that satisfies* $d\tau(t) = 1$ *and* $d\mathbb{M}(t) = \mu(\mathbb{L}(t))dW(t)$ *for all* $1 \leq l \leq L+1$ *where* $\mu(l) := \sqrt{2\mu_l}$, $\mu(L+1) := \sqrt{2}$ *and*

$$q_{(t,x,y,l),(t,x,y,l')} :=$$
$$\begin{cases} (1 - \phi(x,y))\zeta, & l \neq L+1, l' = L+1 \\ \phi(x,y)\psi(t) \sum_{\substack{1 \leq j \leq J: \\ \{l \to l'\} \subseteq R_j}} f^l_j(x,y,\alpha(t \vee 0, x, y)), & l, l' \neq L+1, (x,y) \in (0;1)^2 \\ 0, & \text{otherwise} \end{cases}$$

for all $(x,y) \in \mathbb{R}^2$, *where* ϕ *and* ψ *are as in Definition 1. Moreover,* $(\tilde{z}_K(t))_{t \geq 0}$ *converges in distribution to* $(\tau(t), \mathbb{M}(t), \mathbb{L}(t))_{t \geq 0}$, *as* $K \to \infty$, *and* $(\tau(t), \mathbb{M}(t),$

$\mathbb{L}(t))_{t\geq 0}$ *has a probability density function solving the forward Kolmogorov PDE*

$$\partial_t \rho(t,x,y,l) = \mu_l(\triangle \rho)(t,x,y,l) + \sum_{l'=1}^{L+1} \rho(t,x,y,l')q_{(t,x,y,l'),(t,x,y,l)}$$

Example 7. With $\rho_l(t,x,y) := \rho(t,x,y,l)$, the forward Kolmogorov PDE of the switching diffusion $(\tilde{z}_K(t))_{t\geq 0}$ of Example 6 is

$$\partial_t \rho_1 = \mu_1 \triangle \rho_1 - (1-\phi)\zeta - \phi\rho_1\left(\sigma + \min\left(c, \nu\left(\eta + \frac{\alpha_2}{\alpha_1}\right)\right)\right)$$

$$\partial_t \rho_2 = \mu_2 \triangle \rho_2 - (1-\phi)\zeta + \phi\left(\rho_1 \min\left(c, \nu\left(\eta + \frac{\alpha_2}{\alpha_1}\right)\right) - \gamma\rho_2\right)$$

$$\partial_t \rho_3 = \mu_3 \triangle \rho_3 - (1-\phi)\zeta + \phi\left(\sigma\rho_1 + \gamma\rho_2\right) \tag{11}$$

As before, we skipped the PDE of the spatial coffin state. Notice that the terms $(1-\phi)$ in (11) are zero for $(x,y) \in (\delta; 1-\delta)^2$. Indeed, by choosing $\zeta > 0$ large and $\delta > 0$ small, we are essentially approximating the DBCs, at the price of an increased stiffness in the system arising from large values of ζ. However, the removal of the boundary by means of fast stochastic jumps is a technical trick, and in practice we can introduce back the boundary, imposing DBCs on the forward PDE from Theorem 4. In fact, an application of the main theorem in [1] shows that the PDE solution of Theorem 4 converges, as $\zeta \to \infty$ and $\delta \to 0$, to the solution of the PDE system with DBCs.

For instance, in Example 7, the introduction of DBCs yields the following.

Example 8. With $\rho_l(t,x,y) := \rho(t,x,y,l)$, we approximate the forward Kolmogorov PDE of the switching diffusion $(\tilde{z}_K(t))_{t\geq 0}$ from Example 6 by the PDE

$$\partial_t \rho_1 = \mu_1 \triangle \rho_1 - \rho_1\left(\sigma + \min\left(c, \nu\left(\eta + \frac{\alpha_2}{\alpha_1}\right)\right)\right)$$

$$\partial_t \rho_2 = \mu_2 \triangle \rho_2 - \gamma\rho_2 + \rho_1 \min\left(c, \nu\left(\eta + \frac{\alpha_2}{\alpha_1}\right)\right)$$

$$\partial_t \rho_3 = \mu_3 \triangle \rho_3 + \sigma\rho_1 + \gamma\rho_2, \tag{12}$$

that satisfies the DBCs (which in turn implies that $\rho(0,\cdot,\cdot,\cdot)$ enjoys the DBCs). At this point, we wish to draw readers attention to the striking similarity between the equations (12) and those of Example 5.

4   Case Study

In [16,17] the authors provided numerical evidence for the fact that solving the global PDE system (6) is faster than evaluating the ODE system (3) if K is large. This is because a numerical PDE solver discretizes the square $[0;1]^2$ with respect to the right hand side of the PDEs (the stiffer the PDEs, the finer the mesh), while the size of the ODE system (3), given by $L(K+1)^2$, is inherently connected to the underlying stochastic model. In fact, in [17], the stiff ODE solver ode15s

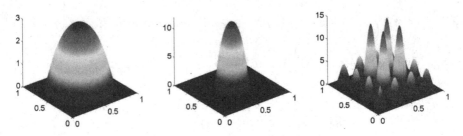

Fig. 1. From left to right: the mollifier functions $\theta_{0.5}$, $\theta_{0.25}$ and the arrival function λ.

of Matlab was crashing due to out of memory errors for models with $L = 10$ and $K \approx 20$, while the corresponding `parabolic` command terminated in less than one second. The very same phenomenon applies to the forward Kolmogorov PDE from Theorem 4 and the forward Kolmogorov ODEs underlying $(z_K(t))_{t \geq 0}$, i.e. the former can be evaluated substantially faster than the latter if K is large.

Instead of lifting the numerical investigations from [16,17] to Kolmogorov equations in a straightforward manner, in particular for what concerns the computational gain, we consider the following two questions related to our running example. First, what is the probability that an agent that starts as a downloader will act as a seed at time T? Second, how likely is it that an agent that started off as seed is still offering the file at time T? Obviously, the first and the second question correspond to the probability value $\int_{[0;1]^2} \rho_1(T, x, y) d(x, y)$ and $\int_{[0;1]^2} \rho_2(T, x, y) d(x, y)$, respectively. In the following, we will compare our method in terms of quality of the approximation and running time with results obtained by stochastic simulation of a model with finite K and N.

Before computing these quantities, we have to chose the time horizon T and the model parameters $\sigma, \gamma, \lambda, c, \eta, \nu$ and $\rho(0, \cdot, \cdot, \cdot), \alpha.(\cdot, \cdot, 0)$. To this end, we fix the mollifier function

$$\theta_\varepsilon(x, y) := \begin{cases} 0 & , \|(x, y) - (0.5, 0.5)\| \geq \varepsilon \\ \frac{\mathcal{I}}{\varepsilon^2} \exp\left(-\dfrac{1}{1 - \varepsilon^{-2}\|(x, y) - (0.5, 0.5)\|^2}\right) & , \|(x, y) - (0.5, 0.5)\| < \varepsilon \end{cases}$$

where $0 < \varepsilon \leq 0.5$ and $\|\cdot\|$ is meant to be the Euclidian norm. θ_ε resembles a bell-shape peaking at the center of the unit square, see Fig. 1. Since $\int_{[0;1]^2} \theta_\varepsilon(x, y) d(x, y) = \int_{[0;1]^2} \theta_{0.5}(x, y) d(x, y)$ for all $\varepsilon > 0$, setting $\mathcal{I} \approx 2.01$ which ensures that $\int_{[0;1]^2} \theta_{0.5}(x, y) d(x, y) = 1.0$ makes θ_ε a probability density function on $[0;1]^2$ for all $0 < \varepsilon \leq 0.5$. In particular, it holds that the probability measure $A \mapsto \int_A \theta_\varepsilon(x, y) d(x, y)$, where $A \in \mathfrak{B}([0;1]^2)$, converges towards the Dirac measure $\delta_{(0.5, 0.5)}$ as $\varepsilon \downarrow 0$.

Armed with this, we randomly choose $T = 0.1$, $\mu_1 = \mu_2 = \mu_3 = 0.1$, $\sigma \equiv 1.0$, $\gamma \equiv 5.0$, $c \equiv 10.0$, $\nu \equiv 1.0$, $\eta \equiv 0.3$ and $\lambda(x, y) := 5.0 \cdot \tilde{\theta}_{0.5}(x, y) \sin(4\pi(x - 0.5))^2 \sin(4\pi(y - 0.5))^2$ where $\tilde{\theta}_{0.5}$ arises from $\theta_{0.5}$ by replacing the Euclidian norm by the supremum norm. This ensures that the decay towards the boundary of $[0;1]^2$ happens not in concentric circles but in terms of concentric squares.

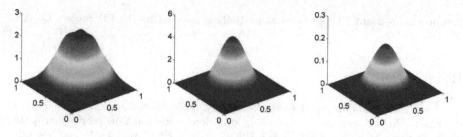

Fig. 2. From left to right: $\alpha_1(\cdot, \cdot, T)$, $\rho(T, \cdot, \cdot, 1)$ and $\rho(T, \cdot, \cdot, 2)$

The presence of the trigonometric functions, instead, leads to a volatile topological profile. The overall shape of λ can be seen in Fig. 1. Instead, the initial values are given by $\alpha_1(x, y, 0) = \theta_{0.5}(x, y)$, $\alpha_2(x, y, 0) = \alpha_1(x, y, 0)/4$, $\alpha_3(x, y, t) \equiv 0$, $\rho(0, x, y, 1) = \theta_{0.25}(x, y)$, $\rho(0, x, y, 2) \equiv 0$ and $\rho(0, x, y, 3) \equiv 0$.

We solved the global PDE system (8) and the forward Kolmogorov PDE (12) by invoking the Matlab command `parabolic` in less than one second. Figure 2 shows the plots of $\alpha_1(\cdot, \cdot, T)$, $\rho(T, \cdot, \cdot, 1)$ and $\rho(T, \cdot, \cdot, 2)$. The probabilities were computed as $\int_{[0;1]^2} \rho(T, x, y, 1) d(x, y) \approx 0.865$ and $\int_{[0;1]^2} \rho_2(T, x, y, 1) d(x, y) \approx 0.038$. Note that the remaining quantity $1.0 - 0.865 - 0.038$ accounts for all other scenarios, i.e. the agent left $[0;1]^2$ before time T, either due to migration or due to the local departure rates σ and γ. Note in particular that this includes also the scenario where the agent changes from state one into state two and abandons $[0;1]^2$ before time T. Compare these two numbers with those obtained by stochastic simulation (average over 1000 runs), for $N = 50$ and $K = 25$, respectively equal to 0.848 ± 0.022 and 0.034 ± 0.011. As we can see the error is very small, as the estimate by fast spatial simulation falls in the 95 % confidence bounds of stochastic simulation. However, the running time of stochastic simulation (using a Java implementation of the GB algorithm [11]) is 2790.58 seconds (about 45 min), versus less than one second to solve PDEs (in Matlab). Hence our method, even for small population and grid size, provides good accuracy and a speedup of 3 orders of magnitude.

5   Conclusion

We introduced fast spatial simulation, combining results on approximation of the behavior of individual agents in large population models with results to approximate the collective dynamics of spatio-temporal population processes by PDEs. The result is an efficient framework to assess properties of random agents moving in space and interacting locally with other agents. Practically, we need to solve a low dimensional PDE, avoiding the burden of stochastic simulation or fluid approximation of grid-based spatial models. In the future, we plan to extend our approach to more complex boundary conditions, possibly using the same trick as in Sect. 3, and applying the framework to more interesting case studies, including the assessment of more complex properties as in [5].

Acknowledgement. This work was partially supported by the EU project QUAN-TICOL, 600708.

References

1. Abate, A., Prandini, M., Lygeros, J., Sastry, S.S.: Approximation of general stochastic hybrid systems by switching diffusions with random hybrid jumps. In: Egerstedt, M., Mishra, B. (eds.) HSCC 2008. LNCS, vol. 4981, pp. 598–601. Springer, Heidelberg (2008)
2. Akyildiz, I.F., Ho, J.S.M., Lin, Y.B.: Movement-based location update and selective paging for PCS networks. IEEE/ACM Trans. Netw. 4(4), 629–638 (1996)
3. Bar-Noy, A., Kessler, I., Sidi, M.: Mobile users: to uptdate or not to update? In: INFOCOM, pp. 570–576 (1994)
4. Bortolussi, L.: Limit behavior of the hybrid approximation of stochastic process algebras. In: Al-Begain, K., Fiems, D., Knottenbelt, W.J. (eds.) ASMTA 2010. LNCS, vol. 6148, pp. 367–381. Springer, Heidelberg (2010)
5. Bortolussi, L., Hillston, J.: Fluid model checking. In: Koutny, M., Ulidowski, I. (eds.) CONCUR 2012. LNCS, vol. 7454, pp. 333–347. Springer, Heidelberg (2012)
6. Bortolussi, L., Hillston, J.: Model checking single agent behaviours by fluid approximation. Inf. Comput. 242, 183–226 (2015)
7. Bortolussi, L., Hillston, J., Latella, D., Massink, M.: Continuous approximation of collective system behaviour: a tutorial. Perform. Eval. 70(5), 317–349 (2013)
8. Boudec, J.Y.L., McDonald, D., Mundinger, J.: A generic mean field convergence result for systems of interacting objects. In: QEST 2007, pp. 3–18 (2007)
9. Darling, R.W.R., Norris, J.R.: Differential equation approximations for Markov chains. Probab. Surv. [electroninc only] 5, 37–79 (2008)
10. Gamal, A., Mammen, J., Prabhakar, B., Shah, D.: Throughput-delay trade-off in wireless networks. In: INFOCOM (2004)
11. Gibson, M.A., Bruck, J.: Efficient exact stochastic simulation of chemical systems with many species and many channels. J. Phys. Chem. A 104(9), 1876–1889 (2000)
12. Ioannidis, S., Marbach, P.: A brownian motion model for last encounter routing. In: INFOCOM (2006)
13. Kowal, M., Tschaikowski, M., Tribastone, M., Schaefer, I.: Scaling size and parameter spaces in variability-aware software performance models. In: ASE 2015 (2015)
14. Qiu, D., Srikant, R.: Modeling and performance analysis of BitTorrent-like peer-to-peer networks. In: SIGCOMM (2004)
15. Ramponi, A.: Mixture dynamics and regime switching diffusions with application to option pricing. Method. Comput. Appl. Probab. 13(2), 349–368 (2009). http://dx.doi.org/10.1007/s11009-009-9155-1
16. Tschaikowski, M., Tribastone, M.: Spatial Fluid Limits for Stochastic Mobile Networks. http://arxiv.org/abs/1307.4566
17. Tschaikowski, M., Tribastone, M.: A partial-differential approximation for spatial stochastic process algebra. In: VALUETOOLS (2014)
18. Yin, G., Zhu, C.: Hybrid Switching Diffusions. Springer, New York (2010)

Efficient Implementations of the EM-Algorithm for Transient Markovian Arrival Processes

Mindaugas Braženas[1], Gábor Horváth[2]([✉]), and Miklós Telek[2,3]

[1] Department of Applied Mathematics,
Kaunas University of Technology, Kaunas, Lithuania
mindaugas.brazenas@yahoo.com
[2] Department of Networked Systems and Services,
Budapest University of Technology and Economics, Budapest, Hungary
{ghorvath,telek}@hit.bme.hu
[3] MTA-BME Information Systems Research Group, Budapest, Hungary

Abstract. There are real life applications (e.g., requests of http sessions in web browsing) with a finite number of events and correlated inter-arrival times. Terminating point processes can be used to model such behavior. Transient Markov arrival processes (TMAPs) are computationally appealing terminating point processes which are terminating versions of Markov arrival processes.

In this work we propose algorithms for creating a TMAP based on empirical measurement data and compare various (series/parallel, CPU/GPU) implementations of the EM method for TMAP fitting.

1 Introduction

Stochastic models with background continuous time Markov chain (CTMC) are widely used in stochastic modeling. Phase type (PH) distributions and Markov arrival processes (MAP) exemplify the flexibility and the ease of application of such models. In this work we cope with terminating stochastic processes [1]. Indeed, Phase type distributions are defined by a terminating (also referred to as transient) background Markov chain, but it generates exactly one event. A transient Markovian arrival process (TMAPs) is a point processes with a finite number of possibly correlated inter event times which is governed by a terminating background Markov chain [8]. Basic properties of TMAPs, such as the distribution of the number of generated arrivals or the time until the last arrival, are presented in [8], further properties and moments based characterization are discussed in [6]. TMAPs can be used in a wide range of application fields from traffic modeling of computer systems to risk analysis, including also population dynamics in biological systems. For instance TMAPs are applied to women's lifetime modeling in several countries in [5].

In this work we consider the parameter estimation of TMAPs to experimental data sets based on the EM method. The EM method has been used successfully

This work was supported by the Hungarian research project OTKA K119750.

S. Wittevrongel and T. Phung-Duc (Eds.): ASMTA 2016, LNCS 9845, pp. 107–122, 2016.
DOI: 10.1007/978-3-319-43904-4_8

for parameter estimation of several models with background Markov chains, e.g., for PH distributions [3], for PH distributions with structural restriction [10], for MAPs [4], for MAPs with structural restrictions [7,9]. The experiences from these previous research results indicate that the inherent redundancy of the stochastic models with background Markov chains makes the parameter estimation of the general models inefficient. In this work we avoid the implementation of the EM based estimation of general TMAPs and immediately apply a similar structural restriction as the one which turned out to be efficient in case of PH distributions [10] and MAPs [7,9]. The formulas of the EM method for TMAP fitting show similarities with the ones for MAP fitting in [7], but there are intricate details associated with the handling of background process termination which require a non-trivial reconsideration of the expectation and the maximization steps of the method.

Apart of the algorithmic description of the EM method for TMAP fitting we pay attention to efficient implementation for both traditional computing devices (CPU) and graphics processing unit (GPU). Both platforms required various implementation optimizations for efficient computing of the steps of the fitting method. Together with the fitting results and the related computation times we present the applied implementation optimization methods and the related considerations.

The rest of the paper is organized as follows. The next section summarizes the basic properties of TMAPs. Section 3 presents the theoretical foundation of the EM method for TMAP fitting and the high level procedural description of the method. Section 4 discusses several implementation versions for CPU as well as for GPU-based computation. Numerical results are provided in Sect. 5 and the paper is concluded in Sect. 6.

2 Transient Markovian Arrival Processes

Transient Markovian Arrival Processes (TMAPs) are continuous time terminating point processes where the inter-arrival times depend on a background Markov chain, hence they can be dependent.

TMAPs can be characterized by an initial probability vector, α, holding the initial state distribution of the background Markov chain at time 0 ($\alpha \mathbb{1} = 1$, where $\mathbb{1}$ is the column vector of ones), and two matrices, D_0 and D_1. Matrix D_0 contains the rates of the internal transitions that are not accompanied by an arrival, and matrix D_1 consists of the rates of those transitions that generate an arrival. However, contrary to non-terminating MAPs, the generator matrix of the background Markov chain of TMAPs, $D = D_0 + D_1$, is transient, that is $D\mathbb{1} \neq 0$ and the non-negative vector $d = -D\mathbb{1}$ describes the termination rates of the background Markov chain. Based on practical considerations we assume that the termination is an observed event (an arrival), which means that a TMAP generates at least one arrival. (If only the "arrival events" are known, which is commonly the case in practice, the TMAPs which do not generate any arrival are not observed. Without knowing how many TMAPs terminated

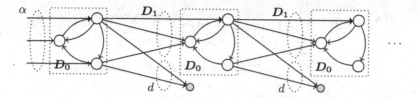

Fig. 1. The structure of the Markov chain representing the number of arrivals and the state of the background process.

without generating any arrival event there is no information to estimate the parameters of those invisible cases.) It also means that the TMAPs considered here are special cases of the ones defined in [8], since we assume that $d_0 = 0$ and our vector d equals to vector d_1 in [8]. The fact that the background Markov chain is transient ensures that the number of events generated by the process is finite. The Markov chain representing the number of arrivals and the state of the background process is depicted in Fig. 1.

Matrix $P = (-D_0)^{-1} D_1$ describes the state transition probabilities embedded at arrival instants. P holds the state transition probabilities of a transient discrete time Markov chain (DTMC) with termination vector $p = 1 - P1$. Note that P is sub-stochastic matrix (it has non-negative elements and $P1 \leq 1$), and $(I - P)^{-1} p = 1$ holds.

In case of TMAPs not only the statistical quantities related to the inter-arrival times are of interest, but also the ones related to the number of generated arrivals.

The number of arrivals $\mathcal{K}$ is characterized by a discrete phase-type (DPH) distribution with initial vector α and transition probability matrix P. Hence, the mean number of arrivals is given by

$$E(\mathcal{K}) = \sum_{k=1}^{\infty} \alpha k P^{k-1} p = \alpha (I - P)^{-2} p = \alpha (I - P)^{-1} 1. \tag{1}$$

If the inter-arrival times are denoted by $\mathcal{X}_1, \mathcal{X}_2, \ldots$, then the joint density function of the inter-arrival times is

$$f(x_1, x_2, \ldots, x_k) = \lim_{\Delta \to 0} \frac{1}{\Delta} P(\mathcal{X}_1 \in (x_1, x_1 + \Delta), \ldots, \mathcal{X}_k \in (x_k, x_k + \Delta))$$

$$= \alpha e^{D_0 x_1} D_1 e^{D_0 x_2} D_1 \cdots e^{D_0 x_k} (D_1 1 + d). \tag{2}$$

If it exists, the nth moment of $\mathcal{X}_{k+1}$ is

$$E(\mathcal{X}_{k+1}^n | \mathcal{X}_{k+1} < \infty) = \frac{E(\mathcal{X}_{k+1}^n I_{\{\mathcal{X}_{k+1} < \infty\}})}{P(\mathcal{X}_{k+1} < \infty)} = \frac{n! \alpha P^k (-D_0)^{-n} 1}{\alpha P^k 1}. \tag{3}$$

The mean of the inter-arrival times $E(\mathcal{X})$ is not as easy to express as for ordinary MAPs, it is obtained from $E(\mathcal{X}) = E\left(\sum_{k=1}^{\mathcal{K}} \mathcal{X}_k\right) / E(\mathcal{K})$, where the numerator is derived as

$$E\left(\sum_{k=1}^{K} \mathcal{X}_k\right) = \sum_{\kappa=1}^{\infty} E\left(\mathcal{I}_{\{K=\kappa\}} \sum_{k=1}^{K} \mathcal{X}_k\right) = \sum_{\kappa=1}^{\infty} \sum_{i=0}^{\kappa-1} \alpha P^i U P^{\kappa-1-i} p$$

$$= \sum_{i=0}^{\infty} \sum_{\kappa=0}^{\infty} \alpha P^i U P^{\kappa} p = \alpha (I - P)^{-1} U (I - P)^{-1} p, \qquad (4)$$

where $U = (-D_0)^{-1}$, and the denominator is given by (1). As a result the mean inter-arrival time is

$$E(\mathcal{X}) = \frac{E\left(\sum_{k=1}^{K} \mathcal{X}_k\right)}{E(K)} = \frac{\alpha (I - P)^{-1} U (I - P)^{-1} p}{\alpha (I - P)^{-2} p} = \frac{\alpha (I - P)^{-1} U \mathbb{1}}{\alpha (I - P)^{-1} \mathbb{1}}. \qquad (5)$$

To discuss the correlation of the inter-arrival times we introduce the notation $\hat{\mathcal{X}}_k = \mathcal{X}_k \mid \mathcal{X}_k < \infty$. Note that $\hat{\mathcal{X}}_1 = \mathcal{X}_1$ due to the modeling assumption of at least one arrival. By this notation from (3) we have

$$E\left(\hat{\mathcal{X}}_{k+1}^n\right) = \frac{n! \alpha P^k U^n \mathbb{1}}{\alpha P^k \mathbb{1}}.$$

The expectation of the product of two subsequent inter-arrival times is

$$E\left(\mathcal{X}_1 \hat{\mathcal{X}}_{k+1}\right) = \frac{E\left(\mathcal{X}_1 \mathcal{X}_{k+1} \mathcal{I}_{\{\mathcal{X}_{k+1}<\infty\}}\right)}{P(\mathcal{X}_{k+1} < \infty)} \qquad (6)$$

$$= \frac{\alpha(-D_0)^{-2} D_1 P^{k-1} (-D_0)^{-2} (D_1 \mathbb{1} + d)}{\alpha(-D_0)^{-1} D_1 P^{k-1} (-D_0)^{-1} (D_1 \mathbb{1} + d)} = \frac{\alpha U P^k U \mathbb{1}}{\alpha P^k \mathbb{1}},$$

where we used that $(-D_0)^{-1} D_1 \mathbb{1} + d = \mathbb{1}$, due to $D_0 \mathbb{1} + D_1 \mathbb{1} + d = 0$. Based on the joint expectation the correlation is

$$Corr(\mathcal{X}_1, \hat{\mathcal{X}}_k + 1) = \frac{E\left(\mathcal{X}_1 \hat{\mathcal{X}}_{k+1}\right) - E(\mathcal{X}_1) E\left(\hat{\mathcal{X}}_{k+1}\right)}{\sqrt{E(\mathcal{X}_1^2) - E^2(\mathcal{X}_1)} \sqrt{E\left(\hat{\mathcal{X}}_{k+1}^2\right) - E^2\left(\hat{\mathcal{X}}_{k+1}\right)}}. \qquad (7)$$

3   An EM Algorithm for TMAPs

In this section an EM algorithm is presented to create a TMAP from measurement data. The measurement data is given by samples $X = (x_k^{(\ell)}, \ k = 1, \ldots, K_\ell, \ \ell = 1, \ldots, L)$. We refer the set of dependent samples for a given ℓ as the ℓth run, where the ℓth run is composed by K_ℓ samples. The aim of the EM algorithm is to find $\Theta = (\alpha, D_0, D_1)$ by which the likelihood of the observations,

$$\mathcal{L}(\Theta|X) = \prod_{\ell=1}^{L} \alpha e^{D_0 x_1^{(\ell)}} D_1 \cdots e^{D_0 x_{K_\ell}^{(\ell)}} d, \qquad (8)$$

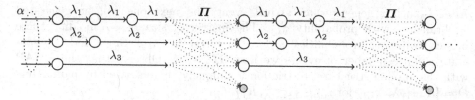

Fig. 2. The special TMAP structure used for fitting.

is maximized. Introducing the run-dependent forward likelihood (row) vectors recursively as

$$
a^{(\ell)}[k] = \begin{cases} \alpha, & k = 0, \\ a^{(\ell)}[k-1]e^{\boldsymbol{D}_0 x_k^{(\ell)}} \boldsymbol{D}_1, & k > 0, \end{cases}
\tag{9}
$$

for $\ell = 1, \ldots, L$ and $k = 0, \ldots, K_\ell - 1$, and backward likelihood (column) vectors as

$$
b^{(\ell)}[k] = \begin{cases} e^{\boldsymbol{D}_0 x_k^{(\ell)}} \boldsymbol{D}_1 b^{(\ell)}[k+1], & k < K_\ell, \\ d, & k = K_\ell, \end{cases}
\tag{10}
$$

for $\ell = 1, \ldots, L$ and $k = K_\ell, \ldots, 1$, the likelihood can be obtained as

$$
\mathcal{L}(\Theta|X) = \prod_{\ell=1}^{L} a^{(\ell)}[k_\ell] \cdot b^{(\ell)}[k_\ell + 1],
\tag{11}
$$

for every $k_\ell = 0, \ldots, K_\ell - 1$.

The forward and backward likelihood vectors play an important role in the presented EM algorithm. However, computing the matrix exponential terms is numerically demanding. To reduce the computational complexity we apply the same structural restriction as in [7,9,10], thus we introduce a special TMAP structure composed of a number of Erlang distributed branches. When a given branch is selected the inter-arrival time is Erlang distributed defined by the parameters (rate and order) of the selected Erlang branch, and after each arrival event a sub-stochastic transition probability matrix determines which Erlang branch to choose for the next inter-arrival given the branch generating the current arrival (see Fig. 2). Due to applied structural restriction the computations of matrix exponential terms, e.g. in (8), are replaced by the computations of scalar exponential terms in the form of (12).

In the proposed special structure the inter-arrival times are generated by one of the R Erlang branches. The order and the intensity parameters of the branches are denoted by r_i, λ_i, for $i \in \{1, \ldots, R\}$, respectively. The density of the inter-arrival times generated by branch i is

$$
f_i(x) = \frac{(\lambda_i x)^{r_i - 1}}{(r_i - 1)!} \lambda_i e^{-\lambda_i x}.
\tag{12}
$$

After branch i generates an arrival event, the next one will be generated by branch j with probability $\pi_{i,j}$. The matrix of size $R \times R$ holding these branch switching probabilities is denoted by $\Pi = [\pi_{i,j}]$. Since TMAPs generate a finite number of events we have $\Pi 1 < 1$. Observe that the TMAP with the applied structural restriction is uniquely characterized by parameters $\Theta = \{\alpha_i, r_i, \lambda_i, \pi_{i,j}, \text{ for } i,j \in \{1, \ldots, R\}\}$.

By this special TMAP structure the forward and backward likelihood vectors can be obtained without computing matrix exponentials, since

$$a_i^{(\ell)}[k] = \sum_{j=1}^{M} a_j^{(\ell)}[k-1] f_j(x_k^{(\ell)}) \pi_{j,i}, \tag{13}$$

$$b_j^{(\ell)}[k] = \sum_{i=1}^{M} f_j(x_k^{(\ell)}) \pi_{j,i} b_i^{(\ell)}[k+1]. \tag{14}$$

The EM algorithm assumes that the data X available for fitting is incomplete, and there is a hidden data Y. In our case, the hidden data $y_k^{(\ell)} \in Y$ is and integer number representing the Erlang branch that generated the kth inter-arrival time of run ℓ, thus $x_k^{(\ell)}$. If the hidden data was known, the logarithm of the likelihood would be easy to express, as

$$\log \mathcal{L}(\Theta | X, Y) = \sum_{\ell=1}^{L} \sum_{k=1}^{K_\ell} \log \left(f_{y_k^{(\ell)}}(x_k^{(\ell)}) \right). \tag{15}$$

Maximizing (15) with respect to λ_i gives

$$\hat{\lambda}_i = \frac{\sum_{\ell=1}^{L} \sum_{k=1}^{K_\ell} r_i \cdot I_{\{y_k^{(\ell)}=i\}}}{\sum_{\ell=1}^{L} \sum_{k=1}^{K_\ell} x_k^{(\ell)} I_{\{y_k^{(\ell)}=i\}}}, \tag{16}$$

where $\hat{\lambda}_i$, and the similar subsequent notations, denotes the optimum assuming Y is known. From [2] we have

$$\hat{\pi}_{i,j} = \frac{\sum_{\ell=1}^{L} \sum_{k=1}^{K_\ell-1} I_{\{y_k^{(\ell)}=i, y_{k+1}^{(\ell)}=j\}}}{\sum_{\ell=1}^{L} \sum_{k=1}^{K_\ell} I_{\{y_k^{(\ell)}=i\}}}. \tag{17}$$

Note that the summation over k in the denominator runs up to K_ℓ, while the one in the numerator runs up to $K_\ell - 1$, thus matrix $\hat{\Pi}$ is sub-stochastic, reflecting the terminating behavior of TMAPs. The maximum likelihood estimation for the initial vector is

$$\hat{\alpha}_i = \frac{1}{L} \sum_{\ell=1}^{L} I_{\{y_1^{(\ell)}=i\}}. \tag{18}$$

The hidden data is, however, unknown. The marginal distribution of the hidden data $y_k^{(\ell)}$ can be derived from the forward and backward likelihood vectors leading to

$$
\begin{aligned}
q_i^{(\ell)}[k] = P(y_k^{(\ell)} = i | X, \Theta) &= \frac{P(y_k^{(\ell)} = i, X | \Theta)}{P(X | \Theta)} \\
&= \frac{\left(a_i^{(\ell)}[k-1] \cdot b_i^{(\ell)}[k]\right) \prod_{m \neq \ell} a^{(m)}[0] \cdot b^{(m)}[1]}{\prod_{m=1}^{L} a^{(m)}[0] \cdot b^{(m)}[1]} \\
&= \frac{a_i^{(\ell)}[k-1] \cdot b_i^{(\ell)}[k]}{\alpha \cdot b^{(\ell)}[1]}, \quad k = 1, \ldots, K_\ell,
\end{aligned}
\tag{19}
$$

where we used $a^{(\ell)}[0] = \alpha$.

To characterize the joint distribution of the branches generating two consecutive inter-arrival times we also need the probabilities

$$
\begin{aligned}
q_{i,j}^{(\ell)}[k] = P(y_k^{(\ell)} = i, y_{k+1}^{(\ell)} = j | X, \Theta) &= \frac{P(y_k^{(\ell)} = i, y_{k+1}^{(\ell)} = j, X | \Theta)}{P(X | \Theta)} \\
&= \frac{a_i^{(\ell)}[k-1] \cdot f_i(x_k^{(\ell)}) \cdot \pi_{i,j} \cdot b_j^{(\ell)}[k+1]}{\alpha \cdot b^{(\ell)}[1]}.
\end{aligned}
\tag{20}
$$

The calculation of $q_i^{(\ell)}[k]$ and $q_{i,j}^{(\ell)}[k]$ form the E-step of the algorithm.

In the M-step new estimates for Θ are obtained based on the distributions of the hidden data. For λ_i from (16) and (19) we get

$$
\lambda_i = \frac{\sum_{\ell=1}^{L} \sum_{k=1}^{K_\ell} r_i \cdot q_i^{(\ell)}[k]}{\sum_{\ell=1}^{L} \sum_{k=1}^{K_\ell} x_k^{(\ell)} q_i^{(\ell)}[k]} = \frac{\sum_{\ell=1}^{L} \frac{\sum_{k=1}^{K_\ell} r_i a_i^{(\ell)}[k-1] b_i^{(\ell)}[k]}{\alpha \cdot b^{(\ell)}[1]}}{\sum_{\ell=1}^{L} \frac{\sum_{k=1}^{K_\ell} x_k^{(\ell)} a_i^{(\ell)}[k-1] b_i^{(\ell)}[k]}{\alpha \cdot b^{(\ell)}[1]}}.
\tag{21}
$$

Similarly, the new estimates for the branch switching probabilities are obtained from (17) and (20) as

$$
\pi_{i,j} = \frac{\sum_{\ell=1}^{L} \sum_{k=1}^{K_\ell - 1} q_{i,j}^{(\ell)}[k]}{\sum_{\ell=1}^{L} \sum_{k=1}^{K_\ell} q_i^{(\ell)}[k]} = \frac{\sum_{\ell=1}^{L} \frac{\sum_{k=1}^{K_\ell - 1} a_i^{(\ell)}[k-1] f_i(x_k^{(\ell)}) \pi_{i,j} b_j^{(\ell)}[k+1]}{\alpha \cdot b^{(\ell)}[1]}}{\sum_{\ell=1}^{L} \frac{\sum_{k=1}^{K_\ell} a_i^{(\ell)}[k-1] b_i^{(\ell)}[k]}{\alpha \cdot b^{(\ell)}[1]}}.
\tag{22}
$$

Finally, probabilities α_i are derived from (18) and (19), yielding

$$
\alpha_i = \frac{1}{L} \sum_{\ell=1}^{L} q_i^{(\ell)}[1] = \frac{1}{L} \sum_{\ell=1}^{L} \frac{\alpha_i b_i^{(\ell)}[1]}{\alpha \cdot b^{(\ell)}[1]}.
\tag{23}
$$

4   Details of the Numerical Algorithm

The EM algorithm presented in Sect. 3 is not straight forward to implement in an efficient way. While the special structure proposed for fitting does reduce the

---

Algorithm 1. Pseudo-code of the proposed EM algorithm

1: **procedure** EM-FIT($x_k^{(\ell)}, \lambda_i, \pi_{i,j}, \alpha_i, r_i$)
2:     $LogLi \leftarrow -\infty$
3:     **for** $iter = 1$ **to** $maxIter$ **do**
4:         Compute and store conditional densities $f_i(x_k^{(\ell)})$ by (12)
5:         **for** $\ell = 1$ **to** L **do**
6:             **for** $k = 1$ **to** K_ℓ **do**
7:                 Compute and store forward likelihood vectors $a^{(\ell)}[k]$ by (13)
8:             **end for**
9:             Compute and store backward likelihood vector $b^{(\ell)}[K_\ell]$ by (10)
10:            **for** $k = K_\ell - 1$ **down to** l **do**
11:                Compute and store backward likelihood vector $b^{(\ell)}[k]$ by (14)
12:            **end for**
13:        **end for**
14:        $oLogLi \leftarrow LogLi$
15:        $LogLi \leftarrow \sum_{\ell=1}^{L} \alpha \cdot b^{(\ell)}[1]$
16:        **if** $iter > 1$ **and** $(LogLi - LogLi) < ln(1 + \epsilon)$ **then**
17:            **return** $(\lambda_i, \pi_{i,j}, \alpha_i)$
18:        **end if**
19:        Compute new estimates for λ_i by (21)
20:        Compute new estimate for $\pi_{i,j}$ by (22)
21:        Compute new estimate for α_i by (23)
22:    **end for**
23:    **return** $(\lambda_i, \pi_{i,j}, \alpha_i)$
24: **end procedure**

---

computational demand of the procedure significantly, the naive implementation (shown in Fig. 1) still contains many numerical pitfalls.

Our aim is to develop an implementation that enables the practical application of the algorithm, thus

- the execution time must be reasonable with large data sets (containing millions of samples),
- the implementation must be insensitive to the order of magnitude of the input data,
- the implementation should exploit the parallel processing capabilities of modern hardware.

These items are addressed in the subsections below.

4.1   Initial Guess for α, λ_i and Π

We use the following randomly generated initial parameters. α is a random probability vector (composed of R uniform pseudo-random numbers in $(0,1)$ divided by the sum of the R numbers). The mean run length of the data set is computed as $\bar{K} = \sum_{\ell=1}^{L} K_\ell / L$, and based on that each row of matrix Π is a random probability vector multiplied by $1 - 1/\bar{K}$ (that is, initially the exit probability is the same, $1/\bar{K}$, in each Erlang branch). The initial values for λ_i are computed based on the mean inter-arrival time $\bar{T} = \frac{\sum_{\ell=1}^{L} \sum_{k=1}^{K_\ell} x_k^{(\ell)}}{\sum_{\ell=1}^{L} K_\ell}$, and it is $\lambda_i = r_i/\bar{T}$. Let $x_{max} = \max_{\ell,k} x_k^{(\ell)}$ and $\lambda_{max} = \max_i \lambda_i$. In order to avoid underflow during the computation of $e^{-x_k^{(\ell)} \lambda_i}$ in (12) we re-scale this initial guess according to the representation limits of the single precision floating point numbers with $8 + 16$ bits, where one of the 8 bits of the mantissa indicates the sign. That is $2^{2^7} \sim e^{88}$ is the representation limit. Accordingly, if $x_{max} \lambda_{max} > 60$ then we re-scale the initial intensity values to $\lambda_i = \frac{60\lambda_i}{x_{max}\lambda_{max}}$, where 60 is a heuristic choice to be far enough from the representation limit (which is 88).

4.2   Improving Numerical Stability of the Forward and Backward Likelihood Vectors Computation

Computing vectors $a^{(\ell)}[k]$ and $b^{(\ell)}[k]$ by applying recursions (13) and (14) directly can lead to numerical overflow. To overcome this difficulty we express these vectors in the normal form

$$a_i^{(\ell)}[k] = \dot{a}_i^{(\ell)}[k] \cdot 2^{\ddot{a}^{(\ell)}[k]},$$
$$b_i^{(\ell)}[k] = \dot{b}_i^{(\ell)}[k] \cdot 2^{\ddot{b}^{(\ell)}[k]}, \tag{24}$$

where $\ddot{a}^{(\ell)}[k]$ and $\ddot{b}^{(\ell)}[k]$ are integer numbers and the values $\dot{a}_i^{(\ell)}[k]$, $\dot{b}_i^{(\ell)}[k]$ are such that $0.5 \le \dot{a}^{(\ell)}[k]\mathbf{1} < 1$ and $0.5 \le \mathbf{1}^T \dot{b}^{(\ell)}[k] < 1$. For a given vector $a_i^{(\ell)}[k]$, $\dot{a}_i^{(\ell)}[k]$ and $\ddot{a}^{(\ell)}[k]$ can be obtained from

$$\dot{a}_i^{(\ell)}[k] = \frac{a_i^{(\ell)}[k]}{2^{\lceil \log_2(a^{(\ell)}[k]\,) \rceil}}, \quad \ddot{a}^{(\ell)}[k] = \left\lceil \log_2(a^{(\ell)}[k]\mathbf{1}) \right\rceil. \tag{25}$$

To avoid the calculation of $a_i^{(\ell)}[k]$ (that can under- or overflow), it is possible to modify the recursion (13) to work with $\dot{a}_i^{(\ell)}[k]$ and $\ddot{a}^{(\ell)}[k]$ directly, leading to

$$\tilde{a}_i^{(\ell)}[k] = \sum_{j=0}^{M} \dot{a}_i^{(\ell)}[k-1] f_j \left(x_k^{(\ell)} \right) \pi_{j,i},$$

$$\dot{a}_i^{(\ell)}[k] = \frac{\tilde{a}_i^{(\ell)}[k]}{2^{\lceil \log_2(\tilde{a}^{(\ell)}[k]\,) \rceil}}, \quad \ddot{a}^{(\ell)}[k] = \ddot{a}^{(\ell)}[k-1] + \left\lceil \log_2(\tilde{a}^{(\ell)}[k]\mathbf{1}) \right\rceil. \tag{26}$$

Hence, in the first step $\tilde{a}_i^{(\ell)}[k]$ is computed, from which in the second step the normalized quantity is derived and the exponent is incremented by the appropriate magnitude. To obtain the normal form of $\dot{a}_i^{(\ell)}[0]$ and $\ddot{a}^{(\ell)}[0]$, we can apply (25). The treatment of the normal form of the backward likelihood vectors $\dot{b}_i^{(\ell)}[k]$ follow the same pattern.

The parameter estimation formulas using the normal form of the forward and backward likelihood vectors are

$$\lambda_i = \frac{\sum_{\ell=1}^{L} \frac{1}{\alpha \dot{b}^{(\ell)}[1]} \sum_{k=1}^{K_\ell} r_i \dot{a}_i^{(\ell)}[k-1] \dot{b}_i^{(\ell)}[k] 2^{\ddot{a}^{(\ell)}[k-1]+\ddot{b}^{(\ell)}[k]-\ddot{b}^{(\ell)}[1]}}{\sum_{\ell=1}^{L} \frac{1}{\alpha \dot{b}^{(\ell)}[1]} \sum_{k=1}^{K_\ell} x_k^{(\ell)} \dot{a}_i^{(\ell)}[k-1] \dot{b}_i^{(\ell)}[k] 2^{\ddot{a}^{(\ell)}[k-1]+\ddot{b}^{(\ell)}[k]-\ddot{b}^{(\ell)}[1]}}, \tag{27}$$

$$\pi_{i,j} = \frac{\sum_{\ell=1}^{L} \frac{1}{\alpha \dot{b}^{(\ell)}[1]} \sum_{k=1}^{K_\ell} \dot{a}_i^{(\ell)}[k-1] f_i\left(x_k^{(\ell)}\right) \pi_{i,j} \dot{b}_i^{(\ell)}[k+1] 2^{\ddot{a}^{(\ell)}[k-1]+\ddot{b}^{(\ell)}[k+1]-\ddot{b}^{(\ell)}[1]}}{\sum_{\ell=1}^{L} \frac{1}{\alpha \dot{b}^{(\ell)}[1]} \sum_{k=1}^{K_\ell} \dot{a}_i^{(\ell)}[k-1] \dot{b}_i^{(\ell)}[k] 2^{\ddot{a}^{(\ell)}[k-1]+\ddot{b}^{(\ell)}[k]-\ddot{b}^{(\ell)}[1]}}, \tag{28}$$

$$\alpha_i = \frac{1}{L} \sum_{\ell=1}^{L} \frac{\alpha_i \dot{b}_i^{(\ell)}[1]}{\alpha \dot{b}^{(\ell)}[1]}. \tag{29}$$

Observe that the exponent of 2 depends only on the difference of $\ddot{a}^{(\ell)}[k]$ and $\ddot{b}^{(\ell)}[k]$ for consecutive k values according to (26), thus the multiplication and the division with large numbers has been avoided.

Finally, the log-likelihood of the whole trace data can be computed as

$$\mathcal{L}(\Theta|X) = \log\left(\prod_{\ell=1}^{L} \alpha b^{(\ell)}[1]\right) = \sum_{\ell=1}^{L} \log\left(\alpha \dot{b}^{(\ell)}[1]\right) + \ddot{b}^{(\ell)}[1] \log(2). \tag{30}$$

4.3   Serial Implementations

For accuracy and performance comparison we have implemented three versions of the algorithm shown in Fig. 1 (with the discussed modifications for numerical stability):

- Java implementation using double precision floating point numbers,
- C++ implementation using double precision floating point numbers,
- C++ implementation using single precision floating point numbers.

4.4   Parallel Implementation

We have adapted the presented algorithm to be executed on GPUs (graphics processing units) by using CUDA library. GPUs are cheap in the sense of computing power, however, their computing cores are much simpler compared to the ones of CPU. Therefore to fully utilize the hardware low level technical details have to be considered such as the thread grouping, the multi-level memory hierarchy, reducing the number of conditional jumps, memory operations, etc.

The entry part of the algorithm (shown in Fig. 2) is executed on the host environment (i.e. processed by CPU) from which the so called kernels (shown in Figs. 3 and 4) are invoked to be executed on GPU device. Upon kernel launching the number of threads in block, the number of blocks in grid and amount of shared memory (in bytes) to be allocated for every block has to be specified.

After kernel launch host process waits until all the threads are processed by the kernel, and then resumes.

The KERNEL-A, shown in Fig. 3, computes the normalized likelihood vectors $\dot{a}^{(\ell)}[k]$, $\dot{b}^{(\ell)}[k]$ and their respective exponents $\ddot{a}^{(\ell)}[k]$, $\ddot{b}^{(\ell)}[k]$. The number of threads in block and grid size can be chosen freely, so that to utilize the specific capabilities of the GPU. However, the threads should be assigned with similar amount of work in order not to waste computing resources.

The KERNEL-B, shown in Fig. 4, computes new parameter estimates $\Theta = (\lambda_i, \pi_{i,j}, \alpha_i)$. Synchronization between threads is necessary before computing the actual parameter estimate after numerator and denominator values are computed. Since thread synchronization is possible only within a block, the number of blocks is determined by the number of $\pi_{i,j}$ estimates, thus R^2. Thread count in block can by chosen freely.

Note that work load for the kernels are different. The data runs are allocated to threads in grid for KERNEL-A. While for KERNEL-B all the runs are allocated among threads for every block.

Even run allocation to threads is a complex problem. A simple greedy solution is to assign runs in descending order (of number of inter-arrival time samples) to the thread, which has been assigned with the smallest number of inter-arrivals.

Global GPU memory accessing operations are slower compared to shared memory. It is a common practice to load frequently used data from global memory into shared one and after calculations write results back into global memory. In our case parameter estimates as well as structure parameters are uploaded in shared memory.

Additionally previously computed likelihood vector values are cached for computing the next ones. Also Erlang branch densities are computed and stored in shared memory just before to be used in subsequent calculations.

Shared memory can be used for communication, since it is visible for all the threads within block. In KERNEL-B threads perform summation across the assigned runs and the intermediate results are written to shared memory to be loaded by one designated thread to compute the final estimate value.

5   Numerical Experiments

We start the section with a general note on the applied special structure. In spite of the natural expectation that the result of the fitting (in terms of likelihood) with the special structure is worse than the one with the general TMAP class of the same size, however, similar to related results in the literature [7,9] our numerical experience is just the opposite. The general TMAP class is redundant [6], and the EM algorithm goes back and forth between different representations of almost equivalent TMAPs. Our special TMAP class has much less parameters, and the benefit of optimizing according to less parameters dominates the drawback coming from reduced flexibility of the special structure.

Algorithm 2. Pseudo-code of the proposed EM algorithm (CUDA)

1: **procedure** EM-FIT-CUDA($x_k^{(\ell)}, \lambda_i, \pi_{i,j}, \alpha_i, r_i$, {run allocation to threads})
2:     Allocate device memory.
3:     Copy data from host to device memory.
4:     $LogLi \leftarrow -\infty$
5:     **for** $iter = 1$ **to** $maxIter$ **do**
6:         Invoke KERNEL-A for computing likelihood vectors and run likelihoods.
7:         $oLogLi \leftarrow LogLi$
8:         Copy run likelihoods from device to host memory.
9:         Compute trace data log-likelihood $LogLi$.
10:         **if** $iter > 1$ **and** $(LogLi - LogLi) < ln(1 + \epsilon)$ **then**
11:             Copy parameter estimates from device to host memory.
12:             **return** $(\lambda_i, \pi_{i,j}, \alpha_i)$
13:         **end if**
14:         Invoke KERNEL-B for computing new parameter estimates.
15:     **end for**
16:     Copy parameter estimates from device to host memory.
17:     Deallocate device memory.
18:     **return** $(\lambda_i, \pi_{i,j}, \alpha_i)$
19: **end procedure**

Hereafter we compare the behavior of the four implementations (three serial ones (Java (double), C++ (double), C++ (single)) and the one for GPU) of the presented EM algorithm. All numerical experiments were made on an average PC with an Intel Core 2 CPU clocked at 2112 MHz with 32 KB L1 cache and 4096 KB L2 cache, and an ASUS GeForce GTX 560 Ti graphics card with a GPU clocked at 900 MHz having 1 GB of RAM and 384 CUDA cores. For the GPU implementation the first kernel is launched with 64 blocks of 32 threads each, the second one is launched with 9 (R^2) blocks of 192 threads each.

Two data sets are considered, in the first one there are 1000000 runs and 8824586 inter-arrival times in total, and in the second one there are 2000000 runs and 14503248 inter-arrival times in total.

Table 1. Execution times and log likelihoods of different implementations

Implementation	With 1 million samples		With 2 million samples	
	Log-likelihood	Execution time	Log-likelihood	Execution time
Java (double)	$-3.65998 \cdot 10^7$	14 m 22 s	$-5.73999 \cdot 10^7$	22 m 15 s
C++ (double)	$-3.65998 \cdot 10^7$	06 m 21 s	$-5.73999 \cdot 10^7$	10 m 28 s
C++ (single)	$-3.65902 \cdot 10^7$	06 m 44 s	$-5.73752 \cdot 10^7$	11 m 04 s
CUDA (single)	$-3.65997 \cdot 10^7$	46 s	$-5.73997 \cdot 10^7$	01 m 13 s

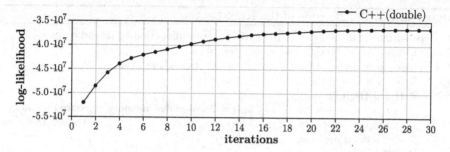

Fig. 3. Log-likelihood of 1000000 sample trace data obtained by C++ (double) procedure.

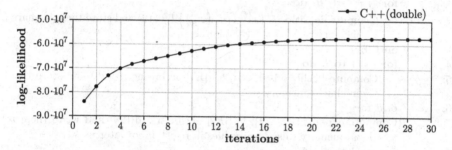

Fig. 4. Log-likelihood of 2000000 sample trace data obtained by C++ (double) procedure.

Based on the experiences in [7] we adopt three Erlang branches ($R = 3$) with 1, 2 and 3 states ($r_1 = 1, r_2 = 2, r_3 = 3$). For fair comparison we have run 30 iterations of the EM algorithm in all cases (after which the algorithm seemed to converge) and the compared results are always initiated with the same initial guesses. The run times and log-likelihoods are compared in Table 1.

Time necessary to allocate/deallocate arrays is not included in run time, because of different C++ and Java memory management policies. However the time for data allocation/deallocation on GPU device is included. The trace data sample distribution among threads is done in advance, thus not included in run time.

After every iteration log-likelihood was computed using double precision floating point variables. Java (double) and C++ (double) implementations gave identical log-likelihoods and are shown in Figs. 3 and 4. Log-likelihoods obtained by running C++ (single) are relatively similar to ones acquired using C++ (double) implementation. Therefore, it is more convenient to plot the difference of log-likelihood obtained from C++ (single) minus C++ (double). The same applies for results obtained by CUDA (single) implementation.

Algorithm 3. Pseudo-code of the KERNEL-A

1: **procedure** KERNEL-A($x_k^{(\ell)}, \lambda_i, \pi_{i,j}, \alpha_i, r_i$, {run allocation to threads})
2:    Identify index ti of thread within the block.
3:    Identify index si of thread within all the threads in grid.
4:    Reference shared memory.
5:    **if** $ti = 0$ **then**
6:        Copy estimates for λ_i, $\pi_{i,j}$ and α_i from global memory to shared memory.
7:        Compute absorption probabilities $\pi_i = 1 - \sum_{j=1}^{R} \pi_{i,j}$ and store them in shared memory.
8:    **end if**
9:    Synchronize threads within block.
10:   **for every** ℓ **assigned to** si **thread do**
11:       **for** $j = 1$ **to** R **do**
12:           Compute densities $f_j\left(x_1^{(\ell)}\right)$, $f_j\left(x_{K_\ell}^{(\ell)}\right)$ by (12) and store them in shared memory.
13:       **end for**
14:       **for** $i = 1$ **to** R **do**
15:           Compute likelihoods $\tilde{a}_i^{(\ell)}[1]$, $\tilde{b}_i^{(\ell)}[K_\ell]$ according to (26) and store them in shared memory.
16:       **end for**
17:       Compute exponents $\ddot{a}^{(\ell)}[1]$, $\ddot{b}^{(\ell)}[K_\ell]$ according to (26) and write them to global memory (also keep values in registers for later use).
18:       **for** $i = 1$ **to** R **do**
19:           Compute normalized likelihoods $\dot{a}_i^{(\ell)}[1]$, $\dot{b}_i^{(\ell)}[K_\ell]$ according to (26), cache them in shared memory and write to global memory.
20:       **end for**
21:       **for** $k = 2$ **to** $K_\ell - 1$ **do**
22:           **for** $j = 1$ **to** R **do**
23:               Compute densities $f_j\left(x_k^{(\ell)}\right)$, $f_j\left(x_{K_\ell - k}^{(\ell)}\right)$ and store them in shared memory.
24:           **end for**
25:           **for** $i = 1$ **to** R **do**
26:               Compute likelihoods $\tilde{a}_i^{(\ell)}[k]$, $\tilde{b}_i^{(\ell)}[K_\ell - k]$ according to (26) and store them in shared memory.
27:           **end for**
28:           Compute exponents $\ddot{a}^{(\ell)}[k]$, $\ddot{b}^{(\ell)}[K_\ell - k]$ according to (26) and write them to global memory (also keep values in registers for later use).
29:           **for** $i = 1$ **to** R **do**
30:               Compute normalized likelihoods $\dot{a}_i^{(\ell)}[k]$, $\dot{b}_i^{(\ell)}[K_\ell - k]$ according to (26), cache them in shared memory and write to global memory.
31:           **end for**
32:       **end for**
33:       Compute ℓth run likelihood and write to global memory.
34:       Compute ℓth run log-likelihood and sum up to register.
35:   **end for**
36:   Write sum of sample log-likelihoods into shared memory.
37:   Synchronize threads within block.
38:   Sum up all the run log-likelihoods within block and write to global memory.
39: **end procedure**

Algorithm 4. Pseudo-code of KERNEL-B

1: **procedure** KERNEL-B$(x_k^{(\ell)}, \lambda_i, \pi_{i,j}, \alpha_i, r_i, \{\text{run allocation to threads}\})$
2:     Identify index (i, j) of the block within grid.
3:     Identify index si of thread within the current block.
4:     Reference shared memory.
5:     **for every** ℓ assigned to si **thread do**
6:         Read ℓth run's likelihood value from global memory and store in register.
7:         Compute part of (28) denominator for (ℓ) and sum up in shared memory.
8:         **if** $K_\ell > 1$ **then**
9:             Compute part of (28) numerator for (ℓ) and sum up in shared memory.
10:        **end if**
11:    **end for**
12:    Synchronize threads within block.
13:    **if** $si = 1$ **then**
14:        Read summed up values for denominator and numerator, compute new estimate $\pi_{i,j}$ by (28) and store in global memory.
15:    **end if**
16:    Synchronize threads within block.
17:    **if** $j = 1$ **then**
18:        **for every** ℓ assigned to si **thread do**
19:            Compute part of (27) numerator and denominator for (ℓ) and sum up in shared memory.
20:        **end for**
21:    **end if**
22:    Synchronize threads within block.
23:    **if** $si = 1$ **then**
24:        Read summed up values for denominator and numerator, compute new estimate λ_i by (27) and store in global memory.
25:    **end if**
26:    Synchronize threads within block.
27:    **if** $j = 1$ **then**
28:        **for every** ℓ assigned to si **thread do**
29:            Compute part of (29) numerator for (ℓ) and sum up in shared memory.
30:        **end for**
31:    **end if**
32:    Synchronize threads within block.
33:    **if** $si = 1$ **then**
34:        Read summed up values for numerator, compute new estimate α_i by (29) and store in global memory.
35:    **end if**
36: **end procedure**

---

6   Conclusions

An EM procedure to estimate special structure TMAP parameters was developed and four of its implementations were tested by fitting reasonable large data sets. The C++ implementations of the fitting procedure indicated that both the single and the double precision floating points versions are stable, and converged to similar limits. Due to the fact that the log-likelihood vectors of independent runs can be computed independently parallel implementation on GPU can speed up the procedure significantly. Log-likelihoods obtained by using CUDA implementation are close to ones obtained computing with serial implementation on CPU.

References

1. Daley, D.J., Vere-Jones, D.: Finite point processes. In: Daley, D.J., Vere-Jones, D. (eds.) An Introduction to the Theory of Point Processes, pp. 111–156. Springer, New York (2003)
2. Anderson, T.W., Goodman, L.A.: Statistical inference about Markov chains. Ann. Math. Stat. **28**, 89–110 (1957)
3. Asmussen, S., Nerman, O., Olsson, M.: Fitting phase-type distributions via the EM algorithm. Scand. J. Stat. **23**, 419–441 (1996)
4. Buchholz, P.: An EM-algorithm for MAP fitting from real traffic data. In: Kemper, P., Sanders, W.H. (eds.) TOOLS 2003. LNCS, vol. 2794, pp. 218–236. Springer, Heidelberg (2003)
5. Hautphenne, S., Latouche, G.: The Markovian binary tree applied to demography. J. Math. Biol. **64**, 1109–1135 (2012)
6. Hautphenne, S., Telek, M.: Extension of some MAP results to transient MAPs and Markovian binary trees. Perform. Eval. **70**(9), 607–622 (2013)
7. Horváth, G., Okamura, H.: A fast EM algorithm for fitting marked Markovian arrival processes with a new special structure. In: Balsamo, M.S., Knottenbelt, W.J., Marin, A. (eds.) EPEW 2013. LNCS, vol. 8168, pp. 119–133. Springer, Heidelberg (2013)
8. Latouche, G., Remiche, M.-A., Taylor, P.: Transient Markov arrival processes. Ann. Appl. Probab. **13**, 628–640 (2003)
9. Okamura, H., Dohi, T.: Faster maximum likelihood estimation algorithms for Markovian arrival processes. In: Sixth International Conference on the Quantitative Evaluation of Systems, QEST 2009, pp. 73–82. IEEE (2009)
10. Thümmler, A., Buchholz, P., Telek, M.: A novel approach for phase-type fitting with the EM algorithm. IEEE Trans. Dependable Secure Comput. **3**(3), 245–258 (2006)

A Retrial Queue to Model a Two-Relay Cooperative Wireless System with Simultaneous Packet Reception

Ioannis Dimitriou$^{(\boxtimes)}$

Department of Mathematics, University of Patras, 26500 Patras, Greece
idimit@math.upatras.gr

Abstract. In this work we analyze a novel queueing system to model cooperative wireless networks with two relay nodes and simultaneous packet reception. We consider a network of three saturated source users, say a central and two background source users, two relay nodes and a common destination node. Source users transmit packets to the destination node with the cooperation of relays, which assist them by re-transmitting their blocked packets. We assume that the central source user forwards its blocked packets to both relay nodes in order to exploit both the spatial diversity they provide, and the broadcast nature of wireless communication. Moreover, each relay node receives also blocked packets from a dedicated background source user. We study a three-dimensional Markov process, investigate its stability condition and show that its steady-state performance is expressed in terms of the solution of a Riemann-Hilbert boundary value problem. Performance metrics are obtained, and numerical results show insights into the system behavior. Some computational issues are also discussed.

Keywords: Retrial queue · Fork-join queue · Boundary value problem · Cooperative communication · Relay nodes · Performance

1 Introduction

Cooperative communication is an effective way to improve the quality of wireless links since it is evidently proved that allows a flexible and robust exchange of data. In a wireless network, each source user increases its quality of service via cooperation with other users that "share" the antennas of their devices and assist the source users to transmit their data to a destination node; e.g., [14, 19, 23–25]. This is so called cooperation with relaying, and the assistant users are called relay nodes.

Such a system operates as follows: Consider a network of a finite number of source users, a finite number of relay nodes, and a destination node. Source users transmit packets to the destination node with the cooperation of the relay(s). If a transmission of a user's packet to the destination fails, the relays store the blocked packet in their buffers and try to re-transmit it to the destination later.

© Springer International Publishing Switzerland 2016
S. Wittevrongel and T. Phung-Duc (Eds.): ASMTA 2016, LNCS 9845, pp. 123–139, 2016.
DOI: 10.1007/978-3-319-43904-4_9

A mechanism must be employed to decide which of the relays will cooperate with sources (i.e., cooperation strategy). This problem gives rise to the usage of a cooperative space diversity protocol [26], under which, each user has a number of "partners" (i.e., relays) that are responsible for transmitting its blocked packets.

The core ideas behind cooperative communications were introduced in [9]. An overview of the rapidly expanding literature can be found in [22], where it was proved that relaying leads to a substantial reduction both on the packet delay, and on the energy consumption of the sources and relay nodes. Thus, further investigation of such systems is of great interest for the research community.

In this work we analyze a novel queueing system to model the impact of using two relay nodes in a network, to assist with relaying packets from a number of users to a destination node, under a cooperative space diversity protocol. We consider three saturated source users, say a central and two background source users, that transmit packets to a destination node, with the cooperation of two relays (i.e., network-level cooperation). Relay nodes have infinite capacity buffers and re-transmit blocked packets of the source users. The cooperation strategy is as follows: When the central source user fails to transmit a packet to the destination node, forwards its blocked packet at both relays (i.e., both relays overhear the transmission due to the wireless multicast advantage of the medium; two "partners"). On the contrary, a background source user cooperates only with a single relay node, and forwards its blocked packet only in that relay node.

Note that the notion of "re-transmission" gives rise to the so-called retrial queues [1–4, 10] (not exhaustive list) that have been proved very useful for modeling communication networks, where a "customer" meeting a busy server tries its luck again after a random time. Our system is modeled as a three-dimensional Markov process, and we prove that its steady-state performance can be expressed in terms of the solution of a Riemann-Hilbert boundary value problem. The study of queueing systems using the boundary value theory had been initiated in [12,13], and a concrete methodological approach was presented in [5]. Important generalizations were given in [2,6–8,17,28] (not exhaustive list).

Contribution of the paper. Besides its practical applicability, our work is also theoretically oriented. We provide for the first time in the related literature, an exact analysis of a model that unifies two fundamental queueing systems: the retrial queue with two orbits and constant retrial policy, and the generalized two-demand model (i.e., fork-join queue). The exact analysis of a typical fork-join queue with c parallel servers is possible only when $c = 2$ (see [15,29]). Moreover, there are also limited results on retrial queues with more than one orbits. In the vast majority of these results, a mean value approach was massively applied due to the complexity of the model [11,20,21]. In this work we present an exact analysis of a very intricate queueing model, and prove that the powerful and quite technical boundary value theory is an adequate technique to handle it.

The paper is organized as follows. In Sect. 2 we present the model in detail, and we derive the balance equations that are used to form the fundamental functional equation. In Sect. 3 we obtain some important results for the following analysis, and investigate stability conditions by considering an associated random walk in the first quadrant. Section 4 is devoted to the formulation and solution of a Riemann-Hilbert boundary value problem, which provides the generating function of the joint queue length distribution of the relays and destination node. Performance metrics are obtained in Sect. 5, and a simple numerical example along with some computational issues are discussed in Sect. 6.

2   The Model

We consider a network with three saturated source users, say S_i, $i = 0, 1, 2$, two relay nodes with infinite capacity queues, say R_1, R_2, and a common destination node D (see Fig. 1). The source users transmit packets towards the destination node with the cooperation of the relay nodes.

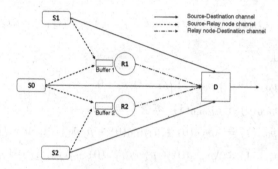

Fig. 1. The model.

User S_i generates packets towards the node D according to a Poisson process with rate λ_i, $i = 0, 1, 2$. Node D can handle at most one packet that forwards outside the network. The service time of a packet at the node D (i.e., the required time to forward the packet outside the network) is exponentially distributed with rate μ (we assume that the acknowledgments of successful or unsuccessful transmissions are instantaneous and error free).

The relays do not generate packets of their own but only re-transmit the packets they have received from the users. A relay node stores a packet in its queue when the direct transmission from a source user to the node D has failed. Specifically, the cooperation strategy to be applied between source users and relay nodes is as follows: If a direct transmission of a user's S_0 packet to the node D fails (i.e., node D is busy (transmitting)), both relay nodes store the blocked packet in their queues and try independently to forward it to the node D later. Moreover, if a direct transmission of a user's S_i, $i = 1, 2$, packet to the node D fails, only node R_i stores it in its queue and is responsible to re-transmit it to the node D later (i.e., user S_i cooperates only with node R_i, $i = 1, 2$). The node R_i tries to re-dispatch a blocked packet to the node D after an exponentially distributed time period with rate μ_i, $i = 1, 2$.

Under such a scheme, the user S_0 exploits both the spatial diversity provided by the relays, and the broadcast nature of wireless communication, where with a single transmission, a number of cooperating relay nodes (i.e., "partners") receive and relay its data [18,23,26]. In another scenario, the user S_0 splits its blocked packet (or job) in two sub-packets (or sub-jobs) and store each sub-packet in each relay node. Moreover, it can be assumed that the user S_0 transmits within the overlapping area created by the intersecting covering regions of both relay nodes, and thus, its blocked packet is forwarded to both relays. On the contrary, user S_i transmits only within the covering region of the node R_i, $i = 1, 2$.

Let $Q_i(t)$ be the number of stored packets in the queue of the relay node R_i, $i = 1, 2$, and $C(t)$ be the number of packets under transmission at the destination node D at time t. Clearly, $X(t) = \{Q_1(t), Q_2(t), C(t); t \geq 0\}$ constitutes a continuous time Markov chain with state space $S = \{0, 1, ...\} \times \{0, 1, ...\} \times \{0, 1\}$. Define the stationary probabilities for $m, n = 0, 1, 2, ..., k = 0, 1$,

$$p_{m,n}(k) = \lim_{t \to \infty} P(Q_1(t) = m, Q_2(t) = n, C(t) = k) = P(Q_1 = m, Q_2 = n, C = k).$$

Then, for $Q_2 = 0$,

$$\lambda p_{0,0}(0) = \mu p_{0,0}(1), \; m = 0, k = 0,$$
$$(\lambda + \mu_1) p_{m,0}(0) = \mu p_{m,0}(1), \; m \geq 1, k = 0,$$
$$(\lambda + \mu) p_{0,0}(1) = \lambda p_{0,0}(0) + \mu_1 p_{1,0}(0) + \mu_2 p_{0,1}(0), \; m = 0, k = 1, \quad (1)$$
$$(\lambda + \mu) p_{m,0}(1) = \lambda p_{m,0}(0) + \mu_1 p_{m+1,0}(0) + \mu_2 p_{m,1}(0)$$
$$+ \lambda_1 p_{m-1,0}(1), \; m \geq 1, k = 1,$$

where $\lambda = \lambda_0 + \lambda_1 + \lambda_2$. For $Q_2 \geq 1$,

$$(\lambda + \mu_2) p_{0,n}(0) = \mu p_{0,n}(1), \; m = 0, k = 0,$$
$$(\lambda + \mu_1 + \mu_2) p_{m,n}(0) = \mu p_{m,n}(1), \; m \geq 1, k = 0,$$
$$(\lambda + \mu) p_{0,n}(1) = \lambda p_{0,n}(0) + \mu_1 p_{1,n}(0) + \mu_2 p_{0,n+1}(0)$$
$$+ \lambda_2 p_{0,n-1}(1), \; m = 0, k = 1, \quad (2)$$
$$(\lambda + \mu) p_{m,n}(1) = \lambda p_{m,n}(0) + \mu_1 p_{m+1,n}(0) + \mu_2 p_{m,n+1}(0)$$
$$+ \lambda_1 p_{m-1,n}(1) + \lambda_0 p_{m-1,n-1}(1)$$
$$+ \lambda_2 p_{m,n-1}(1), \; m \geq 1, k = 1.$$

Define for $|x| \leq 1, |y| \leq 1, k = 0, 1, H^{(k)}(x, y) = \sum_{m=0}^{\infty} \sum_{n=0}^{\infty} p_{m,n} x^m y^n$. Then, using Eqs. (1) and (2) we obtain,

$$(\lambda + \mu_1 + \mu_2) H^{(0)}(x, y) - \mu H^{(1)}(x, y) = \mu_2 H^{(0)}(x, 0) + \mu_1 H^{(0)}(0, y), \quad (3)$$

$$(\lambda xy + \mu_1 y + \mu_2 x) H^{(0)}(x, y) - xy[\lambda_0(1 - xy) + \lambda_1(1 - x)$$
$$+ \lambda_2(1 - y) + \mu] H^{(1)}(x, y) = \mu_2 x H^{(0)}(x, 0) + \mu_1 y H^{(0)}(0, y). \quad (4)$$

Solving (3) with respect to $H^{(1)}(x,y)$ and substituting to (4), we obtain after some algebra the following functional equation,

$$R(x,y)H^{(0)}(x,y) = A(x,y)H^{(0)}(x,0) + B(x,y)H^{(0)}(0,y), \tag{5}$$

where, $\widehat{\lambda}_i = \lambda_i \alpha$, $i = 0,1,2$, $\widehat{\mu}_i = \mu \mu_i$, $i = 1,2$, $\alpha = \lambda + \mu_1 + \mu_2$, $\widehat{\lambda} = \lambda \alpha$,

$$R(x,y) = xy[\widehat{\lambda}_0(1-xy) + \widehat{\lambda}_1(1-x) + \widehat{\lambda}_2(1-y)] - \widehat{\mu}_1 y(1-x) - \widehat{\mu}_2 x(1-y), \tag{6}$$

$$A(x,y) = \mu_2 x[y(\lambda_0(1-xy) + \lambda_1(1-x)) + (\lambda_2 y - \mu)(1-y)],$$

$$B(x,y) = \mu_1 y[x(\lambda_0(1-xy) + \lambda_2(1-y)) + (\lambda_1 x - \mu)(1-x)].$$

Remark 1: Our model can be generalized to incorporate a coordination mechanism between relays that decides, which of the two relays will keep the blocked packet they both have received by the user S_0; [23]. However, since wireless communication is fragile, the coordination between relays may fails. In such a case, both relay nodes will keep the blocked packet of the user S_0 in their queues.

3   General Results

We proceed with the derivation of some general results. Denote for $k = 0,1$,

$$p_{m,.}(k) = \sum_{n=0}^{\infty} p_{m,n}(k), m = 0,1,..., \quad p_{.,n}(k) = \sum_{m=0}^{\infty} p_{m,n}(k), n = 0,1,....$$

Lemma 1. *Let* $\rho_i = \frac{\lambda_i}{\mu} < 1$, $i = 0,1,2$, $\rho = \rho_0 + \rho_1 + \rho_2$. *Then,*

$$\begin{aligned}
H^{(1)}(1,1) &= \tfrac{\rho}{1-\rho_0}, \; H^{(0)}(1,1) = \tfrac{1-\rho_1-\rho_2-2\rho_0}{1-\rho_0}, \\
H^{(0)}(0,1) &= 1 - \tfrac{\rho}{1-\rho_0}(\tfrac{\lambda_0+\lambda_1+\mu_1}{\mu_1}) = 1 - \widehat{\rho}_1, \\
H^{(0)}(1,0) &= 1 - \tfrac{\rho}{1-\rho_0}(\tfrac{\lambda_0+\lambda_2+\mu_2}{\mu_2}) = 1 - \widehat{\rho}_2.
\end{aligned} \tag{7}$$

Proof: For each $m = 0,1,...$, we consider the vertical cut between the states $\{Q_1 = m, C = 1\}$ and $\{Q_1 = m+1, C = 0\}$. Then,

$$(\lambda_0 + \lambda_1)p_{m,.}(1) = \mu_1 p_{m+1,.}(0). \tag{8}$$

Summing for all $m = 0,1,...$, we derive

$$(\lambda_0 + \lambda_1)H^{(1)}(1,1) = \mu_1(H^{(0)}(1,1) - H^{(0)}(0,1)). \tag{9}$$

Note that Eq. (9) is a "conservation of flow" relation, since it equates the flow of jobs into the relay node R_1, with the flow of jobs out of the relay node R_1. Similarly, by repeating the procedure we have

$$(\lambda_0 + \lambda_2)H^{(1)}(1,1) = \mu_2(H^{(0)}(1,1) - H^{(0)}(1,0)). \tag{10}$$

Having in mind that $H^{(1)}(1,1) + H^{(0)}(1,1) = 1$ we conclude in

$$1 - H^{(0)}(0,1) = \tfrac{\lambda_0+\lambda_1+\mu_1}{\mu_1}H^{(1)}(1,1), \; 1 - H^{(0)}(1,0) = \tfrac{\lambda_0+\lambda_2+\mu_2}{\mu_2}H^{(1)}(1,1).$$

Substituting the above equation in (3), with $(x,y) = (1,1)$, we obtain after some algebra Eq. (7). $\qquad\square$

3.1 The Associated Random Walk in Quadrant and Stability Condition

Our model can be seen as a random walk in quarter plane (RWQP) modulated by a two-state Markov process (idle or busy node D). Using its special structure, we convert it to a usual RWQP and investigate its stability condition. Without loss of generality we assume that $\lambda + \mu + \mu_1 + \mu_2 = 1$. Using the notation in [12, 27], the functional Eq. (5) is equivalent to:

$$-h^{(0)}(x,y)\pi^{(0)}(x,y) = h^{(1)}(x,y)\pi_1^{(0)}(x) + h^{(2)}(x,y)\pi_2^{(0)}(y) + h^{(3)}(x,y)p_{0,0}(0), \tag{11}$$

where for $|x| \le 1$, $|y| \le 1$, $k = 0, 1$,

$$\pi^{(k)}(x,y) := \sum_{i=1}^{\infty}\sum_{j=1}^{\infty} p_{i,j}(k)x^{i-1}y^{j-1},$$

$$\pi_1^{(k)}(x) := \pi^{(k)}(x,0) = \sum_{i=1}^{\infty} p_{i,0}(k)x^{i-1}, \quad \pi_2^{(k)}(y) := \pi^{(k)}(0,y) = \sum_{j=1}^{\infty} p_{0,j}(k)y^{j-1},$$

$$h^{(0)}(x,y) = xy\{\widehat{\lambda}_0 xy + \widehat{\lambda}_1 x + \widehat{\lambda}_2 y + \widehat{\mu}_1 x^{-1} + \widehat{\mu}_2 y^{-1} - (\widehat{\lambda} + \widehat{\mu}_1 + \widehat{\mu}_2)\},$$

$$h^{(1)}(x,y) = x\{\lambda_0(\lambda + \mu_1)xy + \lambda_1(\lambda + \mu_1)x + \lambda_2(\lambda + \mu_1)y + \widehat{\mu}_1 x^{-1} \\ -(\lambda(\lambda + \mu_1) + \widehat{\mu}_1)\},$$

$$h^{(2)}(x,y) = y\{\lambda_0(\lambda + \mu_2)xy + \lambda_1(\lambda + \mu_2)x + \lambda_2(\lambda + \mu_2)y + \widehat{\mu}_2 y^{-1} \\ -(\lambda(\lambda + \mu_2) + \widehat{\mu}_2)\},$$

$$h^{(3)}(x,y) = \lambda_0\lambda xy + \lambda_1\lambda x + \lambda_2\lambda y - \lambda^2.$$

Equation (11) is the fundamental form corresponding to a RWQP whose one-step transition probabilities from state (m,n) to $(m+i, n+j)$ are for $-1 \le i, j \le 1$:

$$\widehat{p}_{\{(m,n);(m+i,n+j)\}} = \widehat{p}_{i,j}\delta_{\{m,n>0\}} + \widehat{p}'_{i,j}\delta_{\{m>0,n=0\}} \\ + \widehat{p}''_{i,j}\delta_{\{m=0,n>0\}} + \widehat{p}^{(0)}_{i,j}\delta_{\{m=0,n=0\}},$$

where $\delta_{\{.\}}$ is Kronecker's delta and:

$$\widehat{p}_{1,1} = \widehat{\lambda}_0, \; \widehat{p}_{1,0} = \widehat{\lambda}_1, \; \widehat{p}_{0,1} = \widehat{\lambda}_2, \; \widehat{p}_{-1,0} = \widehat{\mu}_1, \; \widehat{p}_{0,-1} = \widehat{\mu}_2,$$

$$\widehat{p}_{0,0} = 1 - (\widehat{\lambda} + \widehat{\mu}_1 + \widehat{\mu}_2),$$

$$\widehat{p}'_{1,1} = \lambda_0(\lambda + \mu_1), \; \widehat{p}'_{1,0} = \lambda_1(\lambda + \mu_1), \; \widehat{p}'_{0,1} = \lambda_2(\lambda + \mu_1), \; \widehat{p}'_{-1,0} = \widehat{\mu}_1,$$

$$\widehat{p}'_{0,0} = 1 - (\lambda(\lambda + \mu_1) + \widehat{\mu}_1),$$

$$\widehat{p}''_{1,1} = \lambda_0(\lambda + \mu_2), \; \widehat{p}''_{1,0} = \lambda_1(\lambda + \mu_2), \; \widehat{p}''_{0,1} = \lambda_2(\lambda + \mu_2), \; \widehat{p}''_{0,-1} = \widehat{\mu}_2,$$

$$\widehat{p}''_{0,0} = 1 - (\lambda(\lambda + \mu_2) + \widehat{\mu}_2),$$

$$\widehat{p}^{(0)}_{1,1} = \lambda_0\lambda, \; \widehat{p}^{(0)}_{1,0} = \lambda_1\lambda, \; \widehat{p}^{(0)}_{0,1} = \lambda_2\lambda, \; \widehat{p}^{(0)}_{0,0} = 1 - \lambda^2.$$

Following [12], set

$$
\begin{cases}
M = (M_x, M_y) = (\sum_j \widehat{p}_{1,j} - \sum_j \widehat{p}_{-1,j}, \sum_i \widehat{p}_{i,1} - \sum_i \widehat{p}_{i,-1}) \\
\qquad = (\widehat{\lambda}_0 + \widehat{\lambda}_1 - \widehat{\mu}_1, \widehat{\lambda}_0 + \widehat{\lambda}_2 - \widehat{\mu}_2), \\
M^{(1)} = (M_x^{(1)}, M_y^{(1)}) = (\sum_j \widehat{p}'_{1,j} - \sum_j \widehat{p}'_{-1,j}, \sum_i \widehat{p}'_{i,1}) \\
\qquad = ((\lambda_0 + \lambda_1)(\lambda + \mu_1) - \widehat{\mu}_1, (\lambda_0 + \lambda_2)(\lambda + \mu_1)), \\
M^{(2)} = (M_x^{(2)}, M_y^{(2)}) = (\sum_j \widehat{p}''_{1,j}, \sum_i \widehat{p}''_{i,1} - \sum_i \widehat{p}''_{i,-1}) \\
\qquad = ((\lambda_0 + \lambda_1)(\lambda + \mu_2), (\lambda_0 + \lambda_2)(\lambda + \mu_2) - \widehat{\mu}_2).
\end{cases}
$$

Theorem 1 gives necessary and sufficient conditions for the ergodicity of our model.

Theorem 1 [12]. *When $M \neq 0$, a random walk is ergodic if, and only if, one of the following conditions holds,*

1.

$$
\begin{cases}
M_x = \widehat{\lambda}_0 + \widehat{\lambda}_1 - \widehat{\mu}_1 < 0, M_y = \widehat{\lambda}_0 + \widehat{\lambda}_2 - \widehat{\mu}_2 < 0, \\
M_x M_y^{(1)} - M_y M_x^{(1)} < 0 \Leftrightarrow \widehat{\mu}_1 \widehat{\mu}_2 (1 - \rho_0)(\widehat{\rho}_1 - 1) < 0 \Leftrightarrow \widehat{\rho}_1 < 1, \\
M_y M_x^{(2)} - M_x M_y^{(2)} < 0 \Leftrightarrow \widehat{\mu}_1 \widehat{\mu}_2 (1 - \rho_0)(\widehat{\rho}_2 - 1) < 0 \Leftrightarrow \widehat{\rho}_2 < 1;
\end{cases}
$$

2. $M_x < 0, M_y \geq 0, M_y M_x^{(2)} - M_x M_y^{(2)} < 0$;
3. $M_x \geq 0, M_y < 0, M_x M_y^{(1)} - M_y M_x^{(1)} < 0$.

Remark 2: Note that under stability condition, $M_x \geq 0, M_y \geq 0$ cannot hold simultaneously. Without loss of generality we assume here on that $M_x < 0$.

3.2 Analysis of the Kernel

We now provide detailed properties on the branch points, and the branches defined by $R(x, y) = 0$. The kernel $R(x, y)$ can be written as a quadratic polynomial in x (resp. y) with coefficients that are polynomial in y (resp. x). Specifically,

$$
R(x, y) = a(x)y^2 + b(x)y + c(x) = \widehat{a}(y)x^2 + \widehat{b}(y)x + \widehat{c}(y),
$$

where

$$
a(x) = -(\widehat{\lambda}_0 x^2 + \widehat{\lambda}_2 x), \; b(x) = x(\widehat{\lambda} + \widehat{\mu}_1 + \widehat{\mu}_2) - \widehat{\mu}_1 - \widehat{\lambda}_1 x^2, \; c(x) = -\widehat{\mu}_2 x,
$$
$$
\widehat{a}(y) = -(\widehat{\lambda}_0 y^2 + \widehat{\lambda}_1 y), \; \widehat{b}(y) = y(\widehat{\lambda} + \widehat{\mu}_1 + \widehat{\mu}_2) - \widehat{\mu}_2 - \widehat{\lambda}_2 y^2, \; \widehat{c}(y) = -\widehat{\mu}_1 y.
$$

The solutions of $R(x, y) = 0$ for each y, x respectively are given by,

$$
X_{\pm}(y) = \frac{-\widehat{b}(y) \pm \sqrt{D_y(y)}}{2\widehat{a}(y)}, \qquad Y_{\pm}(x) = \frac{-b(x) \pm \sqrt{D_x(x)}}{2a(x)}, \tag{12}
$$
$$
D_y(y) = \widehat{b}^2(y) - 4\widehat{a}(y)\widehat{c}(y), \; D_x(x) = b^2(x) - 4a(x)c(x).
$$

We now focus on the branch points. Denote by x_i, y_i, $i = 1, 2, 3, 4$, the zeros of $D_x(x)$, $D_y(y)$ respectively. Clearly, $b(x) = 0$ has two solutions given by: $x_\pm^b = \frac{\widehat{\lambda} + \widehat{\mu}_1 + \widehat{\mu}_2 \pm \sqrt{(\widehat{\lambda} + \widehat{\mu}_1 + \widehat{\mu}_2)^2 - 4\widehat{\lambda}_1 \widehat{\mu}_1}}{2\lambda_1}$, with $x_-^b < 1 < x_+^b$. Then, it is readily seen from,

$$D_x(-\infty) = +\infty, \ D_x(0) = \widehat{\mu}_1^2 > 0, \ D_x(1) = (\widehat{\lambda}_0 + \widehat{\lambda}_2 - \widehat{\mu}_2)^2 > 0,$$
$$D_x(x_-^b) \leq 0, \ D_x(x_+^b) \leq 0, \ D_x(+\infty) = +\infty,$$

that x_is are real, such that $0 < x_1 \leq x_-^b \leq x_2 < 1 < x_3 \leq x_+^b < x_4 < \infty$. Moreover, $D_x(x) < 0$, $x \in (x_1, x_2) \cup (x_3, x_4)$, $D_x(x) > 0$, $x \in (-\infty, x_1) \cup (x_2, x_3) \cup (x_4, \infty)$. Similarly, we can prove that y_is are also real, and such that $0 < y_1 < y_2 < 1 < y_3 < y_4 < \infty$. Furthermore, $D_y(y) < 0$, $y \in (y_1, y_2) \cup (y_3, y_4)$, $D_y(y) > 0$, $y \in (-\infty, y_1) \cup (y_2, y_3) \cup (y_4, \infty)$.

To ensure the continuity of the two valued function $Y(x)$ (resp. $X(y)$), we consider the following cut planes: $\tilde{C}_x = C_x - [x_3, x_4]$, $\tilde{C}_y = C_y - [y_3, y_4]$, $\widehat{C}_x = C_x - ([x_1, x_2] \cup [x_3, x_4])$, $\widehat{C}_y = C_y - ([y_1, y_2] \cup [y_3, y_4])$, where C_x, C_y the complex planes of x, y, respectively. For $x \in \widehat{C}_x$, the two branches of $Y(x)$ are defined by

$$Y_0(x) = \begin{cases} Y_-(x) \ \text{if} \ |Y_-(x)| \leq |Y_+(x)|, \\ Y_+(x) \ \text{if} \ |Y_-(x)| > |Y_+(x)|; \end{cases} \quad Y_1(x) = \begin{cases} Y_+(x) \ \text{if} \ |Y_-(x)| \leq |Y_+(x)|, \\ Y_-(x) \ \text{if} \ |Y_-(x)| > |Y_+(x)|. \end{cases}$$

Similarly, we can define functions $X_0(y)$, $X_1(y)$, $y \in \widehat{C}_y$ based on $X_\pm(y)$. We proceed with some properties of $Y_0(x)$, $Y_1(x)$:

Lemma 2. *The functions $Y_i(x)$, $x \in C_x$, $i = 0, 1$ are meromorphic. Moreover,*

1. *$Y_0(x)$ has one zero and no poles (i.e. it is analytic in $\widehat{C}_x$). $Y_1(x)$ has two poles and no zeros.*
2. *$|Y_0(x)| \leq |Y_1(x)|$, $x \in \widehat{C}_x$, and equality takes place only on the cuts.*
3. *When $|x| = 1$, $|Y_0(x)| \leq 1$. For $x = 1$, $Y_0(1) = 1$.*
4. *$Y_0'(1) = -\frac{\widehat{\mu}_1 - \widehat{\lambda}_0 - \widehat{\lambda}_1}{\widehat{\mu}_2 - \widehat{\lambda}_0 - \widehat{\lambda}_2}$.*

Similar results can be obtained for $X_0(y)$, $X_1(y)$.

Proof. The proof of $1. - 3.$ is based on Lemma 2.3.4, Theorem 5.3.3 in [12]. 4. is proved by noticing that

$$Y_0(x)Y_1(x) = \frac{\widehat{\mu}_2}{\widehat{\lambda}_0 x + \widehat{\lambda}_2}, \ Y_0(x) + Y_1(x) = \frac{(\widehat{\lambda} + \widehat{\mu}_1 + \widehat{\mu}_2)x - \widehat{\mu}_1 - \widehat{\lambda}_1 x^2}{\widehat{\lambda}_0 x^2 + \widehat{\lambda}_2 x}. \tag{13}$$

Using (13) and taking into account that $Y_0(1) = 1$, we can obtain after some basic algebra the desired result. $\qquad \square$

Define the following image contours: $\mathcal{L} = Y_0[\overrightarrow{x_1, x_2}]$, $\mathcal{L}_{ext} = Y_0[\overleftarrow{x_3, x_4}]$, $\mathcal{M} = X_0[\overleftarrow{y_1, y_2}]$, $\mathcal{M}_{ext} = X_0[\overleftarrow{y_3, y_4}]$, where $[\overrightarrow{u, v}]$ stands for the contour traversed from u to v along the upper edge of the slit $[u, v]$ and then back to u along the lower edge of the slit. Then, we have the following lemma:

Lemma 3. *1. For $x \in [x_1, x_2]$, the algebraic function $Y(x)$ lies on a closed contour $\mathcal{L}$, which is symmetric with respect to the real line and defined by*

$$|y|^2 = \frac{2\widehat{\mu}_2(2\widehat{\lambda}_0 Re(y) + \widehat{\lambda}_1)}{\widehat{\lambda}_0[\widehat{\lambda} + \widehat{\mu}_1 + \widehat{\mu}_2 - 2\widehat{\lambda}_2 Re(y) - \sqrt{\Delta_y}] + 2\widehat{\lambda}_2(2\widehat{\lambda}_0 Re(y) + \widehat{\lambda}_1)}, \quad |y|^2 \le \frac{\widehat{\mu}_2}{\widehat{\lambda}_0 x_1 + \widehat{\lambda}_2}, \quad (14)$$

where $\Delta_y = (\widehat{\lambda} + \widehat{\mu}_1 + \widehat{\mu}_2 - 2\widehat{\lambda}_2 Re(y))^2 - 4\widehat{\mu}_1(2\widehat{\lambda}_0 Re(y) + \widehat{\lambda}_1)$. Moreover, set $\zeta := |Y(x_1)| = \sqrt{\frac{\widehat{\mu}_2}{\widehat{\lambda}_0 x_1 + \widehat{\lambda}_2}}$, the point on $\mathcal{L}$ with the largest modulus. The point $Y_0(x_2) = -\sqrt{\frac{\widehat{\mu}_2}{\widehat{\lambda}_0 x_2 + \widehat{\lambda}_2}}$ is the extreme left point of $\mathcal{L}$.
2. Similarly, for $y \in [y_1, y_2]$, the algebraic function $X(y)$ lies on a closed contour $\mathcal{M}$, which is symmetric with respect to the real line and defined by

$$|x|^2 = \frac{2\widehat{\mu}_1(2\widehat{\lambda}_0 Re(x) + \widehat{\lambda}_2)}{\widehat{\lambda}_0[\widehat{\lambda} + \widehat{\mu}_1 + \widehat{\mu}_2 - 2\widehat{\lambda}_1 Re(x) - \sqrt{\Delta_x}] + 2\widehat{\lambda}_1(2\widehat{\lambda}_0 Re(x) + \widehat{\lambda}_2)}, \quad |x|^2 \le \frac{\widehat{\mu}_1}{\widehat{\lambda}_0 y_1 + \widehat{\lambda}_1}, \quad (15)$$

where $\Delta_x = (\widehat{\lambda} + \widehat{\mu}_1 + \widehat{\mu}_2 - 2\widehat{\lambda}_1 Re(x)))^2 - 4\widehat{\mu}_2(2\widehat{\lambda}_0 Re(x) + \widehat{\lambda}_2)$. Moreover, set $\beta := |X(y_1)| = \sqrt{\frac{\widehat{\mu}_1}{\widehat{\lambda}_0 y_1 + \widehat{\lambda}_1}} > 1$ (see Remark 2), the point on $\mathcal{M}$ with the largest modulus. $X_0(y_2) = -\sqrt{\frac{\widehat{\mu}_1}{\widehat{\lambda}_0 y_2 + \widehat{\lambda}_1}}$ is the extreme left point of $\mathcal{M}$.

Proof: We prove the first part for $Y(x)$ (the proof of 2. is similar). Clearly, $D_x(x) < 0$, $x \in (x_1, x_2)$ and $Y_0(x), Y_1(x)$ are complex conjugates. Moreover,

$$Re(Y(x)) = \frac{x(\widehat{\lambda} + \widehat{\mu}_1 + \widehat{\mu}_2) - \widehat{\mu}_1 - \widehat{\lambda}_1 x^2}{2(\widehat{\lambda}_0 x^2 + \widehat{\lambda}_2 x)}. \quad (16)$$

Since $R(x, Y(x)) = 0$ we have $|Y(x)|^2 = \frac{\widehat{\mu}_2}{\widehat{\lambda}_0 x + \widehat{\lambda}_2} \Leftrightarrow |Y(x)| = \sqrt{\frac{\widehat{\mu}_2}{\widehat{\lambda}_0 x + \widehat{\lambda}_2}}$. Clearly, $|Y(x)|$ is a decreasing function in x. Thus, $|Y(x)| \le |Y(x_1)| = \zeta := \sqrt{\frac{\widehat{\mu}_2}{\widehat{\lambda}_0 x_1 + \widehat{\lambda}_2}}$, which is the extreme right point of $\mathcal{L}$. Solving (16) with respect to x and taking the solution such that $x \in [0, 1]$ yields,

$$\widetilde{x} = \frac{\widehat{\lambda} + \widehat{\mu}_1 + \widehat{\mu}_2 - 2\widehat{\lambda}_2 Re(y) - \sqrt{(\widehat{\lambda} + \widehat{\mu}_1 + \widehat{\mu}_2 - 2\widehat{\lambda}_2 Re(y))^2 - 4\widehat{\mu}_1(2\widehat{\lambda}_0 Re(y) + \widehat{\lambda}_1)}}{2((2\widehat{\lambda}_0 Re(y) + \widehat{\lambda}_1))}. \quad (17)$$

Substituting (17) into $|y|^2 = \widehat{\mu}_2/(\widehat{\lambda}_0 x + \widehat{\lambda}_2)$ (i.e., for $x = \widetilde{x}$) yields (14). $\square$

Finally, for any simple closed contour $\mathcal{U}$, denote by $G_{\mathcal{U}}$ (resp. $G_{\mathcal{U}}^c$) the interior (resp. exterior) domain bounded by $\mathcal{U}$. The next result gives topological and algebraic properties for the associated RWQP, summarized in Lemma 4 (see also Theorem 5.3.2, Corrolary 5.3.5 in [12]). Define,

$$\Delta = \begin{vmatrix} \widehat{p}_{1,1} & \widehat{p}_{1,0} & \widehat{p}_{1,-1} \\ \widehat{p}_{0,1} & \widehat{p}_{0,0} - 1 & \widehat{p}_{0,-1} \\ \widehat{p}_{-1,1} & \widehat{p}_{-1,0} & \widehat{p}_{-1,-1} \end{vmatrix} = \begin{vmatrix} \widehat{\lambda}_0 & \widehat{\lambda}_1 & 0 \\ \widehat{\lambda}_2 & -(\widehat{\lambda} + \widehat{\mu}_1 + \widehat{\mu}_2) & \widehat{\mu}_2 \\ 0 & \widehat{\mu}_1 & 0 \end{vmatrix} = -\widehat{\mu}_1 \widehat{\mu}_2 \widehat{\lambda}_0 < 0.$$

Lemma 4. *(i) The curves $\mathcal{L}$ and $\mathcal{L}_{ext}$ (resp. $\mathcal{M}$ and $\mathcal{M}_{ext}$) are quartic, symmetrical with respect to the real axis, closed and simple. Since $\Delta < 0$, $[y_1, y_2] \subset$*

$G_{\mathcal{L}_{ext}} \subset G_{\mathcal{L}}$ and $[y_3, y_4] \subset G_{\mathcal{L}}^c$. Similar results hold for $\mathcal{M}$, $\mathcal{M}_{ext}$, $[x_1, x_2]$, $[x_3, x_4]$.

(ii) $Y_0(x) : G_{\mathcal{M}} - [x_1, x_2] \to G_{\mathcal{L}} - [y_1, y_2]$, $X_0(y) : G_{\mathcal{L}} - [y_1, y_2] \to G_{\mathcal{M}} - [x_1, x_2]$ are conformal mappings, and since $\Delta < 0$, the values of $Y_0(x)$ (resp. $X_0(y)$) are contained in $G_{\mathcal{L}}$ (resp. $G_{\mathcal{M}}$), whereas the values of $Y_1(x)$ (resp. $X_1(y)$) are contained in $G_{\mathcal{L}_{ext}}^c$ (resp. $G_{\mathcal{M}_{ext}}^c$).

3.3   Intersection Points of the Curves

The analytic continuation of $H^{(0)}(x,0)$ (resp. $H^{(0)}(0,y)$) outside the unit disc is achieved by various methods (e.g., Lemma 2.2.1 and Theorem 3.2.3 in [12]). Note that the common solutions of $R(x,y) = 0$, $A(x,y) = 0$ (resp. $B(x,y)$) are potential singularities for the functions $H^{(0)}(x,0)$, $H^{(0)}(0,y)$. Thus, the study of the intersection points of the curves $R(x,y) = 0$, $A(x,y) = 0$ (resp. $B(x,y) = 0$) is crucial for the analytic continuation of $H^{(0)}(x,0)$, $H^{(0)}(0,y)$.

Intersection points of the curves $R(x,y) = 0$, $A(x,y) = 0$. Let $x \in \widehat{C}_x$ and $R(x,y) = 0$, $y = Y_\pm(x)$. Their intersection points (if any) are the roots of their resultant. We can easily show that the resultant in y of the two polynomials $R(x,y)$ and $A(x,y)$ is

$$Res_y(R, A; x) = (\lambda_0 x + \lambda_2)\widehat{\mu}_2^2 x^2 (x-1) T_y(x),$$

where $T_y(x) = (\lambda(\lambda_0 + \lambda_1) + \lambda_0 \mu_1)(\lambda + \mu_1)x^2 + \lambda\mu_1(\lambda - \mu + \mu_1)x - \mu_1\widehat{\mu}_1$. Note that $T_y(0) = -\mu_1\widehat{\mu}_1 < 0$, $T_y(1) = \widehat{\mu}_1(\lambda + \mu_1)(1 - \rho_0)(\widehat{\rho}_1 - 1) < 0$ (due to the stability condition), and $\lim_{x \to \pm\infty} T_y(x) = +\infty$. Thus, $T_y(x) = 0$ has two roots of opposite sign with $x_* < 0 < 1 < x^*$.

Intersection points of the curves $R(x,y) = 0$, $B(x,y) = 0$. Let $y \in \widehat{C}_y$ and $R(x,y) = 0$, $x = X_\pm(y)$. It is easy to see that

$$R(x,y) = \tfrac{\alpha}{\mu_1} B(x,y) + \lambda\mu y(1 - x) + \widehat{\mu}_2(y - x).$$

Thus, $R(x,y) = 0$, $B(x,y) = 0$, implies that,

$$\lambda_0 x(1 - xy) + \lambda_2 x(1 - y) + (\lambda_1 x - \mu)(1 - x) = 0,$$
$$\lambda\mu y(1 - x) + \widehat{\mu}_2(y - x) = 0. \tag{18}$$

The second equation in (18) gives $x = (\lambda + \mu_2)y/(\lambda y + \mu_2)$, and substituting in the first one yields,

$$L(y) = \tfrac{1-y}{(\lambda y + \mu_2)^2} Z_x(y) = 0,$$

where $Z_x(y) = y^2(\lambda_0(\lambda + \mu_2) + \lambda\lambda_2)(\lambda + \mu_2) + y\lambda\mu_2(\lambda + \mu_2 - \mu) - \mu_2\widehat{\mu}_2$. Note that $Z_x(0) = -\mu_2\widehat{\mu}_2 < 0$, $Z_x(1) = (\lambda + \mu_2)\widehat{\mu}_2(1 - \rho_0)(\widehat{\rho}_2 - 1) < 0$ (due to the stability condition), and $\lim_{y \to \pm\infty} Z_x(y) = +\infty$. Thus, $Z_x(y)$ has two zeros of opposite sign $y_* < 0 < 1 < y^*$, and $Z_x(y) < 0$, $y \in [0, 1]$. Therefore, $L(y) < 0$, $y \in [0, 1)$, which in turn implies that $B(X_0(y), y) \neq 0$, $y \in [y_1, y_2] \subset [0, 1)$, or equivalently $B(x, Y_0(x)) \neq 0$, $x \in \mathcal{M}$.

4 Formulation and Solution of a Boundary Value Problem

For zero pairs (x, y) of $R(x, y) = 0$, $y \in D_y = \{y \in C_y : |y| \leq 1, |X_0(y)| \leq 1\}$,

$$A(X_0(y), y)H^{(0)}(X_0(y), 0) = -B(X_0(y), y)H^{(0)}(0, y). \tag{19}$$

For $y \in D_y - [y_1, y_2]$ the functions $H^{(0)}(0, y)$, $H^{(0)}(X_0(y), 0)$ are both analytic. This entails from (19) that $A(X_0(y), y)$, $B(X_0(y), y)$ must not vanish in $D_y - [y_1, y_2]$, otherwise $H^{(0)}(0, y)$, $H^{(0)}(x, 0)$ would have poles in $|x| \leq 1$, $|y| \leq 1$. Then, the right hand side in (19) can be analytically continued up to the slit $[y_1, y_2]$ and thus,

$$A(X_0(y), y)H^{(0)}(X_0(y), 0) + B(X_0(y), y)H^{(0)}(0, y) = 0, \; y \in [y_1, y_2], \tag{20}$$

or equivalently

$$A(x, Y_0(x))H^{(0)}(x, 0) + B(x, Y_0(x))H^{(0)}(0, Y_0(x)) = 0, \; x \in \mathcal{M}. \tag{21}$$

The function $H^{(0)}(x, 0)$ is holomorphic in $D_x = \{x \in C_x : |x| < 1\}$ and continuous in $\bar{D}_x = \{x \in C_x : |x| \leq 1\}$, but might have poles in $S_x := G_\mathcal{M} \cap (\bar{D}_x)^c$. We also know that (see also Corollary 5.3.5 in [12]) for $x \in S_x$, $|Y_0(x)| \leq 1$, as a consequence of the maximum modulus principle. Hence, from (21) the possible poles of $H^{(0)}(x, 0)$ in S_x are the zeros of $A(x, Y_0(x))$ in this region. Specifically, the only possible zero obtained in Subsect. 3.3 and given by

$$x^* = \frac{-\lambda\mu_1(\lambda + \mu_1 - \mu) + \sqrt{(\lambda\mu_1(\lambda + \mu_1 - \mu))^2 + 4\mu_1\hat{\mu}_1(\lambda(\lambda_0 + \lambda_1) + \lambda_0\mu_1)(\lambda + \mu_1)}}{2(\lambda(\lambda_0 + \lambda_1) + \lambda_0\mu_1)(\lambda + \mu_1)}.$$

Remark 3: Note that the other zero, $x_*(< 0)$ (see Subsect. 3.3), cannot belong to the region S_x. Indeed, it can be easily shown that

$$A(x_*, Y_0(x_*)) = 0 \Leftrightarrow \lambda x_*(1 - Y_0(x_*)) + \mu_1(x_* - Y_0(x_*)) = 0,$$

and since $-1 \leq Y_0(x_*) \leq 1$, then, $x_*(1 - Y_0(x_*)) \leq 0$, $x_* - Y_0(x_*) < 0$, which implies that $x_* \notin S_x$. Thus, we focus only on the positive zero x^*.

If $x^* > \beta$, then $A(x, Y_0(x)) \neq 0$ for $x \in S_x$. If $x^* \in S_x$, then x^* is a zero of $A(x, Y_0(x))$ provided that $|Y_0(x^*)| \leq 1$. Therefore, set $r = 1$, if $x^* \leq \beta$ and $|Y_0(x^*)| \leq 1$, and $r = 0$ elsewhere. If $r = 1$, then $A(x, Y_0(x))$ has a unique zero in S_x given by $x = x^*$. Otherwise, $A(x, Y_0(x))$ does not vanish in S_x. It is easy to prove that when $A(x, Y_0(x))$ vanishes at $x = x^*$, then, this zero has multiplicity equal to one, since it can be shown that $dA(x, Y_0(x))/dx$ does not vanish at $x = x^*$.

For $y \in [y_1, y_2]$, letting $X_0(y) = x \in \mathcal{M}$ and realizing that $Y_0(X_0(y)) = y$ so that $y = Y_0(x)$, we rewrite (20) as $(B(x, Y_0(x)) \neq 0, x \in \mathcal{M}$; see Subsect. 3.3)

$$\frac{A(x, Y_0(x))}{B(x, Y_0(x))}H^{(0)}(x, 0) = -H^{(0)}(0, Y_0(x)), \; x \in \mathcal{M}. \tag{22}$$

Taking into account the possible zero of $A(x, Y_0(x))$ for $x \in S_x$, multiplying both sides of (22) by the imaginary complex number i, and noticing that $H^{(0)}(0, Y_0(x))$ is real for $x \in \mathcal{M}$, since $Y_0(x) \in [y_1, y_2]$, we have

$$Re[iU(x)G(x)] = 0, \ x \in \mathcal{M},$$

$$U(x) = \frac{A(x, Y_0(x))}{(x-x^*)^r B(x, Y_0(x))}, \ G(x) = (x - x^*)^r H^{(0)}(x, 0),$$

(23)

where, $G(x)$ is regular for $x \in G_{\mathcal{M}}$, continuous for $x \in \mathcal{M} \cup G_{\mathcal{M}}$, and $U(x)$ is a non-vanishing function on $\mathcal{M}$. In order to solve the Riemann-Hilbert boundary value problem formulated in (23), we must conformally transform it to the unit circle $\mathcal{C}$. Define the conformal mapping and its inverse respectively by

$$z = f(x) : G_{\mathcal{M}} \to G_{\mathcal{C}}, \ x = f_0(z) : G_{\mathcal{C}} \to G_{\mathcal{M}}.$$

Then, the Riemann-Hilbert problem formulated in (23) is reduced to the following: Determine a function $F(z) := G(f_0(z))$, regular in $G_{\mathcal{C}}$ and continuous in $G_{\mathcal{C}} \cup \mathcal{C}$ satisfying

$$Re[iU(f_0(z))F(z)] = 0, \ z \in \mathcal{C}. \tag{24}$$

Define $\chi = \frac{-1}{\pi}[arg\{U(x)\}]_{x \in \mathcal{M}}$, i.e., the index of the Riemann-Hilbert problem, where $[arg\{U(x)\}]_{x \in \mathcal{M}}$, denotes the variation of the argument of the function $U(x)$ as x moves along the closed contour $\mathcal{M}$ in the positive direction, provided that $U(x) \neq 0$, $x \in \mathcal{M}$. As expected [2,13], under the stability conditions given in Theorem 1, the index $\chi = 0$. Following the lines in [13] (remind from Remark 2 that $M_x < 0$):

Lemma 5. *1. If $M_y < 0$, then $\chi = 0$ is equivalent to*

$$\frac{dA(x, Y_0(x))}{dx}\Big|_{x=1} = \frac{\hat{\mu}_2(\hat{\rho}_1 - 1)}{\hat{\mu}_2 - \hat{\lambda}_0 - \hat{\lambda}_2} < 0 \Leftrightarrow \hat{\rho}_1 < 1,$$

$$\frac{dB(X_0(y), y)}{dy}\Big|_{y=1} = \frac{\hat{\mu}_1(\hat{\rho}_2 - 1)}{\hat{\mu}_1 - \hat{\lambda}_0 - \hat{\lambda}_1} < 0 \Leftrightarrow \hat{\rho}_2 < 1.$$

2. If $M_y \geq 0$, $\chi = 0$ is equivalent to $\frac{dB(X_0(y), y)}{dy}\Big|_{y=1} < 0 \Leftrightarrow \hat{\rho}_2 < 1$.

Thus, under stability conditions (see Theorem 1) the homogeneous Riemann-Hilbert problem (23) has a unique solution given by,

$$H^{(0)}(x, 0) = W(x - x^*)^{-r} \exp[\frac{1}{2i\pi} \int_{|t|=1} \frac{\log\{J(t)\}}{t - f(x)} dt], \ x \in G_{\mathcal{M}}, \tag{25}$$

where W is a constant, and $J(t) = \frac{\overline{U(t)}}{U(t)}$, $U(t) = U(f_0(t))$. Since $1 \in G_{\mathcal{M}}$, W is obtained by setting $x = 1$ in (25) and combining with the value of $H^{(0)}(1, 0)$ found in (7). After some algebra we conclude, for $x \in G_{\mathcal{M}}$, in

$$H^{(0)}(x, 0) = (\frac{1-x^*}{x-x^*})^r (1 - \hat{\rho}_2) \exp[\frac{1}{2i\pi} \int_{|t|=1} \frac{\log\{J(t)\}(f(x) - f(1))}{(t - f(x))(t - f(1))} dt]. \tag{26}$$

We now focus on the determination of the conformal mapping and its inverse. For this purpose, we need a representation of $\mathcal{M}$ in polar coordinates, i.e., $\mathcal{M} = \{x : x = \rho(\phi)\exp(i\phi), \phi \in [0, 2\pi]\}$. This representation can be obtained as follows: Since $0 \in G_{\mathcal{M}}$, for each $x \in \mathcal{M}$, a relation between its absolute value and its real part is given by $|x|^2 = m(Re(x))$ (see (15)), where

$$m(\delta) := \frac{2\widehat{\mu}_1(2\widehat{\lambda}_0\delta + \widehat{\lambda}_2)}{\widehat{\lambda}_0[\widehat{\lambda} + \widehat{\mu}_1 + \widehat{\mu}_2 - 2\widehat{\lambda}_1\delta - \sqrt{\Delta_x(\delta)}] + 2\widehat{\lambda}_1(2\widehat{\lambda}_0\delta + \widehat{\lambda}_2)},$$

and $\Delta_x(\delta) = (\widehat{\lambda} + \widehat{\mu}_1 + \widehat{\mu}_2 - 2\widehat{\lambda}_1\delta)^2 - 4\widehat{\mu}_2(2\widehat{\lambda}_0\delta + \widehat{\lambda}_2)$. Given the angle ϕ of some point on $\mathcal{M}$, the real part of this point, say $\delta(\phi)$, is the solution of $\delta - \cos(\phi)\sqrt{m(\delta)}$, $\phi \in [0, 2\pi]$. Since $\mathcal{M}$ is a smooth, egg-shaped contour, the solution is unique. Clearly, $\rho(\phi) = \frac{\delta(\phi)}{\cos(\phi)}$, and the parametrization of $\mathcal{M}$ in polar coordinates is fully specified. Then, the mapping from $z \in G_C$ to $x \in G_{\mathcal{M}}$, where $z = e^{i\phi}$ and $x = \rho(\psi(\phi))e^{i\psi(\phi)}$, satisfying $f_0(0) = 0$ and $f_0(z) = \overline{f_0(z)}$ is uniquely determined by (see [5], Sect. 1.4.4),

$$f_0(z) = z\exp[\tfrac{1}{2\pi}\int_0^{2\pi}\log\{\rho(\psi(\omega))\}\tfrac{e^{i\omega}+z}{e^{i\omega}-z}d\omega], \; |z| < 1. \tag{27}$$

The angular deformation $\psi(.)$ is uniquely determined as the solution of Theodorsen integral equation

$$\psi(\phi) = \phi - \int_0^{2\pi}\log\{\rho(\psi(\omega))\}\cot(\tfrac{\omega-\phi}{2})d\omega, \; 0 \leq \phi \leq 2\pi, \tag{28}$$

with $\psi(\phi) = 2\pi - \psi(2\pi - \phi)$. Due to the correspondence-boundaries theorem, $f_0(z)$ is continuous in $\mathcal{C} \cup G_C$. Note that the non linear Eq. (28) cannot be solved in closed form but numerically, although a unique solution can be proven to exist. The numerical procedure will be discussed later.

Similarly, we can determine $H^{(0)}(0, y)$ by solving another Riemann-Hilbert boundary value problem on the closed contour $\mathcal{L}$. Then, using the fundamental functional Eq. (5) we obtain $H^{(0)}(x, y)$, and substituting back in (3), the generating function $H^{(1)}(x, y)$ is also uniquely determined.

5   Performance Metrics

In the following we derive formulas for the probability of an empty system and the expected number of packets at each relay node in steady state. Note that since $0 \in G_{\mathcal{M}}$, $P(Q_1 = 0, Q_2 = 0, C = 0) = H^{(0)}(0, 0)$. Clearly,

$$P(Q_1 = 0, Q_2 = 0, C = 0) = \frac{(1-\widehat{\rho}_2)(x^*-1)^r}{(x^*)^r}\exp[\tfrac{-f(1)}{2i\pi}\int_{|t|=1}\tfrac{\log\{J(t)\}}{t(t-f(1))}dt],$$

$$E(Q_1) = \sum_{m=1}^{\infty}m\sum_{n=0}^{\infty}\sum_{k=0}^{1}p_{m,n}(k) = \tfrac{d}{dx}H^{(0)}(x, 1)|_{x=1} + \tfrac{d}{dx}H^{(1)}(x, 1)|_{x=1},$$

$$E(Q_2) = \sum_{n=1}^{\infty}n\sum_{m=0}^{\infty}\sum_{k=0}^{1}p_{m,n}(k) = \tfrac{d}{dy}H^{(0)}(1, y)|_{y=1} + \tfrac{d}{dy}H^{(1)}(1, y)|_{y=1}.$$

Differentiating (5) and (3) with respect to x and setting $(x, y) = (1, 1)$, we obtain respectively after some algebra,

$$\tfrac{d}{dx}H^{(0)}(x, 1)|_{x=1} = \tfrac{\lambda_0+\lambda_1}{M_x}\{\tfrac{\widehat{\mu}_1[(\alpha-\mu_1)H^{(0)}(0,1)-\mu_2H^{(0)}(1,0)]}{M_x} + \mu_2\tfrac{d}{dx}H^{(0)}(x, 0)|_{x=1}\},$$

$$\tfrac{d}{dx}H^{(1)}(x, 1)|_{x=1} = \tfrac{\alpha}{\mu}\tfrac{d}{dx}H^{(0)}(x, 1)|_{x=1} - \tfrac{\mu_2}{\mu}\tfrac{d}{dx}H^{(0)}(x, 0)|_{x=1}.$$

From (26),

$$\tfrac{d}{dx}H^{(0)}(x,0)|_{x=1} = (1 - \widehat{\rho}_2)[\tfrac{-r}{1-x^*} + \tfrac{1}{2\pi i}\int_{|t|=1}\tfrac{\log\{J(t)\}f'(1)}{(t-f(1))^2}dt]. \qquad (29)$$

Then, using the last two equations, we can easily derive

$$E(Q_1) = \tfrac{(\lambda_0+\lambda_1)\widehat{\mu}_1(\mu+\alpha)}{\mu M_x^2}[(\alpha - \mu_1)H^{(0)}(0,1) - \mu_2 H^{(0)}(1,0)]$$
$$+ \tfrac{\widehat{\mu}_1}{M_x}(\tfrac{\lambda_0+\lambda_1+\mu_1}{\mu})\tfrac{d}{dx}H^{(0)}(x,0)|_{x=1}.$$

Similarly,

$$E(Q_2) = \tfrac{(\lambda_0+\lambda_2)\widehat{\mu}_2(\mu+\alpha)}{\mu M_y^2}[(\alpha - \mu_2)H^{(0)}(1,0) - \mu_1 H^{(0)}(0,1)]$$
$$+ \tfrac{\widehat{\mu}_2}{M_y}(\tfrac{\lambda_0+\lambda_2+\mu_2}{\mu})\tfrac{d}{dy}H^{(0)}(0,y)|_{y=1},$$

where by differentiating (19) with respect to y and setting $y = 1$, we obtain

$$\tfrac{d}{dy}H^{(0)}(0,y)|_{y=1} = \tfrac{[(\lambda_0+\lambda_1)X_0'(1)+\lambda_2+\lambda_0-\mu]\mu_2}{(X_0'(1)(\mu-\lambda_0-\lambda_1)-\lambda_2-\lambda_0)\mu_1}\tfrac{d}{dy}H^{(0)}(x,0)|_{x=1}$$
$$+ \tfrac{\widehat{\lambda}[X_0''(1)+2X_0'(1)(1-X_0'(1))]}{2\widehat{\mu}_1\widehat{\mu}_2(1-\rho_0)^2(\widehat{\rho}_2-1)^2}H^{(0)}(1,0),$$

$$X_0'(1) = -\tfrac{\widehat{\mu}_2-\widehat{\lambda}_0-\widehat{\lambda}_2}{\widehat{\mu}_1-\widehat{\lambda}_0-\widehat{\lambda}_1}, \quad X_1'(1) = \tfrac{\widehat{\mu}_1[(\widehat{\lambda}_0+\widehat{\lambda}_1)(\widehat{\mu}_2-\widehat{\lambda}_2)-\widehat{\lambda}_0\widehat{\mu}_1]}{(\widehat{\lambda}_0+\widehat{\lambda}_1)^2(\widehat{\mu}_1-\widehat{\lambda}_0-\widehat{\lambda}_1)},$$

$$X_0''(1) = \tfrac{2[X_0'(1)X_1'(1)(\widehat{\lambda}_0+\widehat{\lambda}_1)^2+\widehat{\lambda}_0(\widehat{\lambda}_0+\widehat{\lambda}_2)-\widehat{\mu}_2(2\widehat{\lambda}_0+\widehat{\lambda}_1)]}{(\widehat{\lambda}_0+\widehat{\lambda}_1)(\widehat{\lambda}_0+\widehat{\lambda}_1-\widehat{\mu}_1)}.$$

6   A Numerical Example

We proceed with a simple numerical example to illustrate the validity of the expressions derived in the previous section. The calculation of $P(Q_1 = 0, Q_2 = 0, C = 0)$, $E(Q_i)$, $i = 1, 2$, requires the evaluation of the integrals in (26), (29) as well as the numerical determination of the mapping $f_0(z)$ (see [5,28]). We now outline how these integrals can be computed: Firstly, we rewrite the integrals (26) and (29), by substituting $t = e^{i\phi}$:

$$H^{(0)}(x,0) = (\tfrac{1-x^*}{x-x^*})^r(1 - \widehat{\rho}_2)\exp[\tfrac{1}{2\pi}\int_0^{2\pi}\tfrac{\log\{J(e^{i\phi})\}(f(x)-f(1))}{(e^{i\phi}-f(x))(e^{i\phi}-f(1))}e^{i\phi}d\phi],$$
$$\tfrac{d}{dx}H^{(0)}(x,0)|_{x=1} = (1 - \widehat{\rho}_2)[\tfrac{-r}{1-x^*} + \tfrac{1}{2\pi}\int_0^{2\pi}\tfrac{\log\{J(e^{i\phi})\}f'(1)e^{i\phi}}{(e^{i\phi}-f(1))^2}d\phi]. \qquad (30)$$

Then, we split the interval $[0, 2\pi]$ into K parts of length $2\pi/K$. For the K points given by their angles $\{\phi_0, ..., \phi_{K-1}\}$, we solve the Theodorsen integral Eq. (28) to obtain iteratively the corresponding points $\{\psi(\phi_1), ..., \psi(\phi_{K-1})\}$ from:

$$\psi_0(\phi_k) = \phi_k, \quad \psi_{n+1}(\phi_k) = \phi_k - \tfrac{1}{2\pi}\int_0^{2\pi}\log\left\{\tfrac{\delta(\psi_n(\omega))}{\cos(\psi_n(\omega))}\right\}\cot[\tfrac{1}{2}(\omega - \phi_k)]d\omega,$$

where $\delta(\psi_n(\omega))$ is determined by $\delta(\psi_n(\omega)) - \cos(\psi_n(\omega))\sqrt{m(\delta(\psi_n(\omega)))} = 0$, using the Newton-Raphson method. For the iteration, we use the stopping criterion: $\max_{k \in \{0,1,...,K-1\}} |\psi_{n+1}(\phi_k) - \psi_n(\phi_k)| < 10^{-6}$. Having obtained $\psi(\phi_k)$ numerically, the values of the conformal mapping $f_0(z)$, $|z| \leq 1$ are given by

$$f_0(e^{i\phi_k}) = e^{i\psi(\phi_k)}\frac{\delta(\psi(\phi_k))}{\cos(\psi(\phi_k))} = \delta(\psi(\phi_k))[1 + i\tan(\psi(\phi_k))], \; k = 0,1,...,K-1.$$

It remains to determine $f(1)$, $f'(1)$. Clearly, $f(1) = \eta$ means $f_0(\eta) = 1$. Thus, $f(1)$ is the unique solution of $f_0(\eta) = 1$ in $[0,1]$, and can be obtained using (27) and the Newton-Raphson method. Furthermore,

$$f'(1) = (\tfrac{d}{dz}f_0(z)|_{z=\eta})^{-1} = \{\tfrac{1}{f(1)} + \tfrac{1}{2\pi}\int_0^{2\pi} \frac{\log\{\rho(\psi(\omega))\}2e^{i\omega}}{(e^{i\omega}-f(1))^2}d\omega\}^{-1},$$

is numerically determined using the trapezium rule.

We now use the above described procedure and set $\mu_2 = 2$, $\mu = 10$, $\lambda_2 = 0.2$, $K = 4000$. The left hand side figure in Fig. 2 shows the impact of λ_0 (i.e., the packet generation rate of the user S_0) on the probability of an empty system for increasing values of λ_1. As expected, $P(Q_1 = 0, Q_2 = 0, C = 0)$ decreases as λ_1 increases. However, we can observe that the values of $P(Q_1 = 0, Q_2 = 0, C = 0)$ deviate significantly for larger λ_0 and especially for small values of λ_1.

In the right hand side figure of Fig. 2 we can observe how the expected length of the relay nodes, $E(Q_1)$, $E(Q_2)$, vary for increasing values of the retrial rate μ_1. Clearly, $E(Q_1)$ decreases for increasing values of μ_1, while $E(Q_2)$ increases, since packets in R_1 retry faster than packets in R_2. Moreover, the impact of λ_0 remains significant, since by increasing its values, $E(Q_1)$, $E(Q_2)$ increase too.

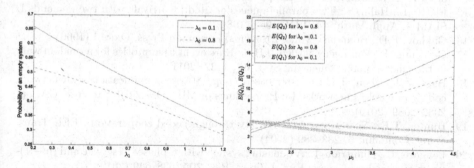

Fig. 2. $P(Q_1 = 0, Q_2 = 0, C = 0)$ for $\mu_1 = 2$ (left). $E(Q_i)$, $i = 1, 2$ for $\lambda_1 = 0.4$ (right).

References

1. Artalejo, J.R., Gomez-Corral, A.: Retrial Queueing Systems: A Computational Approach. Springer, Berlin (2008)

2. Avrachenkov, K., Nain, P., Yechiali, U.: A retrial system with two input streams and two orbit queues. Queueing Syst. **77**, 1–31 (2014)
3. Avrachenkov, K., Yechiali, U.: Retrial networks with finite buffers and their applications to internet data traffic. Probab. Eng. Inf. Sci. **22**, 519–536 (2010)
4. Avrachenkov, K., Morozov, E., Steyaert, B.: Sufficient stability conditions for multi-class constant retrial rate systems. Queueing Syst. **82**(1–2), 149–171 (2016)
5. Cohen, J.W., Boxma, O.: Boundary Value Problems in Queueing Systems Analysis. North Holland Publishing Company, Amsterdam (1983)
6. Cohen, J.W.: Boundary value problems in queueing theory. Queueing Syst. **3**, 97–128 (1988)
7. Boxma, O.: Two symmetric queues with alternating service and switching times. In: Gelenbe, E. (ed.) Performance 1984, pp. 409–431. North-Holland, Amsterdam (1984)
8. Dimitriou, I.: A queueing model with two types of retrial customers and paired services. Ann. Oper. Res. **238**(1), 123–143 (2016)
9. Cover, M., Gamal, A.: Capacity theorems for the relay channel. IEEE Trans. Inf. Theor. **25**(5), 572–584 (1979)
10. Falin, G.I., Templeton, J.G.C.: Retrial Queues. Chapman & Hall, London (1997)
11. Falin, G.I.: On a multiclass batch arrival retrial queue. Adv. Appl. Probab. **20**, 483–487 (1988)
12. Fayolle, G., Iasnogorodski, R., Malyshev, V.: Random Walks in the Quarter-Plane, Algebraic Methods, Boundary Value Problems and Applications. Springer, Berlin (1999)
13. Fayolle, G., Iasnogorodski, R.: Two coupled processors: the reduction to a Riemann-Hilbert problem. Zeitschrift fur Wahrscheinlichkeitstheorie und Verwandte Gebiete **47**, 325–351 (1979)
14. Fiems, D., Andreev, S., Demoor, T., Bruneel, H., Koucheryavy, Y., De Turck, K.: Analytic evaluation of power saving in cooperative communication. In: Proceedings of CFIC 2013, Coibra, Portugal, pp. 1–9. ACM (2013)
15. Flatto, L., Hahn, S.: Two parallel queues created by arrivals with two demands I. SIAM J. Appl. Math. **44**, 1041–1053 (1984)
16. Gakhov, F.D.: Boundary Value Problems. Pergamon Press, Oxford (1966)
17. Guillemin, F., Van Leeuwaarden, J.H.S.: Rare event asymptotics for a random walk in the quarter plane. Queueing Syst. **67**, 1–32 (2011)
18. Haas, Z.J., Chen, T.C.: Cluster-based cooperative communication with network coding in wireless networks. In: Proceedings of MILCOM 2010, San Jose, CA, pp. 2082–2089 (2010)
19. Hunter, T.E., Nosratinia, A.: Diversity through coded cooperation. IEEE Trans. Wirel. Commun. **5**, 283–289 (2004)
20. Langaris, C., Dimitriou, I.: A queueing system with n-phases of service and $(n-1)$-types of retrial customers. Eur. J. Oper. Res. **205**, 638–649 (2010)
21. Moutzoukis, E., Langaris, C.: Non-preemptive priorities and vacations in a multi-class retrial queueing system. Stoch. Models **12**(3), 455–472 (1996)
22. Nosratinia, A., Hunter, T.E., Hedayat, A.: Cooperative communication in wireless networks. Commun. Mag. **42**(10), 74–80 (2004)
23. Papadimitriou, G., Pappas, N., Traganitis, A., Angelakis, V.: Network-level performance evaluation of a two-relay cooperative random access wireless system. Comput. Netw. **88**, 187–201 (2015)
24. Pappas, N., Kountouris, M., Ephremides, A., Traganitis, A.: Relay-assisted multiple access with full-duplex multi-packet reception. IEEE Trans. Wirel. Commun. **14**, 3544–3558 (2015)

25. Sadek, A., Liu, K., Ephremides, A.: Cognitive multiple access via cooperation: protocol design and performance analysis. IEEE Trans. Inf. Theor. **53**(10), 3677–3696 (2007)

26. Sendonaris, A., Erkip, E., Aazhang, B.: User cooperation diversity-Part I: system description. IEEE Trans. Commun. **51**, 1927–1938 (2003)

27. Song, Y., Liu, Z., Zhao, Y.: Exact tail asymptotics – revisit of a retrial queue with two input streams and two orbits. Ann. Oper. Res. 1–24 (2015). doi:10.1007/s10479-015-1945-y

28. Van Leeuwaarden, J.S.H., Resing, J.A.C.: A tandem queue with coupled processors: computational issues. Queueing Syst. **50**, 29–52 (2005)

29. Wright, P.E.: Two parallel processors with coupled inputs. Adv. Appl. Probab. **24**, 986–1007 (1992)

Fingerprinting and Reconstruction of Functionals of Discrete Time Markov Chains

Attila Egri[1], Illés Horváth[2(✉)], Ferenc Kovács[3], and Roland Molontay[4]

[1] Department of Differential Equations,
Budapest University of Technology and Economics, Budapest, Hungary
kleinkoe@math.bme.hu
[2] MTA-BME Information Systems Research Group, Budapest, Hungary
pollux@math.bme.hu
[3] Nokia, Bell-Labs, Murray Hill, USA
ferenc.2.kovacs@nokia.com
[4] Department of Stochastics,
Budapest University of Technology and Economics, Budapest, Hungary
molontay@math.bme.hu

Abstract. We explore various fingerprinting options for functionals of Markov chains with a low number of parameters describing both stationary behaviour and correlation over time. We also present reconstruction methods using lazy Markov chains for the various options. The proposed methods allow for efficient simulation of input data with statistical properties similar to actual real data to serve as realistic input of a data processing system. Possible applications include resource allocation in data processing systems. The methods are validated on data from real-life telecommunication systems.

Keywords: Fingerprinting · Markov processes · Reconstruction · Validation

1 Introduction

Fingerprinting is traditionally used for a number of different purposes [9] including data identification and clustering [2].

In the present paper, we address the use of fingerprinting for simulation; instead of using a large data set, we use a small fingerprint that allows for simulation of data.

The telecommunication systems are complex and their components are not independent from each other, moreover they are tightly coupled. Thus it is difficult to set up a realistic environment for developing and testing different approaches. Usually, the test environments use simulators for simplifying the overall architecture and reducing the costs of the experimentation. The fingerprinting of the component loads allows to simulate and evaluate different scenarios with realistic loads. In this paper, a data processing application is investigated. We assume that the processing complexity of a record depends on some complexity measure of the records (e.g.: record length). The aim is not to

© Springer International Publishing Switzerland 2016
S. Wittevrongel and T. Phung-Duc (Eds.): ASMTA 2016, LNCS 9845, pp. 140–154, 2016.
DOI: 10.1007/978-3-319-43904-4_10

reconstruct the original records themselves, but to reconstruct a series of records that have similar complexity measures.

Functionals of Markov chains arise when a sequence of records within a large data set is assumed to be (or approximated by) a discrete time Markov chain, and a single relevant parameter (*the functional*) is used to describe each record. We are interested in fingerprinting functionals of Markov chains, reconstruction from the fingerprints, and evaluating quality of the reconstruction. We use the term functional in order to emphasize that we are not only interested in the background Markov chain, but also in the values of the function. In our case, the functional will be the aforementioned complexity measure of the records.

We aim to keep the number of parameters in the fingerprint as low as possible. As a result, the problem of reconstruction is severely ill-posed, and the reconstruction will not be perfect; however, with a proper selection of parameters of the functional, the reconstructed Markov chain functional will be close to the original.

The paper is structured as follows: in Sect. 2, we present the preliminaries and the various parameters included in the fingerprints. In Sect. 3, we present reconstruction methods for the various setups of parameters. Section 4 describes the distance measure for MC-functionals used to test the quality of each reconstruction method. Section 5 contains validation of the methods on data from real-life telecommunication systems. Section 6 summarises the results.

As a part of validation, we also look to settle on the best options for raw data fingerprinting. The validation uses a novel distance measure between functionals defined in Sect. 4. We note that a distance measure between functionals is different from a distance measure between actual realizations. For realizations, that is, the actual time series, various similarity measures are present in the literature; for a recent survey, see [8].

For stationary data sequences, other models also exist, e.g. ARMA processes [5,7]. Markov chain functionals offer a more natural approach for models which lack a continuous random noise usually present in ARMA models. Also, in the particular application we are going to test, there is a pre-classification of records (see Subsect. 5) that leads to the Markov model naturally.

2   Preliminaries

Definition 1. *A Markov chain functional is described by a pair* (P, f)*, where* P *is a probability transition matrix of size* $cl \times cl$ *and* f *is a real vector of length* cl*.*

The Markov chain functional should be understood as follows: let X_n be a stationary discrete time Markov chain on cl classes (states) with transition matrix P. f corresponds to a functional on the same cl classes, that is, f_i ($i \in \{1, \ldots, cl\}$) are the values assigned to classes $i \in \{1, \ldots, cl\}$. The Markov chain functional (MC-functional) process is $f(X_t)$. (As a slight abuse of notation, f will be treated as either a function or a vector of the values depending on the situation, so the notation $f(X_n)$ and f_i are both used throughout the paper.)

We assume P is irreducible and aperiodic. Then the stationary distribution π is unique to P, and the behaviour of $f(X_t)$ is completely described by P and f. Of course, as long as f is injective, the process $f(X_t)$ is itself a Markov chain; the distinction is made in order to emphasize the fact that the actual values of f are also relevant. The distance measure detailed in Sect. 4 takes this into account, assigning a positive distance to $(P, f^{(1)})$ and $(P, f^{(2)})$ when $f^{(1)} \neq f^{(2)}$.

MC-functionals are clearly invariant to permutations of the classes, thus we may assume that f is increasing.

Next we consider various parameters of a MC-functional relevant to finger-printing. The parameters are divided into two classes, based on their different roles. *Stationary parameters* describe the behaviour of a MC-functional in the stationary regime, while *correlation parameters* correspond to dynamic behaviour.

2.1 Stationary Parameters

A most natural and relevant parameter is the expected value of f with respect to the stationary distribution, which can be calculated by scalar product:

$$E(f) = \pi^T f$$

where both π and f are written as column vectors.

$$f_{\min} = f(1) \quad \text{and} \quad f_{\max} = f(cl)$$

denote the smallest and largest value of f respectively, and

$$\text{width}(f) = f_{\max} - f_{\min}.$$

The variance of f with respect to the stationary distribution will be denoted by $\sigma_0^2(f)$ and is calculated by

$$\sigma_0^2(f) = \left(\sum_{i=1}^{cl} \pi_i f_i^2 \right) - (E(f))^2.$$

We introduce two more parameters. They are obtained from an order 2 Taylor approximation of π. Consider the vectors

$$v_0 = \begin{pmatrix} 1 \\ 1 \\ \vdots \\ 1 \end{pmatrix}, \quad v_1 = \begin{pmatrix} 1 \\ 2 \\ \vdots \\ cl \end{pmatrix}, \quad v_2 = \begin{pmatrix} (-(cl-1)/2)^2 \\ (-(cl-1)/2+1)^2 \\ \vdots \\ ((cl-1)/2)^2 \end{pmatrix}$$

and the linear subspace

$$a\frac{v_0}{v_0^T 1} + b\frac{v_1}{v_1^T 1} + (1-a-b)\frac{v_2}{v_2^T 1} \qquad (a, b \in \mathbb{R})$$

where $\mathbf{1}$ is the constant column vector 1 of length cl.

We make a linear regression of π from this subspace in the least squares sense with the restrictions that the approximating vector is nonnegative (a or b themselves are allowed to be negative). The solution is denoted by a_π, b_π. We note that the restrictions are linear in (a, b) and the solution itself can also be obtained linearly. a_π and b_π are allowed to be negative.

2.2   Correlation Parameters

The *one-step covariance* of f, denoted by $\mathrm{cov}_1(f)$ is the covariance between $f(X_1)$ (which denotes a real random variable, the value of the functional for the first element of the Markov chain according to stationary initial distribution) and $f(X_2)$ (which is the value of the functional for the second element of the Markov chain, naturally correlated with $f(X_1)$), and can be calculated as

$$\mathrm{cov}_1(f) = E\left((f(X_1) - E(f))(f(X_2) - E(f))\right) =$$
$$= \pi^T \cdot \mathrm{diag}[P(f - E(f)\mathbf{1})](f - E(f)) \tag{1}$$

where diag is the operator that makes a diagonal matrix from a vector.

Another similar parameter is the so-called *asymptotic variance* of f, defined as follows:

$$\sigma^2(f) = \sum_{t=1}^{\infty} E\left((f(X_1) - E(f))(f(X_t) - E(f))\right).$$

The sum is finite since the correlation between X_1 and X_t decays exponentially as $t \to \infty$ (assuming the MC is nondegenerate, that is, the transition matrix P is not completely deterministic). Note that σ^2 and cov_1 capture different aspects of the Markov chain, as the long-term correlation and the short-term correlation may be different. That said, a high value of cov_1 generally corresponds to long runs of the same class in the realization of the Markov chain.

Calculation of the asymptotic variance is straightforward:

$$\sigma^2(f) = \sum_{t=0}^{\infty} \pi^T \cdot \mathrm{diag}[P^t(f - E(f)\mathbf{1})](f - E(f)\mathbf{1})$$
$$= \pi^T \cdot \mathrm{diag}\left[(I - P + \mathbf{1}^T\pi)^{-1}(f - E(f)\mathbf{1}\right](f - E(f)\mathbf{1}).$$

cov_1 is a relatively standard parameter compared to σ^2; for actual data, cov_1 might be more readily available than σ^2.

2.3   Possible Setups of Parameters

To make a meaningful fingerprint for raw data, the following options are viable. From among stationary parameters, select either

- A: $E(f)$ and $\sigma_0^2(f)$, or
- B: $E(f)$ and width(f),
- C: f_{min}, f_{max} and $E(f)$, or
- D: $f_{min}, f_{max}, E(f), a_\pi$ and b_π

while simultaneously select one from the correlation parameters:

- 1: $\sigma^2(f)$, or
- 2: $\text{cov}_1(f)$, or
- 3: nothing.

In the next section, we present reconstruction methods for each of the above combinations.

3   Reconstruction

We aim to reconstruct the Markov chain X_t and the functional f from each of the above descriptions. The reconstruction algorithm has a similar structure for each of the setup of parameters described in Sect. 2.3: it will first reconstruct a function f and a stationary vector π from the stationary parameters, then reconstruct the full MC generator P from the correlation parameter (and f and π).

Note that we will not be able to give a perfect reconstruction of the MC-functional. The main guideline is to reconstruct a MC-functional whose fingerprint is identical to the given fingerprint.

We also note that we use a parameter cl during reconstruction. cl is the number of classes in the reconstructed MC functional. It may be selected as any positive integer $cl \geq 2$; we use $cl = 5$ as a default in Sect. 5.

3.1   Reconstruction of Stationary Behaviour

First we reconstruct f and π from the various stationary parameters. We exclude the trivial (and equivalent) cases when $f_{\max} - f_{\min} = \text{width}(f) = 0$ or $\sigma_0^2(f) = 0$. In such cases, the reconstruction is trivial.

A: reconstruction of f and π from $E(f)$ and $\sigma_0^2(f)$. Essentially, we fit a uniform distribution of proper variance:

$$\pi = \left(\frac{1}{cl}, \ldots, \frac{1}{cl}\right) \quad f = \begin{pmatrix} E(f) - \frac{cl-1}{2}c_0 \\ E(f) - \left(\frac{cl-1}{2} - 1\right)c_0 \\ \vdots \\ E(f) + \left(\frac{cl-1}{2} - 1\right)c_0 \\ E(f) + \frac{cl-1}{2}c_0, \end{pmatrix} \quad \text{where } c_0 = \frac{\sigma_0}{\left(\frac{cl^2-1}{12}\right)^{1/2}}.$$

The resulting f may be negative (on the lower end).

B: reconstruction of f and π from $E(f)$ and $\text{width}(f)$. We fit a uniform distribution of proper width.

$$\pi = \left(\frac{1}{cl}, \ldots, \frac{1}{cl}\right) \quad f = \begin{pmatrix} E(f) - \text{width}(f)/2 \\ E(f) - \text{width}(f)/2 + \frac{1}{cl-1}\text{width}(f) \\ \vdots \\ E(f) - \text{width}(f)/2 + \frac{cl-2}{cl-1}\text{width}(f) \\ E(f) + \text{width}(f)/2 \end{pmatrix}$$

Again, the resulting f may be negative on the lower end.

C: reconstruction of f and π from $f_{\min}, f_{\max}$ and $E(f)$ - the minimax approach. The values of f will be set to partition the interval $f_{\min}, f_{\max}$ in an equidistant manner with cl points:

$$f_i = f_{\min} + (f_{\max} - f_{\min}) \cdot \frac{i-1}{cl-1}, \quad i = 1, \dots, cl.$$

Next we calculate the stationary distribution π. Apart from $f_{\min}$ and $f_{\max}$, only $E(f)$ will be used to calculate π. π will be of the form $\pi = (a, \dots, a, b, \dots, b)$. The reason for this is that we are looking for a π whose smallest value is as large as possible (this property will be important for the next step, the reconstruction of the full P matrix). Such a minimax choice of π is necessarily of the above form. Let the number of a values be j, that is,

$$v_i = a \quad i = 1, \dots, j, \qquad v_i = b \quad i = j+1, \dots, cl.$$

Once the value of j is given, the values of a and b can be determined from the following system of linear equations:

$$E(f) = \sum_{i=1}^{j} a f_i + \sum_{i=j+1}^{cl} b f_i$$
$$1 = aj + b(cl - j)$$

We solve the above system for each choice of $j = 1, \dots, (cl - 1)$ separately, and select the one where $\min(a, b)$ is the highest. Thus π is obtained.

One curious thing about the minimax approach is that the parameter $E(f)$ is used in the reconstruction of π, while the reconstruction of f is trivial. Even this way, numerical examples show that both π and f may differ considerably from the original. This motivates the addition of two more parameters, examined in detail in the next subsection.

D: reconstruction of f and π from $f_{\min}, f_{\max}, E(f)$, a_π and b_π - the order 2 approach. The reconstructed π (denoted by π_r) is obtained simply from a_π and b_π:

$$\pi_r = a_\pi \frac{v_0}{v_0^T 1} + b_\pi \frac{v_1}{v_1^T 1} + (1 - a_\pi - b_\pi) \frac{v_2}{v_2^T 1}$$

where v_0, v_1 and v_2 are the same as in (1).

f is obtained from $f_{\min}, f_{\max}$ and $E(f)$ in a similar order 2 approach; we consider the subspace spanned by the vectors

$$f_0 = \begin{pmatrix} 1 \\ 1 \\ \vdots \\ 1 \end{pmatrix}, \quad f_1 = \begin{pmatrix} 1 \\ 2 \\ \vdots \\ cl \end{pmatrix}, \quad f_2 = \begin{pmatrix} 1^2 \\ 2^2 \\ \vdots \\ cl^2 \end{pmatrix}. \tag{2}$$

We approximate f by a vector of the form

$$f_r = a_1 f_1 + a_2 f_2 + a_3 f_3 \qquad (a_1, a_2, a_3 \in \mathbb{R})$$

with the linear restrictions that

$$f_{\min} = (f_r)_1 \tag{3}$$
$$f_{\max} = (f_r)_{cl} \tag{4}$$
$$(f_r)_1 \leq (f_r)_2 \leq \cdots \leq (f_r)_{cl} \tag{5}$$

The approximation is done via linear regression to minimize for

$$(\pi_r^T f_r - E(f))^2,$$

that is, least squares approximation of $E(f)$ according to weights π_r, subject to the restrictions (3)–(5). Note that π_r is calculated at this point, and given π_r, this is again a linear problem. While the approximation via order 2 Taylor series is a relatively standard idea, it is still somewhat different from most similar applications due to the fact that 2 steps are executed after each other (first p_r, then f_r are calculated); the main gain in doing this in 2 steps is that the calculations throughout involve only linear optimization.

In actual examples, $\pi_r^T f_r = E(f)$ will often be the case; nevertheless, the values of a_1, a_2, a_3 should not be calculated by directly solving $\pi_r^T f_r = E(f)$ along with (3) and (4), because such a solution will not necessarily be monotone, and it offers a considerably worse reconstruction.

3.2   Reconstruction of the Transition Matrix

Next, we reconstruct P from f, π and the various correlation parameters.

1: Reconstruction of P from f, π and $\sigma^2(f)$. Let $\Pi = \mathbf{1}^T \pi$, the matrix whose rows are all equal to π.

We look for P in the form

$$P_c = cI + (1 - c)\Pi, \tag{6}$$

where c is a real constant. As long as $0 \leq c \leq 1$, P_c will be a stochastic matrix naturally with stationary distribution π; however, c may also be negative as long as all elements of P_c are nonnegative. That is,

$$-\frac{\pi_{\min}}{1 - \pi_{\min}} \leq c < 1 \tag{7}$$

where $\pi_{\min}$ denotes the minimal element of π.

A positive c corresponds to what is called a "lazy Markov chain": the MC will typically stay in a single class for many steps before jumping. $c = 0$ corresponds to an iid chain, while $c < 0$ corresponds to a Markov chain where the steps are negatively correlated. Lazy Markov chains often arise during analysis of mixing times; for more on them, see [3]. In our case, lazy Markov chains are used for their easy scalability of correlation.

We will refer to c as the "degree of laziness"; a higher value of c corresponds to the data having longer runs of identical elements. c is always less than 1; $c = 1$ corresponds to a constant sequence (which can be reconstructed trivially). c may be negative, but not "too" negative according to (7).

We calculate the value of c from

$$\sigma^2(f) = \pi \cdot \operatorname{diag}\left[(I - P_c + \mathbf{1}^T \pi)^{-1}(f - E(f)\mathbf{1})\right](f - E(f)\mathbf{1}). \tag{8}$$

We note that

$$(I - P_c + \mathbf{1}^T \pi)^{-1} = \frac{1}{1-c}(I - \Pi) + \Pi,$$

so (8) yields a linear equation for c:

$$\begin{aligned} c &= 1 - \frac{\pi \operatorname{diag}\left[(I - \Pi)(f - E(f)\mathbf{1})\right](f - E(f)\mathbf{1})}{\sigma^2(f) - \pi \operatorname{diag}\left[\Pi(f - E(f)\mathbf{1})\right](f - E(f)\mathbf{1})} \\ &= 1 - \frac{\pi \operatorname{diag}\left[(f - E(f)\mathbf{1})\right](f - E(f)\mathbf{1})}{\sigma^2(f)}, \end{aligned} \tag{9}$$

using the fact $\Pi(f - E(f)\mathbf{1}) = 0$.

In case the solution is smaller than the minimal value $-\frac{\pi_{\min}}{1 - \pi_{\min}}$, we set $c = -\frac{\pi_{\min}}{1 - \pi_{\min}}$.

2: Reconstruction of P from f, π and $\mathrm{cov}_1(f)$. Again,

$$P_c = cI + (1 - c)\Pi$$

where

$$c = \frac{\mathrm{cov}_1(f)}{\pi \operatorname{diag}\left[(f - E(f)\mathbf{1})\right](f - E(f)\mathbf{1})}.$$

Again, we need to check the bounds in (7); in case c would be smaller than the minimal value $-\frac{\pi_{\min}}{1 - \pi_{\min}}$, we set $c = -\frac{\pi_{\min}}{1 - \pi_{\min}}$. In case $c \geq 1$ (which might happen in this case), we set $c = 0.9999$ instead. Allowing $c = 1$ is not a good idea for a number of reasons; it would correspond to a degenerate Markov chain with deterministic transitions. If π is also deterministic, that corresponds to the constant deterministic case which we excluded earlier; π being random while P being deterministic is completely unrealistic. Also, there would be no unique stationary distribution for $P = I$.

3: Reconstruction of P from f, π and no correlation parameter. We simply set

$$P = \Pi.$$

This corresponds to an independent, identically distributed sequence, where the covariance between different values is 0.

3.3 Worked Example

This subsection includes a worked example that illustrates some of the above reconstruction methods. The example originates from the same data that is tested later in Sect. 5, with the length-of-record functional (see Subsect. 5.2).

The original Markov chain functional is described by

$$P = \begin{pmatrix} 0.999457 & 0.000181 & 0. & 0. & 0.000362 \\ 0.000435 & 0.998695 & 0.000435 & 0. & 0.000435 \\ 0. & 0. & 0.998334 & 0.000833 & 0.000833 \\ 0. & 0. & 0.003984 & 0.996016 & 0 \\ 0.002770 & 0.002770 & 0. & 0. & 0.994460 \end{pmatrix}$$

$$\pi = (0.5525, 0.2301, 0.1201, 0.0251, 0.0722)$$
$$f = (117.786, 406.63, 566.855, 1412., 1566.95)$$

Note that the matrix is close to the identity matrix, which corresponds to the fact that the examined data typically contains many consecutive records from the same class. However, for the overall behaviour of the chain (e.g. stationary distribution) the offdiagonal elements are also very important.

The parameters of this Markov chain functional are:

$$E(f) = 375.296 \quad f_{min} = 117.786 \qquad f_{max} = 1566.95 \quad width(f) = 1449.16$$
$$\sigma_0^2(f) = 413.247 \quad \sigma^2(f) = 96952600 \quad cov_1(f) = 170020$$
$$a_\pi = -64.7888 \quad b_\pi = 146.631$$

Reconstruction A gives

$$\pi_r = (0.2, 0.2, 0.2, 0.2, 0.2)$$
$$f_r = (-209.123, 83.0868, 375.296, 667.506, 959.715)$$

and then, for example, A1 gives the reconstructed matrix

$$P_r = \begin{pmatrix} 0.996473 & 0.000882 & 0.000882 & 0.000882 & 0.000882 \\ 0.000882 & 0.996473 & 0.000882 & 0.000882 & 0.000882 \\ 0.000882 & 0.000882 & 0.996473 & 0.000882 & 0.000882 \\ 0.000882 & 0.000882 & 0.000882 & 0.996473 & 0.000882 \\ 0.000882 & 0.000882 & 0.000882 & 0.000882 & 0.996473 \end{pmatrix}$$

In this case, neither π_r or f_r approximate the original π and f well. The simple structure of π_r, f_r and P_r should also be noted. The distance of (P, π, f) and (P_r, π_r, f_r) (defined precisely later in Sect. 4, included here just for illustration) is $d_0 = 1.17262$. Also note that even though P_r is quite close to P in absolute value, the relatively large difference in the offdiagonal elements leads to a quite different stationary distribution.

On the other hand, reconstruction C gives

$$\pi_r = (0.715686, 0.0710785, 0.0710785, 0.0710785, 0.0710785)$$
$$f_r = (117.786, 480.076, 842.366, 1204.66, 1566.95)$$

which is already a considerable improvement. The matrix for C1 is

$$P_r = \begin{pmatrix} 0.999374 & 0.000157 & 0.000157 & 0.000157 & 0.000157 \\ 0.001577 & 0.997954 & 0.000157 & 0.000157 & 0.000157 \\ 0.001577 & 0.000157 & 0.997954 & 0.000157 & 0.000157 \\ 0.001577 & 0.000157 & 0.000157 & 0.997954 & 0.000157 \\ 0.001577 & 0.000157 & 0.000157 & 0.000157 & 0.997954 \end{pmatrix}$$

and the corresponding distance from the original is $d_0 = 0.357345$.
Reconstruction D1 gives

$$\pi_r = (0.540834, 0.262703, 0.0922859, 0.0295832, 0.0745945)$$

$$f_r = (117.786, 372.247, 698.594, 1096.83, 1566.95)$$

$$P_r = \begin{pmatrix} 0.999210 & 0.000452 & 0.000159 & 0.000051 & 0.000128 \\ 0.000931 & 0.998731 & 0.000159 & 0.000051 & 0.000128 \\ 0.000931 & 0.000452 & 0.998438 & 0.000051 & 0.000128 \\ 0.000931 & 0.000452 & 0.000159 & 0.998330 & 0.000128 \\ 0.000931 & 0.000452 & 0.000159 & 0.000051 & 0.998408 \end{pmatrix}$$

For this method, π_r is very close to the original, while the values of f_r are close to the original for coordinates 1 and 2, which have the highest stationary probability in π. The distance here is $d_0 = 0.115363$.

In the above example, D1 offers the best approximation. As test results from Sect. 4 will show, this is more or less typical; however, C1 and C2 also perform well.

4 Measure of Distance Between Functionals of Markov Chains

Assume we have two discrete Markov chain functionals $(P^{(1)}, \pi^{(1)}, f^{(1)})$ and $(P^{(1)}, \pi^{(2)}, f^{(2)})$, where $P^{(1)}$ is a transition matrix on $cl^{(1)}$ classes, π_1 is the corresponding stationary distribution and $f^{(1)}$ a functional on cl_1 classes. Similarly, $P^{(2)}$ is a transition matrix on cl_2 classes, $\pi^{(2)}$ is the corresponding stationary distribution and $f^{(2)}$ a functional on cl_2 classes. We allow cl_1 and cl_2 to be different.

We define a distance measure between $(P^{(1)}, \pi^{(1)}, f^{(1)})$ and $(P^{(2)}, \pi^{(2)}, f^{(2)})$. The following definition is motivated by the L^1-Wasserstein metric in 1 dimension (see e.g. [4]); that said, to the best of our knowledge, it has not been used in 2 dimensions yet, nor has it been used in relation to distances of Markov chain functionals. The main advantage of the distance measure is that it assigns a small value to the distance only if the probabilistic parts $(P^{(1)}, \pi^{(1)})$ and $(P^{(2)}, \pi^{(2)})$ are close *and* the functionals $f^{(1)}$ and $f^{(2)}$ are also close.

First, we consider the joint distribution of $f^{(1)}(X_1)$ and $f^{(1)}(X_2)$, where X_1 and X_2 are the first two elements of a stationary Markov chain with distribution $(P^{(1)}, \pi^{(1)})$. The joint distribution function will be denoted by $F^{(1)}(x, y)$.

Similarly, $F^{(2)}(x, y)$ is the joint distribution function of $f^{(2)}(Y_1)$ and $f^{(2)}(Y_2)$, where Y_1 and Y_2 are the first two elements of a stationary Markov chain with distribution $(P^{(2)}, \pi^{(2)})$.

We assume that the classes of $(P^{(1)}, \pi^{(1)})$ and $(P^{(2)}, \pi^{(2)})$ are ordered such that $f^{(1)}$ and $f^{(2)}$ are increasing. Otherwise, reorder the classes so. Then $F^{(1)}(x, y)$ can be calculated by

$$F^{(1)}(x, y) = \sum_{i: f^{(1)}(i) \leq x} \sum_{j: f^{(1)}(j) \leq y} \pi_i^{(1)} P_{ij}^{(1)}$$

and $F^{(2)}$ by

$$F^{(2)}(x, y) = \sum_{i: f^{(2)}(i) \leq x} \sum_{j: f^{(2)}(j) \leq y} \pi_i^{(2)} P_{ij}^{(2)}.$$

$F^{(1)}$ and the similarly defined $F^{(2)}$ contain all information about the corresponding MC-functionals. We define the following distance metric between $F^{(1)}$ and $F^{(2)}$:

$$d(F^{(1)}, F^{(2)}) = \int_{-\infty}^{\infty} \int_{-\infty}^{\infty} |F^{(1)}(x, y) - F^{(2)}(x, y)| dx dy.$$

$d(\cdot, \cdot)$ has the following properties:

- symmetric: $d(F^{(1)}, F^{(2)}) = d(F^{(2)}, F^{(1)})$,
- satisfies the triangle inequality

$$d(F^{(1)}, F^{(3)}) \leq d(F^{(1)}, F^{(2)}) + d(F^{(2)}, F^{(3)}),$$

- nonnegative: $d(F^{(1)}, F^{(2)}) \geq 0$, and
- $d(F^{(1)}, F^{(2)}) = 0$ only when the functionals $F^{(1)}(X_t)$ and $F^{(2)}(Y_t)$ have the same distribution.

We note that the function $|F^{(1)}(x, y) - F^{(2)}(x, y)|$ is piecewise constant and thus $d(F^{(1)}, F^{(2)})$ can be calculated as the sum of the volume of a finite number of cuboids. Merge sort the values of $f^{(1)}$ and $f^{(2)}$ into a single vector u; then

$$d(F^{(1)}, F^{(2)}) =$$

$$\sum_{i=1}^{|u|-1} \sum_{j=1}^{|u|-1} (u_{i+1} - u_i)(u_{j+1} - u_j) |(F^{(1)}(u_i, u_j) - F^{(2)}(u_i, u_j)|,$$

so actually $F^{(1)}(x, y)$ and $F^{(2)}(x, y)$ needs to be evaluated only for $x, y \in u$. This is algorithmically straightforward.

$d(F^{(1)}, F^{(2)})$ is proportional to the values of $f^{(1)}$ and $f^{(2)}$ in the sense that if we consider $cf^{(1)}$ and $cf^{(2)}$ instead of $f^{(1)}$ and $f^{(2)}$, the distance will be multiplied by c^2. To this end, we introduce some normalization. There are two options for normalizing: either normalize by the mean or the maximal value of the functions:

$$\frac{d(F^{(1)}, F^{(2)})}{E(f^{(1)}) E(f^{(2)})} \quad \text{or} \quad \frac{d(F^{(1)}, F^{(2)})}{\max(f^{(1)}) \max(f^{(2)})}$$

The two normalizations offer different properties. In the case when both $f^{(1)}$ and $f^{(1)}$ have some outlier values (that is, values which are much higher than the mean but with a small stationary probability), the distance normalized by the mean assigns a relatively high penalty to even relatively small differences in either the values or the probabilities of the outliers. On the other hand, normalizing by the maximum assigns too little penalty to even large differences among the non-outlier values.

As a sort of mixed choice, we will use the normalization

$$d_0(F^{(1)}, F^{(2)}) = \frac{d(F^{(1)}, F^{(2)})}{\sqrt{\max(f^{(1)})E(f^{(1)})\max(f^{(2)})E(f^{(2)})}}$$

instead.

The normalization can be problematic when one of the functions has mean 0 or relatively close to 0. This is not an issue when the functionals are nonnegative, which is often the case. We suggest using the normalized version only when the means are nonzero.

An actual threshold value for d_0 is not determined at this point and is the subject of further research. Even without a threshold value, d_0 is certainly applicable for comparison purposes; in the present paper, we use d_0 to compare the quality of reconstruction for the various setups of parameters from Subsect. 2.3.

As a result of the normalization, triangle inequality does not hold for d_0, but it is still symmetric, nonnegative and 0 only when the MC-functionals have identical distribution.

5   Validation on Data from Telecommunication Systems

The experimentation was based on a real telecommunication data processing application that processes records. Records correspond to basic units of communication between two given nodes in the telecommunication network. Each record contains various data fields depending on the type of the package. In the system, the computer time necessary to process a record is roughly proportional to the total length of the record, hence it is natural to select the total length of the record as the complexity functional. This assertion is not validated in the present paper. Depending on the application, another complexity measure might be a better choice.

We focus on the "raw" data itself, not the effect of processing on the data. That is, what is actually obtained by the fingerprinting and reconstruction of the raw data in this case is a way to provide realistic simulated input for the system that matches the stationary and dynamic behaviour of the original data.

5.1   Markov Chain Fitting

A single data set consists of a sequence of records with various fields. For a single record, some of the fields may be empty. The first field is the type of the record, the rest of the fields contain numeric values. For the current study, the following functional will be used: f denotes the total amount of bytes in the fields of a given record.

While there is a prescribed type of each record, the number of types is around 50, which is too high for our purposes. To reduce the number of types, records are clustered according to the value of f. The number of clusters used is $cl = 5$. We used optimal k-means clustering by dynamic programming [10]. (Standard clustering methods can easily handle as many as 10^4 types, and there are methods available for clustering as many as 10^6 types.) After clustering, the values of f are recalculated by averaging the values of the original records; the result is that the value of f is constant within each class. Note that clustering in this case is a purely technical issue; the precise effect of clustering may be subject to further testing.

Once the clustering is done, we fit a (stationary) Markov chain (MC) on raw data. The Markov chain fitting method is rather standard. At this point, only the class is considered relevant for each raw data record, so accordingly, we assume data looks like a sequence whose elements are integers between 1 and cl. The state space of the MC is thus $1, \ldots, cl$. We calculate P as the *empirical transition matrix*, which is a matrix of size $cl \times cl$; this also gives the maximum likelihood estimate for P. For more, see [1].

After obtaining P and f, we calculate the fingerprints and execute the various reconstruction methods. The reconstructed MC-functionals are then compared to the original using the distance metric d_0 defined in Sect. 4.

5.2   Test Results

The data originate from raw data supplied to the Parser unit (see also Subsect. 5). 68 different data sections were tested, each section containing 10000 records.

Test results were executed for the *length-of-record* functional, which corresponds to the total amount of data in the record (in bytes).

The possible setups for reconstruction are denoted by A1, A2, A3, B1, B2, B3, C1, C2, C3, D1, D2 and D3 in accordance with the notations from Subsect. 2.3, so for example A1 means that reconstruction A is used for the reconstruction of the stationary behaviour and reconstruction 1 is used for the reconstruction of the correlation behaviour.

For each setup of parameters, the average distance d_0 of the original and reconstructed Markov chain functionals is included in Table 1. Table 1 shows that reconstructions A and B are quite poor. They nevertheless serve as a basis for comparison.

To give a more complete picture for C1, C2, D1 and D2 (which provided the best means in Table 1), we include histograms in Figs. 1, 2, 3 and 4. In the histograms, the y axis shows the number of samples where the distance falls within the range given by the x axis. The histograms reveal that apart from a few outlier values, the majority of the distances are well below 1.

Table 1. Average distances of original and reconstructed MC functionals for various setups of parameters of reconstruction

Setup	Avg. Dist.	Setup	Avg. Dist.	Setup	Avg. Dist.	Setup	Avg. Dist.
A1	2.79175	B1	4.02120	C1	0.450947	D1	0.423104
A2	2.79179	B2	3.90754	C2	0.452343	D2	0.536618
A3	2.84281	B3	3.90308	C3	0.621195	D3	0.759127

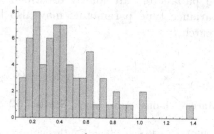

Fig. 3. Histogram for D1

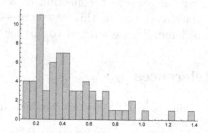

Fig. 4. Histogram for D2

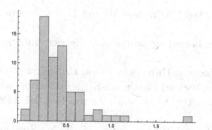

Fig. 1. Histogram for C1

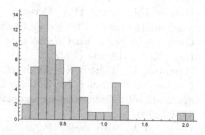

Fig. 2. Histogram for C2

6   Conclusions and Outlook

We have presented a number of different methods to fingerprint and reconstruct Markov chain functionals, and tested the quality of reconstruction for each method on real-life data. The methods handle stationary and long-term behaviour separately, allowing for a modular reconstruction method where the stationary reconstruction and the reconstruction of the transition matrix can be configured separately. We consider the validation method to be interesting in its own right.

Based on the test results, two of the stationary reconstruction methods (A and B) may be dismissed; C (the minimax approach) and D (the order 2 approximation) both provide a considerable improvement. Altogether, C1, C2, D1 and D2 provided the best quality reconstruction. Reconstruction C, the minimax approach proved to be fairly adaptive in a number of cases, while reconstruction D, which is altogether a more standard approach using order 2 approximations, is also efficient. For the data set examined, tests showed that overall D1 provided the best quality reconstruction, but C1 and C2 should not be discarded either (especially since they use less parameters). The difference between C1 and C2 is not significant. C2 may be preferred in cases when cov_1 is available.

Outlook. It may be interesting to extend any of the above methods to include more parameters and examine the trade-off between more parameters included in the description and the improvement in the quality of the reconstruction.

It would also be interesting to extend this approach to multi-dimensional functionals, where a collection of several parameters are jointly described by a Markov chain. In this case, cross-covariance type parameters may also be considered. This is subject to further research.

References

1. Billingsley, P.: Statistical methods in Markov chains. Ann. Math. Statist. **32**(1), 12–40 (1961)
2. Crotti, M., Dusi, M., Gringoli, F., Salgarelli, L.: Traffic classification through simple statistical fingerprinting. ACM SIGCOMM Comput. Commun. Rev. **37**(1), 5–16 (2007)
3. Wilmer, E.L., Levin, D.A., Peres, Y.: Markov Chains and Mixing Times. American Mathematical Soc., Providence (2009)
4. Gibbs, A.L., Su, F.E.: On choosing and bounding probability metrics. Int. Stat. Rev. **70**(3), 419–435 (2002)
5. Hamilton, D.: Time Series Analysis. Princeton University Press, Princeton (1994)
6. Hausser, J., Strimmer, K.: Entropy inference and the James-Stein estimator, with application to nonlinear gene association networks. J. Mach. Learn. Res. **10**, 1469–1484 (2009)
7. Davis, R.A., Brockwell, P.J.: Time Series: Theory and Methods. Springer, Heidelberg (2009)
8. Serrí, J., Arcos, J.L.: An empirical evaluation of similarity measures for time seriesclassification. Know.-Based Syst. **67**, 305–314 (2014)
9. Wagner, N.R.: Fingerprinting. In: Proceedings of the 1983 IEEE Symposium on Security and Privacy, SP 1983, p. 18. IEEE Computer Society, Washington (1983)
10. Wang, H., Song, M.: Ckmeans. 1d. dp: optimal k-means clustering in one dimension bydynamic programming. R J. **3**(2), 29–33 (2011)

On the Blocking Probability and Loss Rates in Nonpreemptive Oscillating Queueing Systems

Fátima Ferreira[1], António Pacheco[2(✉)], and Helena Ribeiro[3]

[1] Departamento de Matemática, CMAT and CEMAT,
Universidade de Trás-os-Montes e Alto Douro, UTAD,
Quinta dos Prados, 5000-801 Vila Real, Portugal
mmferrei@utad.pt

[2] CEMAT and Departamento de Matemática, Instituto Superior Técnico,
Universidade de Lisboa, Av. Rovisco Pais, 1049-001 Lisboa, Portugal
apacheco@math.tecnico.ulisboa.pt

[3] Escola Superior de Tecnologia e Gestão and CEMAT,
Instituto Politécnico de Leiria, Campus 2, Morro do Lena - Alto do Vieiro,
2411-901 Leiria, Portugal
helena.ribeiro@ipleiria.pt

Abstract. We analyze busy period related characteristics for oscillating systems, proposing a method to compute the mean length of busy periods for nonpreemptive oscillating $M^X/G/1/(n,a,b)$ systems. We further evaluate customer loss rates in busy cycles and the long-run blocking probability for such systems using the proposed method and earlier results from the authors for the mean number of customer losses in busy periods.

Keywords: Blocking probability · Busy period · Loss rate · Markov regenerative processes · Nonpreemptive systems · Oscillating queues

1 Introduction

Oscillating $M^X/G/1/\mathbf{n}$ systems, with $\mathbf{n} = (n,a,b)$, are queueing systems with finite capacity, n, and two barriers, a and b with $0 \le a < b \le n$, that impact the way customers are served, as described below.

Customers arrive to the system in batches according to a Poisson process and are served, by a single server, in a first come first served policy. Batch sizes are independent and identically distributed (i.i.d.) random variables with probability mass function $(f_l)_{l \in \mathbb{N}_+}$ and finite mean $\bar{f}$. As regards the customer acceptance policy, we consider what is known as *partial blocking* (see, e.g., [5]) in which, if at arrival of a batch of l customers there are only m, $m < l$, free positions available in the system, then m customers of the batch enter the system and the remaining $l - m$ customers of the batch are blocked.

The sequences of batch sizes and batch interarrival times are independent. However, contrary to the usual classical queuing systems, the customer service times are not independent, oscillating between two phases (1 and 2), reacting

© Springer International Publishing Switzerland 2016
S. Wittevrongel and T. Phung-Duc (Eds.): ASMTA 2016, LNCS 9845, pp. 155–166, 2016.
DOI: 10.1007/978-3-319-43904-4_11

to the congestion of the system. Specifically, the phase in which the system is operating is determined by the evolution of the number of customers in the system, according to the barriers a and b.

When empty, the system operates in phase 1 (a low performance phase) and remains operating in phase 1 until the first subsequent service completion epoch at which the number of customers in the system equals or exceeds the upper barrier b. At that instant, operational phase 1 ends and the system starts to operate in phase 2 (a high performance phase), remaining in such operational phase until the first subsequent epoch at which the number of customers in the system becomes smaller than or equal to the lower barrier a, at which time the system switches back to operate in phase 1, and so on. Service times initiated in phase 1 have duration S_1 with distribution function A_1 and mean μ_1^{-1}, and service times initiated in phase 2 have duration S_2 with distribution function A_2 and mean μ_2^{-1} $(\mu_2 > \mu_1)$.

The state process of an oscillating $M^X/G/1/n$ queue is modelled by the continuous-time bivariate process $(X(t))_{t \geq 0}$, where $X(t) = (X_1(t), X_2(t))$, with $X_1(t)$ denoting the number of customers in system at time t and $X_2(t)$ the phase in which the system is operating under at the same instant. The process has state space

$$E^{(n,a,b)} = \{(i,1) : 0 \leq i \leq b-1\} \cup \{(i,2) : a+1 \leq i \leq n\}$$

and it is a Markov regenerative process (see, e.g., [4]) associated to the renewal sequence of customer post-departure epochs.

Due their relevance in quality control of service provided at low cost, oscillating systems have been addressed by several authors [1–3,5]. In particular, using the potential method, Chydzinsk [1,2] characterized the limit distribution of the number of customers in oscillating $M/G/1$ and $M/G/1/n$ systems. The analysis of queuing systems during busy periods, i.e., during continuous periods of effective use of the server, is relevant from the operator point of view, providing crucial information for system management. In this context, taking profit of the Markov regenerative structure of oscillating $M^X/G/1/n$ systems, Pacheco and Ribeiro [5] derived the distribution of the number of consecutive losses during busy periods and Ferreira *et al.* [3] computed the moments of the number of costumer losses during busy periods.

In [3], the authors found that, when subjected to high traffic intensity, systems with higher group size variability and heavier tailed service time distributions are the ones experiencing smaller mean number of customer losses during busy periods. However, depending on the service time and group size distributions considered, the mean duration of the busy periods can vary considerably, strongly influencing long-run loss rates and blocking probabilities for these systems.

With this in mind, we pursue with the proposal, in Sect. 2, of a recursive method for calculating the mean length of busy periods for oscillating $M^X/G/1/n$ nonpreemptive systems, followed by the derivation, in Sect. 3, of expressions for customer loss rates in busy cycles and the long-run blocking probability for the same systems. The results derived in the paper are applied, in Sect. 4, to several oscillating $M^X/G/1/n$ queues with different service time and group size distributions. Finally, we present some concluding remarks in Sect. 5.

2   Mean Length of Busy Periods

In this section we address the computation of the mean length of busy periods of oscillating $M^X/G/1/n$ systems, hereinafter denoted by $M^X/G_1 - G_2/1/(n, a, b)$ systems. In the study, we consider busy periods started with multiple customers in the system, denoting by $B_{(i,j)}^{(n,a,b)}$ the length of an (i,j)−busy-period of an $M^X/G_1 - G_2/1/(n, a, b)$ queuing system. Specifically, an (i,j)−busy-period is a period that starts at an arrival instant that makes the system stay with i customers and operating in phase j, with a customer initiating service after that arrival instant, and ends at the subsequent epoch at which the system becomes empty.

We start by considering oscillating systems with lower barrier equal to zero, $M^X/G_1 - G_2/1/(n, 0, b)$. We note that, for such systems, when a busy period starts with the system operating in phase 2, the system remains in phase 2 during the entire busy period and its duration does not depend on the value of the upper barrier, b. As a consequence, and taking into account that an $M^X/G_1 - G_2/1/(n, 0, 1)$ system behaves as a regular $M^X/G_2/1/n$ system, it follows that

$$B_{(i,2)}^{(n,0,b)} =_{\mathrm{d}} B_{(i,2)}^{(n,0,1)} =_{\mathrm{d}} B_i^{(n)}, \quad 1 \le b \le i \le n, \tag{1}$$

with $=_{\mathrm{d}}$ denoting the equality in distribution and $B_i^{(n)}$ the length of a busy period initiated with i customers in the regular $M^X/G/1/n$ system with service time distribution A_2. Therefore, the mean length of an $(i,2)$−busy-period in a $M^X/G_1 - G_2/1/(n, 0, b)$ is obtained from

$$E[B_{(i,2)}^{(n,0,b)}] = E[B_{(i,2)}^{(n,0,1)}] = E[B_i^{(n)}], \quad 1 \le b \le i \le n, \tag{2}$$

using the procedures derived in Pacheco and Ribeiro [6] to compute $E[B_i^{(n)}]$.

Moreover, if the busy period starts in phase 1, the mean length of the $(i,1)$−busy-period $(i < b)$ can be obtained from Theorem 1, presented below, by conditioning on the number of customers that arrive to the system during the service of the first customer served in such $(i,1)$−busy-period.

Theorem 1. *The mean length of an $(i,1)$−busy-period of an $M^X/G_1 - G_2/1/(n, 0, b)$ system, with $b > 1$, is such that*

$$E[B_{(i,1)}^{(n,0,b)}] = \xi_i - \tau_i \frac{\xi_0}{\tau_0}, \quad 0 \le i \le b-1, \tag{3}$$

with $(\xi_{b-1}, \tau_{b-1}) = (0, 1)$ and (ξ_{i-1}, τ_{i-1}), $1 \le i \le b-1$, defined recursively for $i = b-1, b-2, \ldots, 1$, by

$$\xi_{i-1} = \frac{\xi_i - \sum_{l=1}^{b-i-1} r_l \, \xi_{l+i-1} - \zeta_i}{r_0} \tag{4}$$

$$\tau_{i-1} = \frac{\tau_i - \sum_{l=1}^{b-i-1} r_l \, \tau_{l+i-1}}{r_0} \tag{5}$$

with

$$\zeta_i = E[S_1] + \sum_{l=b-i}^{n-i-1} r_l\, E[B_{(l+i-1,2)}^{(n,0,b)}] + \left(1 - \sum_{l=0}^{n-i-1} r_l\right) E[B_{(n-1,2)}^{(n,0,b)}] \qquad (6)$$

and r_l denoting the probability that l customers arrive to the system during a period whose duration has distribution function A_1.

Proof. Let C denote the number of customer arrivals during the service of the first customer served in the $(i,1)$−busy-period. By conditioning on the number of customer arrivals during that service, the mean length of an $(i,1)$−busy-period, $1 \le i < b$, verifies

$$[B_{(i,1)}^{(n,0,b)}|C=l]=_d \begin{cases} (S_1|C=l) \oplus B_{(l+i-1,1)}^{(n,0,b)} & 0 \le l \le b-i-1 \\ (S_1|C=l) \oplus B_{(l+i-1,2)}^{(n,0,b)} & b-i \le l \le n-i-1 \\ (S_1|C=l) \oplus B_{(n-1,2)}^{(n,0,b)} & l \ge n-i \end{cases} \qquad (7)$$

where $\oplus$ denotes the addition of independent random variables. As a result,

$$\begin{aligned} E[B_{(i,1)}^{(n,0,b)}] = & \sum_{l=0}^{b-i-1} r_l\, E\left[(S_1|C=l) \oplus B_{(l+i-1,1)}^{(n,0,b)}\right] \\ & + \sum_{l=b-i}^{n-i-1} r_l\, E\left[(S_1|C=l) \oplus B_{(l+i-1,2)}^{(n,0,b)}\right] \\ & + \sum_{l \ge n-i} r_l\, E\left[(S_1|C=l) \oplus B_{(n-1,2)}^{(n,0,b)}\right] \end{aligned} \qquad (8)$$

and, by separating the term $l = 0$ in the previous equation from the remaining terms, it follows that

$$r_0\, E[B_{(i-1,1)}^{(n,0,b)}] = E\left[B_{(i,1)}^{(n,0,b)}\right] - \sum_{l=1}^{b-i-1} r_l\, E\left[B_{(l+i-1,1)}^{(n,0,b)}\right] - \zeta_i, \quad 1 \le i < b, \qquad (9)$$

with ζ_i defined in (6), since $E[S_1] = \sum_{l=0}^{\infty} r_l E\left[S_1|C=l\right]$.

The expression (9) allows us to express the mean length of a $(j,1)$−busy-period as a function of the mean length of a $(b-1,1)$−busy-period, in the form

$$E[B_{(j,1)}^{(n,0,b)}] = \xi_j + \tau_j E[B_{(b-1,1)}^{(n,0,b)}], \quad 0 \le j \le b-1, \qquad (10)$$

with $(\xi_{b-1}, \tau_{b-1}) = (0,1)$ and ξ_{i-1} and τ_{i-1}, $1 \le i \le b-1$, defined recursively for $i = b-1, b-2, \dots, 1$ by (4) and (5), respectively. We will prove the previous statement using backward induction and resorting to (9). As (10) trivially holds for $j = b-1$ with $(\xi_{b-1}, \tau_{b-1}) = (0,1)$, it remains to show that (10) holds for $j = i-1$, with $1 \le i \le b-1$ and ξ_{i-1} and τ_{i-1} defined by (4) and (5), respectively, provided that (10) holds for $j = i, i+1, \ldots, b-1$, with ξ_j and τ_j satisfying (4) and (5), respectively, a statement which follows directly from (9) and (10) by substituting in (9) $E[B_{(j,1)}^{(n,0,b)}]$, $j = i, i+1, \ldots, b-1$, by the equivalent expression [in view of (10)] $\xi_j + \tau_j E[B_{(b-1,1)}^{(n,0,b)}]$.

In view of the previous, to conclude the proof of the theorem it suffices to show that $E[B_{(b-1,1)}^{(n,0,b)}] = -\frac{\xi_0}{\tau_0}$, which results simply by taking into account the boundary condition $0 = E[B_{(0,1)}^{(n,0,b)}] = \xi_0 + \tau_0 E[B_{(b-1,1)}^{(n,0,b)}]$.

We now address the computation of the mean length of busy periods of oscillating systems with positive lower barrier, i.e., an $M^X/G_1-G_2/1/(n,a,b)$ system with $a > 0$. We resort to the Markov regenerative structure of oscillating systems to relate the length of an $(i,j)-$busy-period, in systems with fixed capacity and barriers, with the length of busy periods of systems with smaller or equal capacity and barriers, and initiated with fewer customer.

In fact, supposing that we take out of consideration one of the customers initially present in the $M^X/G_1-G_2/1/(n,a,b)$ system and assuming that such a customer will start being served only when being alone in the system, the Markov regenerative property guarantees that in any $(i,j)-$busy-period starting at time zero $(X(0) = (i,j), X(0-) \neq (i,j)$, with $i > 1)$ the amount of time the system takes to reach state $(1,1)$ from state (i,j) has the same distribution as the duration of an $(i-1,j)-$busy-period of an $M^X/G_1 - G_2/1/(n-1,a-1,b-1)$ system. As a consequence, for $(i,j) \in E^n \backslash \{(0,1),(1,1)\}$ and $a \geq 1$,

$$B_{(i,j)}^{(n,a,b)} =_{\mathrm{d}} B_{(i-1,j)}^{(n-1,a-1,b-1)} \oplus B_{(1,1)}^{(n,a,b)}. \tag{11}$$

By conditioning on the number of customers that arrive to the system during the service of the first customer served in the busy period, Theorem 2, presented below, establishes relations that can be used to construct a recursive procedure [on the system capacity and barriers] to compute the mean length of an $(i,j)-$busy-period in $M^X/G_1 - G_2/1/(n,a,b)$ systems, with $0 < a < n-1$.

Theorem 2. *The mean length of an $(i,j)-$busy-period of an $M^X/G_1-G_2/1/(n, a,b)$ system, with $0 < a < n-1$, is such that, for $(i,j) \in E^{(n,a,b)} \backslash \{(0,1),(1,1)\}$,*

$$E\left[B_{(i,j)}^{(n,a,b)}\right] = E\left[B_{(i-1,j)}^{(n-1,a-1,b-1)}\right] + E\left[B_{(1,1)}^{(n,a,b)}\right] \tag{12}$$

and

$$r_0\, E\left[B_{(1,1)}^{(n,a,b)}\right] = E\left[S_1\right] + \sum_{l=1}^{b-2} r_l\, E\left[B_{(l-1,1)}^{(n-1,a-1,b-1)}\right] + r_{b-1}\, E\left[B_{(b-2,1+1_{\{a<b-1\}})}^{(n-1,a-1,b-1)}\right]$$
$$+ \sum_{l=b}^{n-2} r_l\, E\left[B_{(l-1,2)}^{(n-1,a-1,b-1)}\right] + \left(1 - \sum_{l=0}^{n-2} r_l\right) E\left[B_{(n-2,2)}^{(n-1,a-1,b-1)}\right] \tag{13}$$

with $1_{\{z\}}$ denoting the indicator function of the statement z.

Proof. We begin the proof by noting that (12) results from the linearity of the expected value operator applied to (11). Conditioning on the number of customer arrivals during the service of the first customer served in the busy period (C), we get

$$[B^{(n,a,b)}_{(1,1)}|C=l] \overset{.}{=}_d \begin{cases} (S_1|C=l) & l=0 \\ (S_1|C=l) \oplus B^{(n,a,b)}_{(l,1)} & 1 \le l \le b-2 \\ (S_1|C=l) \oplus B^{(n,a,b)}_{(l,1+1_{\{a<b-1\}})} & l=b-1 \\ (S_1|C=l) \oplus B^{(n,a,b)}_{(l,2)} & b \le l \le n-2 \\ (S_1|C=l) \oplus B^{(n,a,b)}_{(n-1,2)} & l \ge n-1 \end{cases}. \tag{14}$$

Thus, from the total probability law,

$$\begin{aligned} E\left[B^{(n,a,b)}_{(1,1)}\right] &= r_0 E\left[S_1|C=0\right] + \sum_{l=1}^{b-2} r_l E\left[(S_1|C=l) \oplus B^{(n,a,b)}_{(l,1)}\right] \\ &+ r_{b-1} E\left[(S_1|C=b-1) \oplus B^{(n,a,b)}_{(b-1,1+1_{\{a<b-1\}})}\right] \\ &+ \sum_{l=b}^{n-2} r_l E\left[(S_1|C=l) \oplus B^{(n,a,b)}_{(l,2)}\right] \\ &+ \sum_{l \ge n-1} r_l E\left[(S_1|C=l) \oplus B^{(n,a,b)}_{(n-1,2)}\right] \end{aligned} \tag{15}$$

which, from the linearity of the expected value and the fact that $E[S_1] = \sum_{l=0}^{\infty} r_l E$
$[S_1|C=l]$, leads to

$$\begin{aligned} E\left[B^{(n,a,b)}_{(1,1)}\right] &= E[S_1] + \sum_{l=1}^{b-2} r_l E\left[B^{(n,a,b)}_{(l,1)}\right] + r_{b-1} E\left[B^{(n,a,b)}_{(b-1,1+1_{\{a<b-1\}})}\right] \\ &+ \sum_{l=b}^{n-2} r_l E\left[B^{(n,a,b)}_{(l,2)}\right] + \sum_{l \ge n-1} r_l E\left[B^{(n,a,b)}_{(n-1,2)}\right]. \end{aligned} \tag{16}$$

Now, by using (12), the mean length of an $(1,1)-$busy-period is such that

$$\begin{aligned} E\left[B^{(n,a,b)}_{(1,1)}\right] &= E[S_1] + \sum_{l=1}^{b-2} r_l E\left[B^{(n-1,a-1,b-1)}_{(l-1,1)} \oplus B^{(n,a,b)}_{(1,1)}\right] \\ &+ r_{b-1} E\left[B^{(n-1,a-1,b-1)}_{(b-2,1+1_{\{a<b-1\}})} \oplus B^{(n,a,b)}_{(1,1)}\right] \\ &+ \sum_{l=b}^{n-2} r_l E\left[B^{(n-1,a-1,b-1)}_{(l-1,2)} \oplus B^{(n,a,b)}_{(1,1)}\right] \\ &+ \sum_{l \ge n-1} r_l E\left[B^{(n-1,a-1,b-1)}_{(n-2,2)} \oplus B^{(n,a,b)}_{(1,1)}\right] \end{aligned} \tag{17}$$

from which, by the linearity of the expected value and since $1 - \sum_{l \ge 1} r_l = r_0$, (13) results.

Based on the previous results, for (n, a, b) and (i, j) fixed, with $0 \leq a < b \leq n$ and $a < n - 1$, we compute $E[B_{(i,j)}^{(n,a,b)}]$, $(i, j) \in E^{(n,a,b)}$, using the following procedure:

1. Compute $E[B_{(i,2)}^{(n-a,0,b-a)}]$, $b - a \leq i \leq n - a$, using the procedures derived in Pacheco and Ribeiro [6] to compute the equivalent expression $E[B_i^{(n-a)}]$ for a regular system with service distribution A_2.

2. Compute $E[B_{(i,1)}^{(n-a,0,b-a)}]$, $1 \leq i < b - a$, using the procedure described in Theorem 1.

3. For $k = 1, 2, \ldots, a$, compute $E[B_{(i,j)}^{(n-a+k,k,b-a+k)}]$, $(i, j) \in E^{(n-a+k,k,b-a+k)}$, using the procedure described in Theorem 2.

Steps 1 and 2 address the computation of the mean length of busy periods in the system $M^X/G_1 - G_2/1/(n - a, 0, b - a)$ with lower barrier equal to zero, first for busy periods initiated in phase 2 in Step 1, followed by busy periods initiated in phase 1 in Step 2. In Step 3, the mean length of busy periods in the system $M^X/G_1 - G_2/1/(n - a + k, k, b - a + k)$ are computed using the procedure described in Theorem 2, with k taking successively the values $1, 2, \ldots, a$. As such, at the end of the recursion in Step 3, the mean length of busy periods in the system $M^X/G_1 - G_2/1/(n, a, b)$ are obtained.

Finally, we address the mean length of busy periods of oscillating systems with lower barrier $a = b - 1 = n - 1$, $M^X/G_1 - G_2/1/(n, n - 1, n)$ systems. In such queues, whenever a busy period starts operating in phase 1, it remains operating in such phase during the complete duration of the busy period. Therefore, for $1 \leq i < n$, $B_{(i,1)}^{(n,n-1,n)} =_d B_i^{(n)}$ and $E[B_{(i,1)}^{(n,n-1,n)}] = E[B_i^{(n)}]$ where $B_i^{(n)}$ denotes the mean length of a busy period initiated with i customers in the $M^X/G/1/n$ system with service time distribution A_1. In addition, when the busy period starts operating in phase 2, which occurs only when $l \geq n$ customers arrive to the empty system,

$$E[B_{(n,2)}^{(n,n-1,n)}] = E\left[S_2 \oplus B_{(n-1,1)}^{(n,n-1,n)}\right] = E[S_2] + E\left[B_{n-1}^{(n)}\right].$$

3   Blocking Probability

In this section we combine results from the previous section with earlier results from Ferreira et al. [3] for the mean number of customer losses in busy periods, in order to analyse the long-run blocking probability and loss rates in $M^X/G_1 - G_2/1/n$ nonpreemptive systems.

Taking advantage of the Markov regenerative structure of such systems, the long-run blocking probability, P_b, is computed from

$$P_b = \frac{T^n}{\lambda \bar{f}} \tag{18}$$

where λ denotes the batch arrival rate of customers, $\bar{f}$ the mean batch size, and T^n the long-run customer loss rate.

Note that, as $T^{\mathbf{n}}$ equals the ratio between the mean number of customer losses in a general busy cycle (or busy period), $E[L^{\mathbf{n}}]$, and the mean length of a general busy cycle,

$$T^{\mathbf{n}} = \frac{E[L^{\mathbf{n}}]}{E[B^{\mathbf{n}}] + \lambda^{-1}} \tag{19}$$

with $E[B^{\mathbf{n}}]$ denoting the mean length of a general busy period. Here, by general busy period (busy cycle) we mean a busy period (busy cycle) that starts with an arbitrary number of customers in the system with the distribution of a customer batch size.

Capitalizing on the results derived in the previous section, we compute the mean length of a general busy period by conditioning on the number of customers present in the batch that starts the busy period, obtaining

$$E[B^{\mathbf{n}}] = \sum_{i=1}^{b-1} f_i \, E[B_{(i,1)}^{\mathbf{n}}] + \sum_{i=b}^{n} f_i \, E[B_{(i,2)}^{\mathbf{n}}] + E[B_{(n,2)}^{\mathbf{n}}] \left(1 - \sum_{i=1}^{n} f_i\right). \tag{20}$$

Similarly, using the results derived in [3] for the mean number of customer losses in (i,j)−busy-periods, the mean number of customer losses during a general busy period is obtained from

$$E[L^{\mathbf{n}}] = \sum_{i=1}^{b-1} f_i \, E[L_{(i,1)}^{\mathbf{n}}] + \sum_{i=b}^{n} f_i \, E[L_{(i,2)}^{\mathbf{n}}] + \sum_{i \geq n+1} f_i \left(i - n + E[L_{(n,2)}^{\mathbf{n}}]\right)$$

with $L_{(i,j)}^{\mathbf{n}}$ denoting the number of customer losses in an (i,j)−busy-period, or, equivalently,

$$E[L^{\mathbf{n}}] = \sum_{i=1}^{b-1} f_i \, E[L_{(i,1)}^{\mathbf{n}}] + \sum_{i=b}^{n} f_i \, E[L_{(i,2)}^{\mathbf{n}}] \\
+ \left(1 - \sum_{i=1}^{n} f_i\right) \left(E[L_{(n,2)}^{\mathbf{n}}] - n\right) + \bar{f} - \sum_{i=1}^{n} i f_i. \tag{21}$$

Note that, similarly to the expression for the long-run customer loss rate given in (19), the (long-run) customer loss rate in (i,j)-busy-cycles is simply given by

$$\frac{E[L_{(i,j)}^{\mathbf{n}}]}{E[B_{(i,j)}^{\mathbf{n}}] + \lambda^{-1}}. \tag{22}$$

4   Numerical Illustration

In this section we compute mean lengths of busy periods and the long-run blocking probability for $M^X/G(\mu_1) - G(\mu_2)/1/\mathbf{n}$ systems, illustrating the sensitivity of such measures with respect to different traffic intensities and different service distributions and group sizes.

To this purpose, several different batch size and service time distributions have been considered. Specifically, it is assumed that customers arrive in batches with the following size distributions, with common mean $\bar{f}$: $D(\bar{f})$ – deterministic with constant value $\bar{f}$; $Geo(1/\bar{f})$ – geometric with success probability $1/\bar{f}$; $1 + B(c, (\bar{f}-1)/c)$ – shifted binomial, a binomial with c trials and success probability $(\bar{f}-1)/c$, added of one unit; and $U\{1, \cdots, 2\bar{f}-1\}$ – discrete uniform on the set $\{1, 2, \cdots, 2\bar{f}-1\}$. We also consider the following service time distributions, with mean μ^{-1}: $M(\mu)$ – exponential with rate μ; $U(0, 2/\mu)$ – uniform on the interval $(0, 2/\mu)$; $GP(\kappa, \theta, \beta)$ – generalized Pareto with parameters κ, θ and β, with $\theta = (\kappa - 1)/(\mu\beta)$; and $SP(\kappa, \theta)$ – shifted Pareto with parameters κ and θ, with $\theta = (\kappa - 1)/\kappa\mu$.

The results derived in the previous sections and the recursion proposed in [3] to compute the mean number of customer losses in busy periods have been computed with algorithms implemented in MATLAB.

Figure 1 illustrates the evolution of the mean length of busy periods as a function of the traffic intensity, for several service time and batch size distributions. As expected, the curves displayed show that the mean length of busy periods increases as the traffic intensity increases, with the rate of increase being small for small traffic intensities and turning progressively larger for larger traffic intensities. For small traffic intensities, the mean length of busy periods is similar for systems with the service time and batch size distributions considered. Differences on the values of the mean length of busy periods become larger for larger traffic intensities.

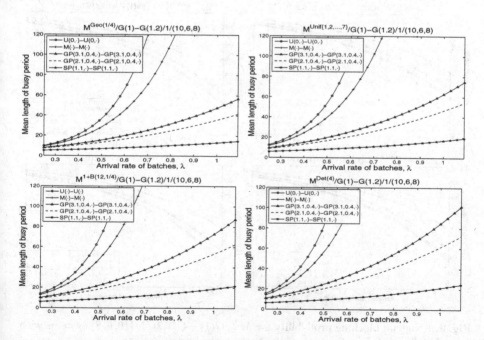

Fig. 1. Mean length of busy periods of $M^X/G(1) - G(1.2)/1/(10, 6, 8)$ systems with mean batch size 4, as a function of the batch arrival rate.

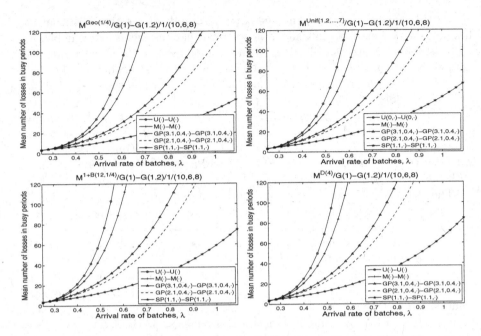

Fig. 2. Mean number of customer losses in $M^X/G(1) - G(1.2)/1/(10,6,8)$ systems with mean batch size 4, as a function of batch arrival rate.

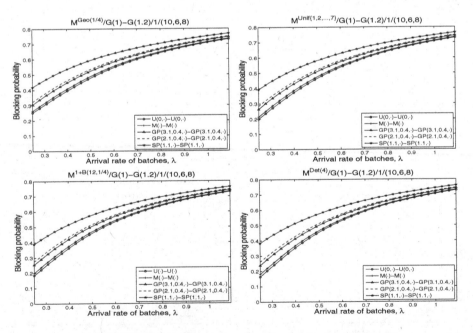

Fig. 3. Long-run blocking probability for $M^X/G(1) - G(1.2)/1/(10,6,8)$ systems with mean batch size 4, as a function of the batch arrival rate.

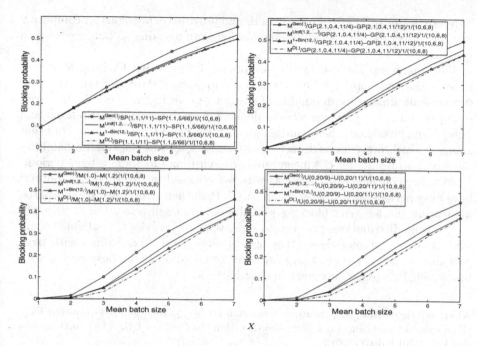

Fig. 4. Long-run blocking probability for $M^X/G(1) - G(1.2)/1/(10,6,8)$ systems with arrival rate 0.20, as a function of the mean batch size.

We have also observed that systems with heavy tailed service time distributions ($SP(1.1, \cdot)$, $GP(2.1, 0.4, \cdot)$, and $GP(3.1, 0.4, \cdot)$) and batch size distributions with greater variability ($Geo(1/2)$) have smaller mean length of busy periods. Moreover, as pointed out in [3] (cf. Fig. 2), the evolution of the mean number of customer losses in a busy cycle as a function of the batch arrival rate follows the same trend as the mean length of busy periods.

Curves for long-run customer blocking probabilities, which can be computed using (18), are given in Figs. 3 and 4. As can be observed in Fig. 2, systems with heavy tailed service time distributions are the ones that present larger mean number of customer losses in a busy period – as well as smaller mean duration of busy periods. We can also observe that long-run blocking probabilities are quite sensitive to the service time distribution, but less sensitive to the batch size distribution for fixed mean batch size.

5 Conclusions

In this work we have investigated nonpreemptive oscillating queueing systems, which are systems for which the server changes the service type (service time distribution or service rate) only at the beginning of service, depending on the number of customers waiting in the queue. We derived a recursive procedure to compute the mean length of busy periods and provided expressions for computing customer loss rates in busy cycles and the long-run blocking probability for such systems.

The analysis performed by the authors in [3] found that for large load traffic scenarios, systems with heavy tailed service time distributions are the ones experiencing smaller mean number of customer losses in busy periods. However, as illustrated in the previous section, depending on the service distributions and group size considered, the mean length of busy periods varies in a significant manner, which prevents to draw conclusions for long-run blocking probabilities directly from results for the mean number of customer losses in a busy period.

We have concluded that systems with heavy tailed service time distributions have busy periods with smaller mean length than their light tailed counterparts, influencing the long-run blocking probability. Although heavy tailed customer service time distributions give rise to some very long service times, which could lead to many customer losses, they also produce short service times with large probability and that could lead to a faster termination of the busy period with heavy tailed customer service time distributions.

Acknowledgments. This research was financed by Portuguese funds through FCT (Fundação para a Ciência e a Tecnologia), within the Projects UID/MAT/00013/2013 and UID/Multi/04621/2013.

References

1. Chydzinski, A.: The $M/G-G/1$ oscillating queueing system. Queueing Syst. **42**(3), 255–268 (2002)
2. Chydzinski, A.: The oscillating queue with finite buffer. Perform. Eval. **57**(3), 341–355 (2004)
3. Ferreira, F., Pacheco, A., Ribeiro, H.: Moments of losses during busy-periods of regular and nonpreemptive oscillating $M^X/G/1/n$ systems. Annals of Operations Research (2016, in press)
4. Kulkarni, V.G.: Modeling and Analysis of Stochastic Systems. Chapman and Hall, London (1995)
5. Pacheco, A., Ribeiro, H.: Consecutive customer losses in regular and oscillating $M^X/G/1/n$ systems. Queueing Syst. **58**(2), 121–136 (2008)
6. Pacheco, A., Ribeiro, H.: Moments of the duration of busy-periods of $M^X/G/1/n$ systems. Probab. Eng. Inf. Sci. **22**, 1–8 (2008)

Analysis of a Two-Class Priority Queue with Correlated Arrivals from Another Node

Abdulfetah Khalid[✉], Sofian De Clercq, Bart Steyaert, and Joris Walraevens

Department of Telecommunications and Information Processing (EA07),
Ghent University - UGent, Sint-Pietersnieuwstraat 41, B-9000 Gent, Belgium
{abdulfetahhadi.khalid,sofian.declercq,
bart.steyaert,joris.walraevens}@UGent.be

Abstract. Exact analysis of tandem priority queues is a difficult problem. In this paper, we model the output process of the first stage as a three-state Markov chain and analyze the second stage. The arrival process of this second stage is the superposition of this output process and an uncorrelated arrival process. We calculate the joint probability generating function of the number of high- and low-priority packets in the second stage and show that two implicit functions appear in this expression. We demonstrate how to deal with these implicitly defined functions in the calculation of moments.

Keywords: Priority queue · Generating functions · Markov chain · Correlated arrivals

1 Introduction

HOL (Head-of-the-Line) priority scheduling is one of the main scheduling types in network buffers to diversify the delays of traffic streams with different delay requirements [5]. When delay-sensitive high-priority packets (packets of voice and video streams, interactive gaming ...) are present in the buffer, they are transmitted. Best-effort low-priority packets can thus only be transmitted when no high-priority traffic is present.

Isolated priority queues with uncorrelated arrival processes have been studied abundantly in the past [6–8,11]. Analysis of networks of priority queues is much more complicated. Even the simplest network, a tandem queue, combined with priority has not been solved yet. Therefore, we attempt another approach for the analysis of such a tandem priority queue. We consider one of the simplest settings, namely a discrete-time tandem priority queueing system with constant (single-slot) service times, uncorrelated arrivals of both classes to the two stages, and single servers in both stages. The idea is to approximate the output process of the first stage of the tandem queue, use this output process as input (arrival) process of the second stage, and analyze the second stage.

The first step was already taken in a previous paper [9]. We modeled the output process of a discrete-time priority queue with single-slot service times as

S. Wittevrongel and T. Phung-Duc (Eds.): ASMTA 2016, LNCS 9845, pp. 167–178, 2016.
DOI: 10.1007/978-3-319-43904-4_12

a three-state discrete-time Markov chain, with the three states representing 'no output', 'departure of a high-priority packet' and 'departure of a low-priority packet' respectively. Transition probabilities were calculated such that the mean residence times in these states correspond to mean idle and busy periods of both classes.

In this paper, we take the next step. We analyze the number of high- and low-priority packets in a priority queue with an arrival process which is the superposition of an uncorrelated arrival process (coined *external arrivals* in the remainder) and a correlated arrival process as discussed in the previous paragraph (*internal arrivals*). We show that two implicitly defined functions appear in the final expression for the joint probability generating function of the stationary number of packets of both priorities. We also demonstrate that we need to calculate one constant numerically for the calculation of the moments of the number of packets of both priorities in the system. For reference, in the same queue without the second arrival process, only one implicitly defined function is encountered and expressions of the moments of the number of packets in the system can be found completely explicitly in terms of the parameters of the arrival process [10].

Other studies of discrete-time queues with correlated arrivals have been published. Khamisy and Sidi [4] study a priority queue with a correlated arrival process according to a two-state Markovian batch arrival process. Our model is in a way simpler, since the time-correlation in our model only applies at maximum one arrival. On the other hand, we will need a three-state Markov chain instead of two states in [4]. In [2], authors analysed a queue with an arrival process that is a superposition of on-off sources. The numbers of class-1 and class-2 arrivals however are independent, which is not the case in our model. An entirely different approach to correlated arrivals are train arrivals [3,12].

In the next section, we specify the model and notations. In Sect. 3, we calculate the joint probability generating function of the number of low- and high-priority packets in the system and discuss some numerical aspects. Finally, we demonstrate the formulas by means of some numerical examples in Sect. 4 and conclude the paper.

2   Mathematical Model

We consider a single-server queue with infinite buffer space. Time is assumed to be slotted. Packets of two classes arrive to the system, namely packets of class 1 and packets of class 2. The packet stream is furthermore a superposition of two arrival streams.

We denote the number of arrivals of type j of the first arrival stream (external arrivals) during slot k by $a_{j,k}$ ($j = 1, 2$). The numbers of arrivals in this stream are assumed to be i.i.d. from slot to slot and are characterized by the (common) joint probability generating function (PGF)

$$A(z_1, z_2) \triangleq \mathrm{E}\left[z_1^{a_{1,k}} z_2^{a_{2,k}}\right].$$

The marginal PGFs of the numbers of per-slot arrivals of class 1 and class 2 are denoted and given by $A_1(z) = A(z, 1)$ and $A_2(z) = A(1, z)$ respectively. The arrival rate $\lambda_{j,\text{ext}}$ of class-j packets ($j = 1, 2$) is given by $A'_j(1)$.

A second arrival process (internal arrivals) is governed by a three-state discrete-time Markov chain with state space $\{0, 1, 2\}$. We define t_k as the state of the Markov chain in slot k. When the Markov chain is in state 0 no packets arrive in that slot. When the Markov chain is in state j (1 or 2), exactly one class-j packet arrives. We define the transition probabilities as

$$e_i(j) \triangleq \Pr[t_{k+1} = j | t_k = i],$$

$i, j = 0, 1, 2$. We summarize the total arrival process in the generating function matrix $B(z_1, z_2)$:

$$B(z_1, z_2) \triangleq \left[\mathrm{E} \left[z_1^{a_{1,k} + \mathbb{1}_{\{t_k=1\}}} z_2^{a_{2,k} + \mathbb{1}_{\{t_k=2\}}} \mathbb{1}_{\{t_k=j\}} | t_{k-1} = i \right] \right]_{i,j}$$

$$= A(z_1, z_2) \begin{pmatrix} e_0(0) & e_0(1)z_1 & e_0(2)z_2 \\ e_1(0) & e_1(1)z_1 & e_1(2)z_2 \\ e_2(0) & e_2(1)z_1 & e_2(2)z_2 \end{pmatrix}, \tag{1}$$

with $\mathbb{1}_{\{X\}}$ the indicator function of X. The mean internal arrival rates of class j equals $\lambda_{j,\text{int}} = e(j)$, with $e(j)$ the stationary probability that the Markov chain is in state j.

We note here in passing that the case $e_0(1) = e_2(1)$ is an important special case, as it is the case for our motivating example (cf. [9]) and it will lead to some (numerical) simplifications.

We define λ_1 and λ_2 as the total arrival rate of packets of class 1 and 2 respectively, i.e., $\lambda_j \triangleq \lambda_{j,\text{int}} + \lambda_{j,\text{ext}}$, $j = 1, 2$. Similarly we define λ_T as the overall arrival rate, i.e., $\lambda_T \triangleq \lambda_1 + \lambda_2$.

The server serves the packets at the rate of one packet per slot, i.e., all service times equal one slot. It is assumed that class-1 packets have service priority over class-2 packets and we adopt a FCFS discipline within a class.

3   Analysis

We denote the number of packets of class j in the system at the beginning of slot k by $u_{j,k}$ ($j = 1, 2$). The triplet $(t_{k-1}, u_{1,k}, u_{2,k})$ is a first-order Markov chain, and is therefore suitable to analyze the queueing system. We assume that $\lambda_T < 1$, such that this Markov chain reaches a steady state.

3.1   System Equations

Figure 1 shows the time axis for the buffer in stage 2, with t_{k-1} the class of internal packet arriving during slot $k - 1$ and $u_{j,k}$ the buffer occupancy of class

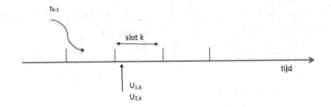

Fig. 1. Time axis for the buffer analysis in stage 2

j during slot k Note that $t_{k-1} = 0$ if there is no internal arrival. The buffer occupancy of both classes are characterized by the following system of equations.

$$u_{1,k+1} = [u_{1,k} - 1]^+ + a_{1,k} + \mathbb{1}_{\{t_k=1\}};$$
$$u_{2,k+1} = [u_{2,k} - \mathbb{1}_{\{u_{1,k}=0\}}]^+ + a_{2,k} + \mathbb{1}_{\{t_k=2\}}. \tag{2}$$

These equations are understood as follow: if $t_k = j$ ($j = 1, 2$) an extra packet of class j arrives. If $u_{1,k} > 0$, a class-1 packet is served because of the priority scheduling; otherwise a class-2 packet is served, if any.

3.2   Functional Equations

We define the partial joint generating functions

$$P_j(z_1, z_2) \triangleq \sum_{m=0}^{\infty} \sum_{n=0}^{\infty} p(j, m, n) z_1^m z_2^n, \quad j = 0, 1, 2,$$

with $p(j, m, n) = \lim_{k\to\infty} \Pr[t_{k-1} = j, u_{1,k} = m, u_{2,k} = n]$, and the row vector $\mathbf{P}(z_1, z_2) = [P_j(z_1, z_2)]_{j=0,1,2}$.

Transforming the system equations (2) to generating functions and letting $k \to \infty$, we find

$$\mathbf{P}(z_1, z_2) = \left[\frac{\mathbf{P}(z_1, z_2) - \mathbf{P}(0, z_2)}{z_1} + \frac{\mathbf{P}(0, z_2) - \mathbf{P}(0, 0)}{z_2} + \mathbf{P}(0, 0) \right] B(z_1, z_2).$$

Solving in $\mathbf{P}(z_1, z_2)$ yields

$$\mathbf{P}(z_1, z_2) = \left[\frac{z_1 - z_2}{z_2} \mathbf{P}(0, z_2) + \frac{z_1(z_2 - 1)}{z_2} \mathbf{P}(0, 0) \right] B(z_1, z_2)(z_1 I - B(z_1, z_2))^{-1}, \tag{3}$$

with I the 3×3 identity matrix.

3.3   Calculation of Unknowns

Expression (3) contains three unknown functions $P_0(0, z_2)$, $P_1(0, z_2)$ and $P_2(0, z_2)$ and three unknown constants $P_0(0, 0)$, $P_1(0, 0)$ and $P_2(0, 0)$. However,

$P_1(0, z_2) = P_2(z_1, 0) = 0$ since at least one class-j packet is present in the system if the state of the Markov chain was equal to j in the previous slot. This eliminates the unknown function $P_1(0, z_2)$ and the unknown parameters $P_1(0, 0)$ and $P_2(0, 0)$.

We now show how to calculate the other unknowns. It can be proved that $f(z_1) \triangleq \det(z_1 I - B(z_1, z_2))$ has three zeroes inside the complex unit disk for each z_2 inside the complex unit disk, see the appendix. One of them is 0. We denote the two other zeroes by $Y_1(z_2)$ and $Y_2(z_2)$. Since PGFs are analytic inside the complex unit disk, $Y_1(z_2)$ and $Y_2(z_2)$ are zeroes of the numerators of $P_j(z_1, z_2)$. Using the notation $C(z_1, z_2)$ for the cofactor matrix of $z_1 I - B(z_1, z_2)$, i.e., $C^T(z_1, z_2) = f(z_1)(z_1 I - B(z_1, z_2))^{-1}$, this leads to the vector equations

$$\left[(Y_i(z_2) - z_2)\mathbf{P}(0, z_2) + Y_i(z_2)(z_2 - 1)\mathbf{P}(0, 0) \right] B(Y_i(z_2), z_2) C^T(Y_i(z_2), z_2) = \mathbf{0},$$
(4)

for $i = 1, 2$.

Since $P_i(0, 0) = 0$, $i \neq 0$, we have that $P_0(0, 0)$ equals the probability that the system is empty and thus $P_0(0, 0) = 1 - \lambda_T$. This leaves the calculation of the unknown functions $P_0(0, z_2)$ and $P_2(0, z_2)$ only. It can be shown that the three scalar equations for each i in (4) are dependent. Hence we only use the first for our system of two equations ($i = 1, 2$):

$$Y_i(z_2) P_0(0, z_2) + X_i(z_2) P_2(0, z_2) = Y_i^2(z_2) \frac{1 - z_2}{Y_i(z_2) - z_2} P_0(0, 0),$$

where we used the notations

$$X_i(z_2) \triangleq A(Y_i(z_2), z_2) \sum_{j=0}^{2} \frac{C_{0j}(Y_i(z_2), z_2)}{C_{00}(Y_i(z_2), z_2)} e_2(j) z_j \Big|_{z_0=1,\, z_1=Y_i(z_2)},$$

$$C_{ij}(z_1, z_2) \triangleq \left(C(z_1, z_2) \right)_{ij},$$

and that $f(Y_i(z_2), z_2) = 0$ can be written as

$$A(Y_i(z_2), z_2) \sum_{j=0}^{2} C_{0j}(Y_i(z_2), z_2) e_0(j) z_j \Big|_{z_0=1,\, z_1=Y_i(z_2)} = Y_i(z_2) C_{00}(Y_i(z_2), z_2).$$

Thus we can solve for the two unknown functions, yielding

$$P_0(0, z_2) = (1 - z_2) \frac{\frac{Y_1^2(z_2) X_2(z_2)}{Y_1(z_2) - z_2} - \frac{Y_2^2(z_2) X_1(z_2)}{Y_2(z_2) - z_2}}{Y_1(z_2) X_2(z_2) - Y_2(z_2) X_1(z_2)} P_0(0, 0)$$
(5)

$$P_2(0, z_2) = (1 - z_2) \frac{\frac{Y_2^2(z_2) Y_1(z_2)}{Y_2(z_2) - z_2} - \frac{Y_1^2(z_2) Y_2(z_2)}{Y_1(z_2) - z_2}}{Y_1(z_2) X_2(z_2) - Y_2(z_2) X_1(z_2)} P_0(0, 0).$$
(6)

Not surprisingly these equations are symmetric in $Y_1(z_2)$ and $Y_2(z_2)$.

3.4 Results

All unknown partial PGFs and constants have been calculated and this results in expressions for $P_i(z_1, z_2)$, $i = 0, 1, 2$ that depend on $A(z_1, z_2)$, the transition probabilities $e_i(j)$ and the implicitly defined functions $Y_i(z_2)$, $i = 1, 2$. In general, the latter functions have to be calculated numerically for each z_2.

From the expressions of the $P_i(z_1, z_2)$, the joint generating function of the stationary buffer occupancies u_1 and u_2 of class 1 and class 2 respectively can be calculated as

$$U(z_1, z_2) \triangleq \mathrm{E}[z_1^{u_1} z_2^{u_2}]$$
$$= P_0(z_1, z_2) + P_1(z_1, z_2) + P_2(z_1, z_2).$$

From this joint PGF, marginal PGFs of the total, class-1 and class-2 buffer occupancies can be calculated as

$$U_T(z) \triangleq \mathrm{E}[z^{u_1 + u_2}] = U(z, z),$$
$$U_1(z) \triangleq \mathrm{E}[z^{u_1}] = U(z, 1),$$
$$U_2(z) \triangleq \mathrm{E}[z^{u_2}] = U(1, z).$$

Although the resulting expressions are too large to show, we can provide some insight.

The expression for $U_T(z)$ does not depend on the implicitly defined functions $Y_i(z)$, and is therefore explicitly known in terms of the arrival processes and parameters. This can also be understood from expression (3). By putting $z_1 = z_2$ in this expression, the unknown functions $P_i(0, z_2)$ disappear and it is these functions that introduced the implicitly defined functions in the final expressions. In fact, $U_T(z)$ is the PGF of the buffer occupancy in a single-class buffer with a superposition of an independent arrival process and a three-state Markov chain with 1 off-state and 2 on-states.

The expression for $U_1(z)$ does depend on the implicitly defined functions through the constants $Y_1(1)$ and $Y_2(1)$ only. This is logically sound as $Y_i(z_2)$ appears in the expression of $U(z_1, z_2)$ and we put $z_2 = 1$ to obtain an expression for $U_1(z)$. We can prove that one of these two constants equals 1 and the other does not (see Appendix). Assume for instance that $Y_1(1) = 1$. Then, $U_1(z)$ can be calculated completely in terms of the arrival processes and parameters apart from one constant that has to be calculated numerically ($Y_2(1)$). We note that $U_1(z)$ is, in fact, the PGF of the buffer occupancy in a single-class buffer with a superposition of an independent arrival process and a three-state Markov chain with 2 off-states and 1 on-state.

Finally, the expression for $U_2(z)$, which is of primary interest to us, contains both implicitly defined functions $Y_1(z)$ and $Y_2(z)$, as expected.

Moments of the total buffer occupancy and the buffer occupancies of class 1 and class 2 can be calculated by means of the moment generating property of PGFs, i.e., by taking derivatives in 1. The mean buffer occupancy, for instance, is calculated by taking the first derivative of the corresponding marginal PGF in 1.

As for the PGFs, the moments of the total buffer occupancy (the class-1 buffer occupancy respectively) are found in terms of the arrival parameters (the arrival parameters and $Y_2(1)$ respectively). The moments of the class-2 buffer occupancy are also found in terms of the arrival distributions and parameters, and $Y_2(1)$. The reason that no other constants but $Y_2(1)$ appear in this expression is that $Y_1(1) = 1$ and that the derivatives of $Y_i(z)$ in 1 can be expressed in terms of $Y_i(1)$. With some abuse of notation, we obtain, for instance,

$$Y_i'(1) = - \left. \frac{\frac{\partial f(z_1,z_2)}{\partial z_2}}{\frac{\partial f(z_1,z_2)}{\partial z_1}} \right|_{z_1=Y_i(1),z_2=1} ,$$

by considering f as a function of z_1 and z_2 (instead of z_1 only). We conclude that all moments of the buffer occupancies can be calculated explicitly in terms of the arrival distributions apart from one constant that has to be calculated numerically.

4 Numerical Results

We now demonstrate our results for some numerical examples.

We assume that the external (independent) arrival process is a two-class Bernoulli process, i.e.,

$$A(z_1, z_2) = (1 - \lambda_{1,\text{ext}} - \lambda_{2,\text{ext}}) + \lambda_{1,\text{ext}} z_1 + \lambda_{2,\text{ext}} z_2 .$$

In choosing this easy arrival process we end up with two similar Bernoulli arrival processes: the external one is a time-independent process and the internal one a time-dependent one. This will therefore demonstrate the effect of this time-dependence fairly.

The internal arrival process is characterized by the 9 transition probabilities $e_i(j)$, $i, j = 0, 1, 2$, of which 6 can be independently chosen ($\sum_{j=0}^{2} e_i(j) = 1$). We first assume $e_0(1) = e_2(1)$, cf. the discussion in Sect. 2. The advantage is that $Y_2(1)$ disappears from the expression of $E[u_1]$ since the 2 off-states can be merged to 1 off-state in this particular case. Therefore, we end up – for the class-1 buffer occupancy – with a single-class system with a superposition of an independent arrival process and a two-state Markov chain with 1 off-state and 1 on-state. Therefore, the mean total buffer occupancy and the mean class-1 buffer occupancy can be calculated explicitly. Since the mean class-2 buffer occupancy is the difference of these two, this measure can also be calculated explicitly in the case $e_0(1) = e_2(1)$. Again, we stress that this is an important special case.

We are then left with 5 independent parameters. To demonstrate our results more intuitively, we map these to 5 other parameters, namely to the traffic intensities of both classes $\lambda_{1,\text{int}}$ and $\lambda_{2,\text{int}}$ and to L_i, $i = 0, 1, 2$ defined as

$$L_i \triangleq \frac{1}{1 - e_i(i)}.$$

The L_i are in fact the mean residence times in state i (the residence time in state i is shifted geometrically distributed with parameter $e_i(i)$) and serve therefore as measures for the time-correlation: the larger L_i, the more slots on average the Markov chain resides in state i. The $\lambda_{i,\text{int}}$, on the other hand, are measures for the frequency of being in the different states.

A difficulty with applying this mapping is that not every 'logical' combination of $(\lambda_{1,\text{int}}, \lambda_{2,\text{int}}, L_0, L_1, L_2)$ leads to values in $[0,1]$ for each of the 9 transition probabilities. In the remainder, we filtered out the ones that do, but this restricts the range of values for the parameters somewhat. We focus on the influence of correlation in the arrival process.

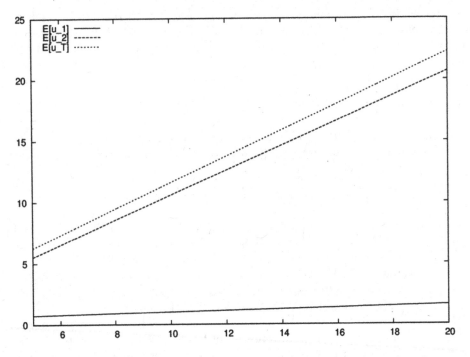

Fig. 2. Mean values of system contents versus the mean residence time L

In Fig. 2, we show the influence of the lengths L_i on the mean buffer occupancy. We depict the mean total class-1 and class-2 buffer occupancies as a function of L, with $L_i = L$, $i = 0, 1, 2$. We assume a perfect balance of class-1 and class-2 arrival rates, with $\lambda_{1,\text{int}} = \lambda_{2,\text{int}} = 0.3, \lambda_{1,\text{ext}} = \lambda_{2,\text{ext}} = 0.15$. We firstly observe that a larger mean residence time L in the states leads to higher mean buffer occupancies of both classes. The reason is that for long periods at least 1 packet per slot arrives (internal arrivals) while occasionally a second packet arrives in that slot (external arrivals). Therefore, the buffer occupancy can not decrease during these (long) periods. Secondly, it is clear that the low-priority class is affected much more. In fact, its slope is about the same as the slope of the mean total buffer occupancy. This is again explained by these long consecutive periods of arrivals: class-1 is only 'punished' by long periods of its own class while class 2 suffers from long periods of both (which can even directly follow eachother).

Figure 3 depicts the influence of the fraction of correlated and independent arrivals. The mean buffer occupancies are shown as function of $\lambda_{T,\text{int}}$ for $\lambda_T = 0.9$, $L_i = 5$ and perfect balance between class-1 and class-2 arrivals. We notice that time-correlation is not necessarily the dominant factor: time-correlation increases when more internal packets arrive, while the mean buffer occupancies decrease. The main reason is that the variance of the number of arrivals in a slot changes as well: when $\lambda_{T,\text{int}} = 0$ or $\lambda_{T,\text{ext}} = 0$ (the two extremes of the plot; the left extreme cannot be shown because of the remark above), at most one packet can arrive in the system per slot and no queueing happens. When $\lambda_{T,\text{int}}$ and $\lambda_{T,\text{ext}}$ are more balanced, the probability of two arrivals in a slot is significantly higher. This leads to a maximum of the buffer occupancies somewhere in the middle.

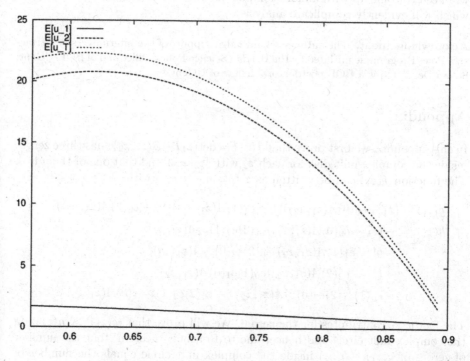

Fig. 3. Mean values of system contents versus $\lambda_{T,\text{int}}$

5   Conclusions and Future Work

We analysed a discrete-time two-class priority queue with an arrival process that is a superposition of an independent process and a time-dependent process. The latter is modeled by a discrete-time Markov chain with three states, no arrivals when in state 0 and one class-1 or class-2 arrivals when in state 1 or state 2 respectively. This model is motivated by a priority tandem queue, where our system can be regarded as the second stage of the tandem and the two arrival processes as models for external and internal (output from the first stage) arrivals from the network. We analysed the buffer occupancy in this system and demonstrated some results. We also identified an important special case that simplifies the analysis and numerical work.

In future work, we will analyse sojourn times. Note that we can define several different sojourn times, not only from different classes but also either of a packet of the external or the internal arrival stream. Furthermore, since the three-state Markov chain is only a very coarse model for the output process of the first stage, we need to test whether the results of our model are accurate approximations of the second stage performance measures of a priority tandem queue. If not, we will need to come up with a more complex model for the internal arrival process which will evidently complicate analysis.

Acknowledgement. The authors acknowledge support of the Interuniversity Attraction Poles Programme initiated by the Belgian Science Policy Office and of BOF-UGent. Sofian De Clercq is a BOF postdoctoral fellow of UGent.

Appendix

In this appendix, we first prove that $f(z_1) = \det(z_1 I - B(z_1, z_2))$ has three zeroes inside the complex unit disk for each z_2 with $|z_2| < 1$ and that one of them is 0. The function $f(z_1)$ can be written as $z_1(g(z_1) - h(z_1))$ with

$$g(z_1) = (z_1 - e_0(0)A(z_1, z_2))(1 - e_1(1)A(z_1, z_2))(z_1 - e_2(2)A(z_1, z_2)), \quad (7)$$

$$\begin{aligned} h(z_1) = &- (z_1 - e_0(0)A(z_1, z_2))e_1(2)e_2(1)z_2 A(z_1, z_2)^2 \\ &- (1 - e_1(1)A(z_1, z_2))e_0(2)e_2(0)z_2 A(z_1, z_2)^2 \\ &- (z_1 - e_2(2)A(z_1, z_2))e_0(1)e_1(0)A(z_1, z_2)^2 \\ &- e_0(1)e_1(2)e_2(0)z_2 A(z_1, z_2)^3 - e_0(2)e_2(1)z_2 e_1(0)A(z_1, z_2)^3. \end{aligned}$$

The factor z_1 contributes for the zero 0. We will prove that $|g(z_1)| > |h(z_1)|$ on the complex unit circle and hence, due to Rouché's theorem, that the number of zeroes of $g(z_1) - h(z_1)$ inside the complex unit circle equals the number of zeroes of $g(z_1)$. Finally, we prove that the latter has two zeroes, again by means of Rouché's theorem.

We have that

$$
\frac{|h(z_1)|}{|g(z_1)|} < \frac{e_1(2)e_2(1)|A(z_1,z_2)|^2}{|1 - e_1(1)A(z_1,z_2)||z_1 - e_2(2)A(z_1,z_2)|}
$$

$$
+ \frac{e_0(2)e_2(0)|A(z_1,z_2)|^2}{|z_1 - e_0(0)A(z_1,z_2)||z_1 - e_2(2)A(z_1,z_2)|}
$$

$$
+ \frac{e_0(1)e_1(0)|A(z_1,z_2)|^2}{|z_1 - e_0(0)A(z_1,z_2)||1 - e_1(1)A(z_1,z_2)|}
$$

$$
+ \frac{e_0(1)e_1(2)e_2(0)|A(z_1,z_2)|^3}{|z_1 - e_0(0)A(z_1,z_2)||1 - e_1(1)A(z_1,z_2)||z_1 - e_2(2)A(z_1,z_2)|}
$$

$$
+ \frac{e_0(2)e_2(1)e_1(0)|A(z_1,z_2)|^3}{|z_1 - e_0(0)A(z_1,z_2)||1 - e_1(1)A(z_1,z_2)||z_1 - e_2(2)A(z_1,z_2)|}. \tag{8}
$$

If we further use that, for $|z_1| = 1$, $|A(z_1,z_2)| \leq 1$, that $|z_1 - e_j(j)A(z_1,z_2)| \geq 1 - e_j(j)$ $(j = 0,2)$, that $1 - e_1(1)A(z_1,z_2) \geq 1 - e_1(1)$ and that $\sum_{j=1}^{2} e_i(j) = 1$, we find that the RHS of (8) is less than or equal to 1. Therefore, $|h(z_1)| < |g(z_1)|$ for $|z_1| = 1$ and the number of zeroes of $g(z_1) - h(z_1)$ in the complex unit circle equals the number of zeroes of $g(z_1)$. The latter function exists of three factors, see (7). Using $|z_1 - e_j(j)A(z_1,z_2)| \geq 1 - e_j(j)$ $(j = 0,2)$ and $1 - e_1(1)A(z_1,z_2) \geq 1 - e_1(1)$ once more, we find that each of the factors has as many zeroes inside the complex unit disk as its first terms. Therefore, $g(z_1)$ - and $g(z_1) - h(z_1)$ - has two zeroes.

Next, we prove that the limit for $z_2 \to 1$ of one of the two zeroes equals 1 while the other one does not. If we substitute z_2 by 1 in $f(z_1)$, Rouché's theorem can no longer be used directly, because $f(z_1)$ is not necessarily analytic for $z_2 = 1$ and $|z_1| = 1$. The result is however still valid, cf. [1], i.e., $f(z_1)$ has two zeroes inside and on the complex unit disk. One of them is equal to 1, since $f(1) = 0$ for $z_2 = 1$. Finally, it is easily proved that $f'(1) \neq 0$ for $z_2 = 1$ which means $z_1 = 1$ is a single zero of f for $z_2 = 1$. The other root is therefore different from 1 for $z_2 \to 1$. It will however lie in $]-1,1[$ which simplifies a numerical search.

References

1. Adan, I., Van Leeuwaarden, J.S.H., Winands, E.M.M.: On the application of Rouché's theorem in queueing theory. Oper. Res. Lett. **34**, 355–360 (2006)
2. Ali, M.M., Song, X.: A performance analysis of a discrete-time priority queueing system with correlated arrivals. Perform. Eval. **57**, 307–339 (2004)
3. Choi, B.D., Choi, D.I., Lee, Y., Sung, D.K.: Priority queueing system with fixed-length packet-train arrivals. IEE Proc.-Commun. **145**, 331–336 (1998)
4. Khamisy, A., Sidi, M.: Discrete-time priority queues with two-state Markov modulated arrivals. Stoch. Models **8**, 337–357 (1992)
5. Roberts, J.: Internet traffic, QoS, and pricing. Proc. IEEE **92**, 1389–1399 (2005)
6. Sleptchenko, A., Selen, J., Adan, I., van Houtum, G.J.: Joint queue length distribution of multi-class, single-server queues with preemptive priorities. QUESTA **81**, 379–395 (2015)
7. Takagi, H.: Queueing analysis: a foundation of performance evaluation. (1991–1993)

8. Tarabia, A.: Two-class priority queueing system with restricted number of priority customers. AEÜ **61**, 534–539 (2007)
9. Walraevens, J., Fiems, D., Wittevrongel, S., Bruneel, H.: Calculation of output characteristics of a priority queue through a busy period analysis. EJOR **198**, 891–898 (2009)
10. Walraevens, J., Steyaert, B., Bruneel, H.: Performance analysis of a single-server ATM queue with a priority scheduling. C&OR **30**, 1807–1829 (2003)
11. Walraevens, J., Steyaert, B., Bruneel, H.: Analysis of a discrete-time preemptive resume priority buffer. EJOR **186**, 182–201 (2008)
12. Walraevens, J., Wittevrongel, S., Bruneel, H.: A discrete-time priority queue with train arrivals. Stoch. Models **23**, 489–512 (2007)

Planning Inland Container Shipping: A Stochastic Assignment Problem

Kees Kooiman[1,2], Frank Phillipson[1(✉)], and Alex Sangers[1]

[1] TNO, The Hague, The Netherlands
frank.phillipson@tno.nl
[2] Erasmus University, Rotterdam, The Netherlands

Abstract. In this paper a case study is presented concerning the collection and delivery of containers between a container terminal and inland ports by barges making a round trip along these ports. The goal is to maximise the number containers arriving on time, transported by barge. To account for uncertain release times of the containers a simulation algorithm is used when making online decisions whether to assign a container to a barge or not. Using a realistic problem instance it is shown that the proposed simulation algorithm is the better performing one among several other benchmark solution methods.

Keywords: Inland container shipping · Stochastic assignment problem · Simulation algorithm · Multimodal transportation planning · Self-optimising

1 Introduction

The Netherlands consists of about 5,046 km of inland waterways, of which 4,800 km is accessible for goods transport. Rotterdam, one of the largest ports of the world, has central container terminals for transport from and to European cities inland. At the other side of these terminals large deep sea vessels come in and leave, transporting huge numbers of containers to and from other parts of the world. At inland ports containers are prepared to be transported by barges to these terminals. Barges arrive at the ports and containers are stacked inside as well as on the deck. The problem, inspired by the work of [1,2], elaborated in this paper is that of assigning containers to barges. In this problem we consider a container terminal from which containers have to be dispatched by barge to a series of inland ports. At the same time these barges collect other containers at these inland ports in order to deliver them at the container terminal. The barges make a round trip along a series of inland ports starting and ending at the container terminal. The barges have fixed capacity in terms of the number of containers they can carry. The round trip along the inland ports is repeated by a number of barges in a fixed schedule. A choice has to be made to which barge a container is assigned such that the maximum number of containers will be able to be delivered by barge. The release times of the containers are stochastic

© Springer International Publishing Switzerland 2016
S. Wittevrongel and T. Phung-Duc (Eds.): ASMTA 2016, LNCS 9845, pp. 179–192, 2016.
DOI: 10.1007/978-3-319-43904-4_13

both at the container terminal and at the inland ports. All containers have a (non-stochastic) due time, the time they have to be at their destination. If it is not possible to get the container at the destination by barge before the due time, it will be assigned to a truck. There is an infinite amount of trucks available at a higher price than assigning a container to a barge.

Not much attention is given to this problem in literature. Most literature concerns the empty container management (ECM) problem, see [11] for an overview. [1,2] do not incorporate uncertainty in their approach. [4] propose a DSS that facilitates the creation of schedules for barges by means of a heuristic approach. Our approach extends to their DSS. An other barge optimisation problem is the barge rotation planning, as presented in [3,5]. The paper of [10] give an overview of current topics and research opportunities in synchromodal container transportation. An important topic here is methods for real-time network planning. We elaborate on this topic and add uncertainty to it.

The rest of this paper is structured as follows. In Sect. 2 a detailed problem description of the Container collect & deliver problem is given. In Sect. 3 we discuss several solution methods. In Sect. 4 a case study is introduced and the results of the solution methods on this case study are shown in Sect. 5. At the end a conclusion is given including some directions for future research.

2   Container Collect and Deliver Problem

In this section we describe the Container collect & deliver problem in detail. First we will sketch the main difficulty in a short example.

2.1   Example

To illustrate the Container collect & deliver problem we first take an example of two ports. The first port has 10 containers with an early due time, and 10 containers with a late due time. All these containers will be available for transport upon arrival of the first barge. The second port has 10 containers with an early due time. The expected release times of these containers are such that part of these containers will be available for transport upon arrival of the first barge whereas the others will not. When we have a barge with a capacity of 20 containers starting at the container terminal and arriving at port one, we should definitely take the 10 containers with an early due time as they will not arrive on time with a later barge. At the time the barge loads in these 10 containers there are 10 spots left on the barge. The decision has to be made how many of these spots have to be reserved for containers on the second port, for which it is uncertain, though, whether they will be released on time or not. Ideally the number of empty spots equals the number of containers available for shipment on the second port upon arrival of the first barge. If too many empty spots are reserved the barge will not use its total capacity what could turnout to be sub optimal. If too few empty spots are reserved more containers will have to go by truck than necessary on the second port.

2.2 Problem Formulation

When the problem becomes bigger with due times fluctuating a lot and release times uncertain the problem gets more complicated. Taking an optimal decision can get very difficult. A mathematical model can be used as a decision support tool. To make a manageable model several assumptions need to be made:

Assumption 1. *The barges have a known fixed capacity.*

Assumption 2. *Containers can only be unloaded at the port of destination or at the container terminal at the end of the round trip.*

Assumption 3. *All containers have a known fixed due time and stochastic release times with a known probability distribution. The mean is different for each container but the standard deviation is equal for all containers.*

In our proposed solution method release times of the containers can have different distribution types and parameters. In this study we assume a normal distribution with equal standard deviation for the sake of keeping it simple. In real life application the choice of the distribution should ideally be based on historical data.

Assumption 4. *All containers are of equal type.*

Assumption 5. *The travel schedule of the barges is fixed and given and allows for the handling of containers.*

The container collect & deliver problem is summarised in the following IP formulation.

Sets:

$T :=$ set of time stamps indexed by t,

$T_i := T_i \subset T$, subset of time stamps t barge i is at a port,

$Z :=$ set of containers indexed by j,

$Z_C := Z_C \subset Z$, containers that need to be collected at the inland ports,

$Z_D := Z_D \subset Z$, containers that need to be delivered at the inland ports,

$N :=$ set of barges indexed by i.

Non-stochastic parameters:

$OQ_{ij} :=$ arrival time of barge i at port holding container $j \in Z_C$ and at destination port of container $j \in Z_D$,

$d_j :=$ due time of container $j \in Z_C$ at terminal and container $j \in Z_D$ at destination port,

$a_i :=$ arrival time at terminal of barge i,

$b_i :=$ departure time at terminal of barge i,

$c_i :=$ container capacity of barge i,

$$
\omega_{ijt} := \begin{cases} 1 & \text{if at time stamp } t \in T_i, \text{ barge } i \text{ has container } j \\ & \text{on deck (if container } j \text{ would be assigned to} \\ & \text{barge } i \text{ after leaving the port),} \\ 0 & \text{otherwise.} \end{cases}
$$

Stochastic parameters:

$$
\tau_j := \text{release time of container } j.
$$

Decision variables:

$$
x_{ij} := \begin{cases} 1 & \text{if container } j \text{ is assigned to barge } i, \\ 0 & \text{otherwise.} \end{cases}
$$

The optimization problem, given a realisation of τ is:

$$
max\, \mathrm{K} = \sum_{j \in Z} \sum_{i \in N} x_{ij}, \tag{1}
$$

subject to

$$
\sum_{i \in N} \tau_j x_{ij} \leq \sum_{i \in N} OQ_{ij} x_{ij} \qquad \text{for } j \in Z_C, \tag{2}
$$

$$
\sum_{i \in N} d_j x_{ij} \geq \sum_{i \in N} a_i x_{ij} \qquad \text{for } j \in Z_C, \tag{3}
$$

$$
\sum_{i \in N} d_j x_{ij} \geq \sum_{i \in N} OQ_{ij} x_{ij} \qquad \text{for } j \in Z_D, \tag{4}
$$

$$
\sum_{i \in N} \tau_j x_{ij} \leq \sum_{i \in N} b_i x_{ij} \qquad \text{for } j \in Z_D, \tag{5}
$$

$$
\sum_{i \in N} x_{ij} \leq 1 \qquad \text{for } j \in Z, \tag{6}
$$

$$
\sum_{j \in Z} \omega_{ijt} x_{ij} \leq c_i \text{ for } i \in N \qquad \text{for } t \in T_i, \tag{7}
$$

$$
x_{ij} \in \{0,1\} \qquad \text{for } i \in N, \text{ for } j \in Z. \tag{8}
$$

The objective is to maximise the number of containers travelling by barge (1). We have a set of constraints that secures that a container that needs to be picked up at a port can only be assigned to a barge if the container is released before or at the same time as the barge departs from the port (2). Constraints (3) make sure a container that needs to be picked up at a port can only be assigned to a barge when the due time of the container is later than or equal to the arrival of the barge at the terminal. If this is not the case the container will be too late. We have a set of constraints that secures that a container that needs to be delivered at a port can only be assigned to a barge if the container is released before or at the same time as the barge departs from the terminal (4). Constraints (5) make sure a container that needs to be delivered at a port can only be assigned to a barge when the due time of the container is later than or equal to the arrival of the barge at the delivery port. If this is not the case the container will be too late. Constraints (6) secure a container can only be assigned to one barge. Constraints (7) restrict the total number of containers assigned to a barge by a maximum capacity for each time stamp the barge is at a port. The constraints (8) define x_{ij} as binary variables.

3   Solution Methods

Given the complexity of our model in the stochastic case it is impossible to obtain an analytical expression for the expected number of containers travelling by barge, i.e., Eq. (1). As a consequence usual (non-stochastic) IP algorithms cannot directly be applied. As soon as all (future) release times are known, though, our problem does reduce to standard IP instances. Having stochastic release times with a known distribution we can simulate complete sets of release times by drawing from this distribution. For each such drawing we could in principle apply an IP algorithm to find the optimal container collect & deliver solution for this set of release times.

Now suppose we have to decide on whether to assign a (given, released) container to one of a number of alternative barges. Inspired by the procedure used in [9], we could proceed as follows. We draw repeatedly from.the distribution of all containers yet to be released. For each draw we have a complete set of release times, so that we can derive the optimal IP solution. In this solution the container under consideration is assigned to one of the available barges or to a truck. If the number of drawings is large enough the results will converge to their mean. It can be checked from these means which alternative gives the highest expected overall result for shipment of the container under consideration. An exact solution can be approximated with any precision, provided the number of replications is sufficiently large (law of large numbers). Since solving each IP problem is non-trivial in terms of the run time involved, the process is hardly feasible in practice, especially with large numbers of replications (details follow below). Thus we have to reduce the computational burden considerably as compared to this near optimal solution.

The computational burden originates from two sources: (i) simulation of the stochastics in the model, and (ii) finding exact IP solutions for a given drawing. This gives three different ways to simplify the approach to finding solutions to our problem. The most drastic one is to skip both features by neglecting the stochastics altogether as well as recurring to a simple decision rule in assigning containers to barges. In Subsect. 3.1 we present three such rules. A more sophisticated approach is to take into account the stochastics in the model by simulation, as in the near optimal solution, but skipping run time consuming optimal IP solutions for each drawing and replacing them by a more simple greedy algorithm. This approach is presented in Subsect. 3.3. A third approach is mentioned here only for reasons of comparison: we stick to the IP part of the near-optimal solution but skip the stochastics by applying the IP algorithm to the set of expected values of the release times of all containers. This approach is presented in Sect. 3.2.

3.1 Rule Based Decision Making

One way to proceed is to use a simple decision rule when deciding whether a container should be dispatched or not. When a barge arrives at a port all released containers at that port can be treated according to this rule. Containers that are not able to make it in time with a barge will always go by truck. Containers that can only make it in time travelling with this barge will always be assigned to this barge. If multiple barges can take this container we define:

$$Cap := \text{total container capacity of barge,}$$
$$rCap := \text{remaining container capacity of barge,}$$
$$nBar := \text{number of barges that can still deliver the container on time,}$$
$$vCon := \text{number of unassigned containers that still need to be visited}$$
$$\text{during the roundtrip of the barge, i.e. only at upstream ports,}$$
$$uCon := \text{total number of unassigned containers.}$$

We apply one of the following three alternative decision rules:

1. **FCFS:**
   Assign container to barge if $rCap \geq 1$,
2. **Barges2Come:**
   Assign container to barge if $(rCap/Cap) \geq (1/nBar)$,
3. **Containers2Come:**
   Assign container to barge if $(rCap/Cap) \geq (vCon/uCon)$.

The FCFS rule is a simple first-come, first-serve process of assigning containers to a barge. The Barges2Come rule is based on the knowledge that there are alternative barges in the pipeline to consider. The more alternative barges available in the pipeline, the less urgent it is to assign the container. The Containers2Come rule looks at the fraction of the containers still to be shipped which is upstream of the barge under consideration. If less containers are in front of the barge the remaining capacity is less valuable which makes it easier to assign the container in consideration to the barge.

3.2 Planning Solution

A planning solution can be calculated by making a planning on time stamp $t = 0$ with the static IP formulation based on the expected release times, i.e. assuming that actual release times are all equal to their expected values. This solution method is equal to the two stage stochastic programming recourse model as described, i.e., in [8]. If an assignment in the planning turns out to be infeasible because the actual release time is later than the departure of the assigned barge the container is assigned to a truck. This gives us an optimal planning only in the case where all actual release times turn out very close to their mean values (and a rather bad assignment in other cases as we will see in Sect. 5.

3.3 Simulation Algorithm

When a barge is available for loading/unloading at the terminal or at an inland port, we consider all containers that could be shipped with this barge. If the due time of a container is such that it cannot arrive on time when shipped by barge it goes by truck. If it can only be on time when shipped with the barge under consideration, it will be shipped if the barge still has empty spots; if not it goes by truck. If a container under consideration can reach its destination on time when travelling by alternative barges we proceed as follows. From the known distribution of release times we repeatedly draw sets of release times of all containers yet to be released, say n times. This gives n complete sets of known release times of all containers yet to be assigned. For each of these sets we determine a full assignment (all containers yet to be assigned, all barges yet to come), *conditional on the choice we make with respect to the container under consideration*. For each choice and each set this gives the total number of containers transported by barges. By averaging these outcomes over the n sets we obtain an estimate of the expected total number of containers shipped by barges per alternative assignment of the container under consideration. The container is then assigned to the alternative with the highest result. This process is repeated for all containers available for shipment by the barge. The barge is loaded accordingly, and departs for the next stop, where the process starts all over again.

Since the full IP solution of the yet to be assigned containers is too time consuming we replace it by a faster assignment rule, labelled greedy assignment. According to this rule containers are assigned to the barge with the highest remaining capacity out of all barges that are able to deliver the container on time. For all barges the remaining capacity is tracked during its round trip. If a container is assigned to a barge the remaining capacity is lowered by 1 as long as the container stays on the barge. The remaining capacity of a barge is determined as the current minimum of the remaining capacities the remainder if the round trip of this barge.

The order in which yet unassigned containers are being processed in the greedy assignment is as follows. First the containers that need to be delivered at the inland ports are assigned and thereafter the containers that have to be collected at the inland ports. Containers that have to be delivered are treated in the order of the port of destination, i.e., containers for port 1 are processed prior to containers for port 2, etc. Once all the containers that need to be delivered have been assigned this way we turn to the containers to be collected. These are treated in the same order, i.e., by their port of origin. For all containers that need to be delivered or collected at the same port we do not use a specific order to assign them, this is done at random. Although an order could be based upon the due time of the containers involved we choose not to do so, since we consider it unlikely to give a notable improvement in the solution.

4   Case Description

The round trip used in this case study consists of 8 Dutch ports; Dordrecht, Geertruidenberg, Den Bosch, Veghel, Roermond, Gennep, Nijmegen, Wageningen (see Fig. 1). In Table 1 the number of containers that need to be collected and delivered at these inland ports is stated, the total number of containers to be processed is 1,000 in this case. The round trip starts and ends at the European Container Terminal (ECT) in Rotterdam. In our problem we have 6 barges that have a respective capacity of; 78, 100, 97, 88, 45 and 76 which totals 484. The time window is 120 h and the round trip schedule is given in Table 2.

Table 1. Number of containers to be collected and delivered at inland ports. Dordrecht(1), Geertruidenberg(2), Den Bosch(3), Veghel(4), Roermond(5), Gennep(6), Nijmegen(7), Wageningen(8).

Port	1	2	3	4	5	6	7	8	Sum
Collect	80	56	69	87	67	57	78	102	596
Deliver	88	75	54	56	28	36	44	23	404

Table 2. Barge schedule in hours

	Barge					
	1	2	3	4	5	6
ECT	0	15	38	56	80	98
Dordrecht	2	17	40	58	82	100
Geertruidenberg	5	20	43	61	85	103
Den Bosch	7	22	45	63	87	105
Veghel	11	26	49	67	91	109
Roermond	13	28	51	69	93	111
Gennep	14	29	52	70	94	112
Nijmegen	17	32	55	73	97	115
Wageningen	19	34	57	75	99	117
ECT	20	35	58	76	100	118

4.1   Creating Due Times and Expected Release Times

In this section we describe the data set that we generated to study solution methods for the Container collect & deliver problem. Let due denote the due time of a container, exp the expected release time of the container, and tt the travel time of a container to its destination. For the containers that need to be collected at the inland ports a due time is uniformly chosen between 20 and 140, i.e., $due =\sim unif(20, 140)$. Containers with a due time before 20 can never

Table 3. The distribution of containers by number of optional barges available. Based on due times and expected release times.

| Collect Containers | | Deliver Containers | |
# Containers	Optional barges	# Containers	Optional barges
79	0	46	0
168	1	117	1
142	2	99	2
88	3	59	3
61	4	54	4
34	5	16	5
24	6	13	6
596		404	

be on time and will always go by truck so we neglect them in our problem instance. Due times later than 140 will be outside our time window and will only act as noise, so we leave these out as well. After a due time is chosen a corresponding expected release time is drawn uniformly between -20, and the minimum of the due time and 120 minus the travel time of the container, i.e., $exp =\sim unif(-20, min(due, 120) - tt)$. The upper bound entails that the container can always go with the last barge if released before or on the expected release time, provided the due time allows this. The lower bound of -20 makes sure that a fair amount of containers will be released when the first barge leaves. The left hand part of Table 3 shows how many optional barges are available per container to be collected. It appears that most containers have only 1 or 2 barges they can travel with. Despite our choice of the generation of expected release times and due times 79 containers cannot travel by barge at all. This is due to the fact that barges only arrive at intervals of 15–20 h at ports so that a container released just after a barge has left has to wait too long before the next barges arrives, given the due time for this container. It should be realised

Fig. 1. Inland ports and waterways

that actual release times differ from the average release times the table is based upon. So with actual release times rather than these expected values a number of these 79 containers might still be available for transport by barge. On the other hand it can also happen that the actual release time is such that a due time can no longer be met by barge transport. If this is the case the container will go by truck.

For the 404 containers that need to be delivered at the inland ports we apply a similar procedure. In particular we generate due times according to $due =\sim unif(tt, 120 + tt)$, and expected release times according to $exp =\sim unif(-20, min(due, 98) - tt)$. The right hand side of Table 3 presents the number of barges available per container to be delivered. The pattern is similar to the pattern generalised for containers to be collected.

5   Results

At the end of the time window all release times of the containers are known. An optimal ex post *offline* solution can be calculated by solving the static IP formulation. Obviously this ideal solution is not realistic since in practice the release times are not completely known when a container has to be assigned. However, it can be used as an upper bound for our solution methods. For the container collect & deliver problem we simulate the release times for the 1,000 containers with a normal distribution with different standard deviations (1, 3, 6, 9 & 15) in hours and the chosen average release times. With these release times results are calculated for the optimal, planning and rule based solutions. For the simulation algorithm per standard deviation we choose different numbers of simulations (3, 9, 27, 81 & 243) for each container assignment decision at a port. To check how steady the solutions perform we repeat the complete process 25 times and look at the average outcomes and their standard deviation.

Table 4 gives an overview of the performance of the simulation algorithm for the container collect & deliver problem. The results shown in the table are the mean difference (and standard deviation) between the simulation algorithm and the other types of solution methods. These results are calculated out of the results of the 25 repeated simulations with release time standard deviation equal to 9. The number of simulations in the simulation algorithm is equal to 243. In Table 4 we see that the difference between the simulation algorithm and the ideal solution is negative meaning that the ideal solution has overall better results as expected. The planning solution performs worse than the simulation algorithm. The larger the release time standard deviation the worse the planning solution performs (not shown in the table). This is as expected since the planning solution generates more unfeasible assignments when uncertainty gets larger. Note that the results presented mean that with the Simulation Algorithm you save 95 % of the extra truck movements in comparison with the Planning solution.

The rule based algorithms all perform worse than the simulation algorithm. As stated before this is due to the fact that the rule based algorithms do not take into consideration the distribution of the yet unknown release times. All

Table 4. Mean and standard deviation of the difference in results between the simulation algorithm and other solution methods.

Type	Mean	S.D.
Ideal Solution	-2.84 (-0.35%)	2.03
Simulation Algorithm	0 (0.0%)	0
Planning Solution	50.64 (6.24%)	3.83
FCFS	132.08 (16.27%)	4.82
Barges2Come	86.56 (10.66%)	5.88
Containers2Come	55.84 (6.88%)	7.91

differences between our simulation algorithm and the four alternative solution methods are significant at the 95 % level (t-test with 25° of freedom).

Table 5 shows the effect of choosing different parameters on the performance of the simulation algorithm. The results display a weak but systematic pattern first, results tend to deteriorate with increasing uncertainty (standard deviation). Second, results tend to improve with increasing numbers of simulations. Third, with increasing numbers of simulations the results seem to converge slower with increasing uncertainty. These observations are as expected as increasing uncertainty is likely to induce more false online decisions. Also we need more simulations to properly represent the probability distribution of future demands. It should be added, though, that some of the differences observed in the table are hardly significant.

Table 5. Average number of containers travelling by barge by number of simulations and standard deviation of the release time distribution. Standard deviation of the averages in parenthesis.

	S.D. 1	S.D. 3	S.D. 6	S.D. 9	S.D. 15
Sim 3	809.8 (0.65)	806.1 (0.75)	800.0 (0.96)	796.9 (1.22)	785.0 (1.44)
Sim 9	819.3 (0.46)	813.8 (0.82)	808.4 (1.07)	803.4 (1.45)	791.5 (1.34)
Sim 27	822.7 (0.26)	818.4 (0.70)	813.7 (0.81)	808.9 (1.18)	797.0 (1.68)
Sim 81	823.1 (0.26)	820.4 (0.54)	815.7 (0.83)	811.7 (1.12)	800.2 (1.58)
Sim 243	823.4 (0.26)	820.7 (0.49)	816.5 (0.71)	812.3 (1.13)	801.8 (1.48)

It is interesting to compare the results of Table 5 to certainty equivalence outcome. That is we impose a standard deviation of 0 so that in all simulations actual release times will equal average release times. As Table 6 shows the results are worse than the results with stochastic release times. This seems strange at first sight. Indeed, by adding noise into the system the solution appears to improve considerably which seems counter intuitive. The table shows that this effect manifests itself already with rather small standard deviations. With a standard deviation as small as 0.3 the number of containers transported by barge increases on average by no less than 9 containers in the instance displayed. This raises the question how to explain this pattern of dependency of the outcomes on the amount of uncertainty.

We conjecture that the clue lies in the comparison with simulated annealing optimisation processes, where we have a (natural) tendency towards an optimum combined with noise in the problem conditions. In our problem we also perturb the initial conditions (i.e., release times) and we systematically favour solutions improving the optimum. Table 6 shows that even with a standard deviation as small as 0.2 a substantial gain is obtained. Apparently a standard deviation of 0.1 is too small to generate any effect whatsoever.

Table 6. Average number of containers travelling by barge by standard deviation of the release time distribution. Standard deviation of the averages in parenthesis.

S.D.	# Containers
0	805.0 (0)
0.1	805.0 (0)
0.2	808.4 (0.37)
0.3	814.4 (0.54)
0.5	812.8 (0.50)
1.0	809.8 (0.65)

The gain obtained by introducing some uncertainty in the system is counteracted by a second effect of uncertainty, that is less predictable release times and therefor a higher probability of online decisions that turn out to be wrong ex post. This second effect of uncertainty apparently becomes dominant at higher standard deviations, as Tables 5 and 6 show.

A different but probably equivalent way to explain the observed pattern of dependency of the outcomes on the standard deviation employed is the following. Although the distribution of release times is symmetrical the implied distribution of the objective under the algorithm employed is far from symmetrical. In fact it is a very complicated non-linear function of the release times embodying the stochastics in the system. Apparently the skewness of the distribution of the objective is such that favourable outcomes dominate unfavourable. In other words the expected value of the optimal solution is larger than the certainty equivalent solution based on the average value of the release times.

With 243 simulations of the release times of the yet to release containers the simulation algorithm takes about 3 min on a standard laptop for the container collect & deliver problem. To put the runtime of the simulation algorithm into perspective we should realise that the runtime reported refers to the complete set of all containers to be shipped. In actual applications, when a specific barge collides with a specific port or terminal we only have to decide about a few containers whether to assign them or not. So the simulation burden is limited and the runtime of the simulations at each port is almost negligible, even with a large number of replications. Indeed, in practice a transport manager could easily run the simulation algorithm on a small handheld computer or even a smart phone, within less than a few seconds.

6   Conclusion and Future Research

In this case study we have demonstrated that an explicit simulation approach to deal with uncertain information in an online assignment problem of reasonable size is feasible. In the problem instance studied the algorithm employed performs close to ideal, while its runtime is small enough to be perfectly executable on small computing devices. By exploiting knowledge of the distribution of uncertain future demands it outperforms more basic decision rules neglecting such information.

Further research along this line can proceed in three ways. First the transport model can be improved upon by evaluation of the assumptions of the model underlying this case study. In particular the fixed time window employed in this exercise is grossly unrealistic. In practical situations a continuous stream of containers to be transported will have to be served by a scheme of round trips of barges which goes on indefinitely. This calls for a rolling time window instead of a fixed one, such that at each time stamp a steady number of containers to be transported will be present. Presumably this calls for a fairly substantial redefinition of our model framework, which we have not tried yet. In addition the model should allow for transport of containers between inland ports, e.g., empty containers to balance shortages and surpluses at different ports. This does not seem too hard to implement.

Second, in terms of run time there seems to be room for improvement of the greedy algorithm employed. As an example we could make use of the available information about due times of the containers yet to be released in determining the order in which the containers are assigned to barges in the greedy algorithm.

Third, the treatment of the stochastics in the model can be improved. In our case study we assumed a known distribution of release times with known constant parameters. In practice this might be unrealistic. More sophisticated statistical methods could be investigated allowing for the uncertainty in the kind of distribution and/or the true value of its parameters. In a decision context a Bayesian approach might be more appropriate than the classical framework underlying our exercise, see e.g. [12]. Another way to improve the treatment of the stochastics in the model is to employ a more sophisticated simulation technique. Note that the simulation is basically meant to get an estimate of the expected value of the objective function of the problem, i.e., Eq. (1). This expectation is a complicated mathematical integral, so the problem boils down to a the numerical problem of evaluating this integral. Advanced simulation methods have been applied in other fields of statistics to reduce the computational burden of this evaluation, e.g., importance sampling, see [7], and diverse Monte Carlo integration techniques, see [6].

Acknowledgements. We thank Dr. T.A.B. Dollevoet, Erasmus University Rotterdam, for his useful input and comments during this research.

References

1. Bandeira, D.L., Becker, J.L., Borenstein, D.: A DSS for integrated distribution of empty and full containers. Decis. Support Syst. **47**(4), 383–397 (2009)
2. Braekers, K., Janssens, G.K., Caris, A.: Challenges in managing empty container movements at multiple planning levels. Transp. Rev. **31**(6), 681–708 (2011)
3. Caris, A., Macharis, C., Janssens, G.K.: Decision support in intermodal transport: a new research agenda. Comput. Ind. **64**(2), 105–112 (2013)
4. Fazi, S., Fransoo, J.C., Van Woensel, T.: A decision support system tool for the transportation by barge of import containers: a case study. Decis. Support Syst. **79**, 33–45 (2015)
5. Feng, F., Pang, Y., Lodewijks, G.: Application of hybrid algorithm to joint decision making in hinterland barge transport planning. In: Proceedings CIE45 (2015)
6. Geyer, C.J.: Practical Markov chain Monte Carlo. Stat. Sci. **7**(4), 473–483 (1992)
7. Glynn, P.W., Iglehart, D.L.: Importance sampling for stochastic simulations. Manag. Sci. **35**(11), 1367–1392 (1989)
8. Haneveld, W.K.K., Stougie, L., Van der Vlerk, M.H.: Simple integer recourse models: convexity and convex approximations. Math. Program. **108**(2–3), 435–473 (2006)
9. Phillipson, F.: Planning nurses in maternity care: a stochastic assignment problem. J. Phys. Conf. Ser. **616**(1), 012006 (2015)
10. van Riessen, B., Negenborn, R.R., Dekker, R.: Synchromodal container transportation: an overview of current topics and research opportunities. In: Corman, F., Voß, S., Negenborn, R.R. (eds.) ICCL 2015. LNCS, vol. 9335, pp. 386–397. Springer, Heidelberg (2015). doi:10.1007/978-3-319-24264-4_27
11. Song, D.P., Dong, J.X.: Empty container repositioning. In: Lee, C.-Y., Meng, Q. (eds.) Handbook of Ocean Container Transport Logistics, pp. 163–208. Springer, Heidelberg (2015)
12. Zellner, A.: An Introduction to Bayesian Inference in Econometrics. Wiley, New York (1996)

A DTMC Model for Performance Evaluation of Irregular Interconnection Networks with Asymmetric Spatial Traffic Distributions

Daniel Lüdtke[1]([✉]) and Dietmar Tutsch[2]

[1] German Aerospace Center (DLR), Simulation and Software Technology,
Lilienthalplatz 7, 38118 Braunschweig, Germany
daniel.luedtke@dlr.de
[2] Automation/Computer Science, University of Wuppertal,
Rainer-Gruenter-Str. 21, 42119 Wuppertal, Germany
tutsch@uni-wuppertal.de

Abstract. Several mathematical models have been proposed to evaluate the performance of interconnection networks used for high-speed connections for supercomputers, switches and routers for local and wide area networks, as well as networks on a chip. Often these models are based on state space reduction by exploiting symmetries of the network and requiring uniform traffic patterns. If an interconnection network is built for a specific application with non-uniform spatial traffic distribution, models that are more general are needed. This paper proposes a mathematical model for performance evaluation of application-specific interconnection networks based on inhomogeneous discrete time Markov chains (DTMC). It supports store and forward routing, irregular network topologies, and asymmetric spatial traffic distributions. The model is described in a generalized way so that it can support arbitrary switching element sizes within the network and its input buffers.

1 Introduction

Interconnection networks with point-to-point links are widely used for high-speed connections in different domains, for example, as networks on a chip (NoC), in supercomputers, and in switches and routers for local and wide area networks. In the space domain, SpaceWire networks [9] are more often used to provide high communication bandwidth onboard spacecraft.

Most interconnection networks have regular or symmetric topologies, because supercomputers or network switches are usually designed for a large variety of applications. However, for application-specific networks, like specialized systems on a chip employing a network on a chip or a spacecraft with many specialized network nodes, irregular topologies are often beneficial.

Additionally, many performance evaluation approaches presume uniform traffic distributions. When planning and optimizing an interconnection network for a specific application, non-uniform traffic distributions, which match the traffic patterns of the application in question, need to be considered.

© Springer International Publishing Switzerland 2016
S. Wittevrongel and T. Phung-Duc (Eds.): ASMTA 2016, LNCS 9845, pp. 193–209, 2016.
DOI: 10.1007/978-3-319-43904-4_14

For instance, in the research project OBC-NG (Onboard Computer — Next Generation) [6], which was carried out at the German Aerospace Center (DLR), a new reconfigurable distributed computer architecture was developed. The goal of this project and its successor, ScOSA (Scalable On-board computing for Space Avionics), is to provide high performance onboard computing power to support complex future space missions. ScOSA utilizes a reconfigurable SpaceWire network that allows the reconfiguration of the spacecraft computer for different mission phases as well as for error mitigation. Each of these configurations are predetermined for which optimization algorithms are needed to find optimal partitioning and mapping of tasks for each mission phase. These optimizations need to consider the available bandwidth and possible congestion in the network.

The project considers two approaches for the stochastic performance evaluation: simulation and an analytical model. CINSim (Component-based Interconnection Network Simulator) [7], a stochastic discrete event simulation framework, is used as simulator. Additionally, a mathematical model, based on inhomogeneous discrete-time Markov chains (DTMC) is developed, to evaluate different network configurations. This model is presented in this paper. It supports direct networks, where terminals are connected at each switch or switching element as well as indirect networks like Multistage Interconnection Networks (MIN). A discrete-time Markov chain, compared to a continuous-time Markov chain, was chosen to model the synchronous behavior of the switch implementation.

The remainder of this paper is structured as follows: the next section gives a brief overview of related work. Section 3 introduces the DTMC model. Followed by some results, where the model results are compared to simulation. Section 5 presents the conclusions and an outlook to ongoing and future work is given.

2   Related Work

Performance evaluation models for interconnection networks, especially MINs, have been proposed for many years. Dias and Jump [3] proposed a model for MINs with a buffer at each switch input. Yoon et al. [12] described a model that supports arbitrary buffer lengths and arbitrary switch dimensions. Youn and Mun [13] established a model that confined packet movement from one switch to another in one clock cycle. Atiquzzaman and Akhtar [2] proposed a model to investigate hot spot traffic in MINs. Bin and Atiquzzaman [14] developed a model for output buffered MINs with asymmetric traffic patterns. Tutsch and Hommel [11] proposed a model, which considers multicast traffic within a MIN. This model reduces the state space by exploiting symmetries in the network topology and assuming uniformly distributed traffic.

All mentioned models are based on Markov chains. In recent years, analytical models have been proposed that are based on queuing networks. Moadeli et al. [8] developed a performance model for the spidergon topology for NoCs. Kiasari et al. [5] proposed an analytical latency model for NoCs for arbitrary topologies. An analytical model for MINs is proposed by Amiri-Zarandi et al. [1]. Hamid et al. [4] introduce an analytical model for a multi-core multi-cluster architecture and Sabbaghi-Nadooshan and Patooghy [10] present a performance model for de Bruijn inspired mesh-based NoCs.

3   Model Description

This section describes the network abstraction, traffic model, and the DTMC model itself.

3.1   Network and Traffic Model

The network model presented in this work is based on CINSim's component-based interconnection network model. It supports combinations of the components source, destination, switch, and buffer. With these atomic components, a great variety of packet-switched interconnection network architectures can be modeled and investigated, especially irregular network topologies.

Sources generate traffic according to a geometric load distribution $r(x)$, with $x \in \{0, 1\}$ (on/off source), which means that a packet is generated at each time step with the probability $r(x)$, and the global spatial distribution ℓ_{sd}^g, which defines the probability that a packet at source s is targeted to destination d. Destinations consume packets immediately. Packets move from sources via buffers and switches to their destinations. Switches are components to realize dynamically changing connections between its inputs and outputs. Inputs and outputs are connected according to the requested output of the packet. If multiple inputs contain packets destined to the same output, one of the packet is randomly selected. Switches have input buffers that store packets. It is required that each switch input is connected to a buffer. Input buffers and destinations can only have one preceding component. Switches work in a timeless manner in this model. Only moving through a buffer costs at least one time step. The buffer size can be set individually for each input buffer. Figure 1 shows a small interconnection network with the described components: sources on the left, input buffers with a capacity of three packets connected to a switch, and two destinations on the right.

The proposed model employs unicast store-and-forward routing, i.e. a packet is the smallest flow control unit in the model. It is based on the local backpressure scheme, which means that packets that enter the network will eventually reach their destination. Packet drops happen only at the sources. *Local* backpressure means that packets can only proceed to the next buffer component during the following time step, if buffer space is available at the current time step. The model implements shortest-path routing. If several shortest paths are available, a path is randomly selected. All network components operate synchronously.

To keep the modeling complexity limited, we assume that all network components operate synchronously, driven by a global internal clock. This assumption is quite realistic for a wide range of applications. For instance, network switches or networks on chips often operate synchronously.

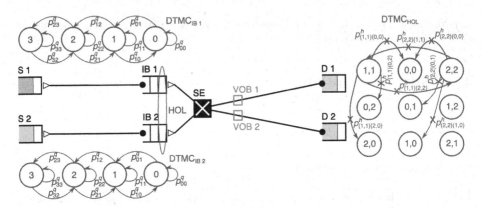

Fig. 1. A 2 × 2 network with two sources on the left (S), two input buffers (IB), a switch (switching element, SE), and two destinations (D) on the right. The two virtual output buffers (VOB) and the corresponding Markov chains (see Sect. 3.2) are also shown: $DTMC_{IB1}$ and $DTMC_{IB2}$ for the input buffers and $DTMC_{HOL}$ for the Head-of-Line (HOL) Markov chain (only unfeasible state transitions are drawn).

3.2   DTMC Model

For an adequate modeling of non-uniform spatial traffic distributions, the global spatial distribution ℓ^g has to be considered at each switch in combination with the actual load distribution of each source. For this, probabilities need to be determined about the packet rate for each input/output combination at every switch. Hence, a local spatial traffic distribution ℓ^l_{io} is derived, which represents the probability that a packet at a switch is transmitted from switch input i to switch output o. *Local* means that this distribution refers to a single switch in contrast to the global spatial distribution ℓ^g. To achieve this, besides the queue length of each input buffer, the Head-of-Line places (HOL) of all buffers are modeled to represent the target outputs of the switch. Additionally, to simplify the equations, virtual output buffers are introduced at the switches. These buffers indicate from which input a packet has arrived in the current time step to adapt ℓ^l_{io}. Since the virtual output buffers have no representation in real hardware, the presence of a packet there does not consume time.

It is not feasible to model even small networks with a single Markov chain due to the state space explosion. In particular, no symmetries can be exploited compared to most models mentioned in the related work section due to the requirement to support arbitrary topologies. For example, a simple model of a single 3 × 3 switch with four buffer places at each input leads to 24,303 states and 261,081 state transitions. Larger networks cannot be modeled in such a way, since the number of states increases exponentially. Thus, a decomposition technique is used that was also applied in previous work (e.g. [11]). However, the proposed model in this paper does not exploit topological symmetries or requires uniformly distributed spatial traffic distributions.

Several Markov chains are setup to reflect different behavioral aspects of the system. Hence, the number of states and state transitions grows only linearly.

The connection between individual Markov chains is achieved by dependencies in the state transition probabilities of the individual Markov chains, i.e. the transient state probabilities of an individual Markov chain are considered in the state transition probabilities of dependent other Markov chains. These state transition probabilities change every time step until the network reaches steady-state, leading to inhomogeneous DTMCs.

Figure 1 shows, next to the network, the corresponding Markov chains and virtual output buffers. For the HOL DTMC, only the unfeasible state transitions are shown, all other transitions are feasible.

State Space of HOL DTMCs. All HOL buffer places in a switch with i_{max} inputs and input buffers, respectively, as well as o_{max} outputs and virtual output buffers, respectively, are represented by a HOL Markov chain with a finite state space $S^h(i_{max}, o_{max})$. A combined Markov chain for all HOL buffer places is chosen to adequately model the blocking behavior of a switch. The states of $S^h(i_{max}, o_{max})$ are denoted by i_{max}-tuples of an $(o_{max} + 1)$-set. Each HOL place of an input buffer i has the possible states $s_i^h \in \{0, 1, \ldots, o_{max}\}$. The first element of each state represents the state of the HOL place of the first input buffer. A '0' represents the case, where no packet is present; a '1', where a packet wants to move to output 1 etc. s_i^h denotes the target output of a packet. For instance, $s_3^h = 2$ means that the HOL packet in the third input buffer of a switch is destined to the second output of the switch.

$S^h(i_{max}, o_{max})$ of the HOL Markov chain for a switch is defined by the i_{max}-permutation with repetition (PR) of the set of outputs s_i^h:

$$S^h(i_{max}, o_{max}) := PR(i_{max}, s_i^h) = \{(0, \ldots, 0), \ldots, (o_{max}, \ldots, o_{max})\}.$$

The number of possible states is given by $\left|S^h(i_{max}, o_{max})\right| = (o_{max} + 1)^{i_{max}}$. Hence, the state probability vector ν^h of the head-of-line Markov chain has $\left|S^h(i_{max}, o_{max})\right|$ components and $\left|S^h(i_{max}, o_{max})\right|^2$ elements in its corresponding transition probability matrix $\mathbf{P}^h$.

Initially, it is assumed that all buffers are empty. Thus, $\nu_{(0,0)}^h(0) = 1$ in case of the 2×2 example. To calculate the state probability vector of the second time step $n = 1$, the equation $\nu^h(1) = \mathbf{P}^h(0) \cdot \nu^h(0)$ has to be solved according to the Chapman-Kolmogorov equation.

State Space of Queue Length DTMCs. The finite state space S^q of the Markov chain that represents the queue length of input buffer i is constructed by $S^q(m_{max}(i)) := \{0, 1, \ldots, m_{max}(i)\}$, where $m_{max}(i)$ denotes the number of buffer places of input buffer i. For example, an input buffer with the capacity for four packets has the state space $S^q(4) = \{0, 1, 2, 3, 4\}$.

The number of possible states is given by $\left|S^q(m_{max}(i))\right| = m_{max}(i) + 1$. Hence, the queue length DTMC state probability vector ν^q has $\left|S^q(m_{max}(i))\right|$ components and a transition probability matrix $\mathbf{P}^q$ with $\left|S^q(m_{max}(i))\right|^2$ elements. $\nu_0^q(n)$ denotes the state probability that the buffer is empty at time step n and $\nu_{m_{max}(i)}^q(n)$ represents the state probability that the buffer is full, respectively.

In the following, it is assumed that $m_{\max}(i) > 1$. With only one buffer place in an input buffer, a packet cannot simultaneously be sent and received during one time step due to the local backpressure scheme. With $m_{\max}(i) = 1$, the normalized throughput of this connection would maximally be 0.5.

State Space of Virtual Output Buffers. As previously mentioned, virtual output buffers are established to simplify the modeling of the spatial distribution within the network. Packets are only transferred to these buffers if they can be transferred to the next switch or network output in the following time step. This reduces the modeling of the output buffers to simple probabilities that can be derived from the HOL states. The probabilities are updated directly during the same iteration step. Nevertheless, the states of the virtual output buffers in a switch are also denoted as a state probability vector $\boldsymbol{\nu}^{o}$. The state space of a virtual output buffer is defined by the number of the input buffers of the switch in question: $S^{o}(i_{\max}) := \{0, 1, \ldots, i_{\max}\}$. The number of possible states is given by $|S^{o}(i_{\max})| = i_{\max} + 1$. Hence, the state probability vector $\boldsymbol{\nu}^{o}$ has $|S^{o}(i_{\max})|$ components.

State Transition Probabilities for HOL DTMCs. The state transition probabilities for the HOL DTMCs are shown in Fig. 2. They depend on the competition of packets at the HOL position of the input buffers for an output. Since only a single packet can win such a competition, the losing packets stay at their input buffer. Conflicts are resolved randomly.

Each entry of the state transition matrix $\mathbf{P}^{h}(n)$ for the HOL DTMCs at time step n is calculated by (1). All state transitions from state s to state t at time step n that are not feasible are set to zero. Infeasible state transitions describe situations, for instance, if two HOL packets destined to the same switch output, could have been successfully sent, which is not possible because only one packet could have won the conflict. For instance, a 5×5 switch has 7,776 possible HOL states and 60,466,176 entries in its $\mathbf{P}^{h}(n)$ matrix. 22,221,176 state transitions are feasible. Thus, 38,245,000 entries in $\mathbf{P}^{h}(n)$ are always zero.

(2) determines if a state transition from s to t is feasible. It is feasible if for all HOL packets that are destined to the same switch output o (conflicting inputs, ci (o, s^{h}), see (4)) only a single input changes from s to t ($|\Delta i(o, s, t)| \le 1$, see (5)). This has to be true for all outputs with more than one HOL as target ("courted outputs", $o \in$ co (s^{h}), see (3)).

The probability for a feasible state transition is the product of the probabilities of a state change from state s to t of the HOL packet for each input. Two general cases have to be distinguished.

First, we consider inputs that are not part of any conflict set at state s. This includes all empty input buffers and HOL packets at s, which have no other competitors for its output (non-conflicting inputs $i \in$ nci (s), see (6)). The probabilities for these inputs are calculated individually with $p_{st}^{\text{noConf}}(i, n)$, defined in (7).

$$p_{st}^{h}(n) = \begin{cases} 0, & \text{if feasible}(s,t) = 0 \\ \prod_{i \in \text{nci}(s)} p_{st}^{\text{noConf}}(i,n) \cdot \prod_{o \in \text{co}(s)} p_{st}^{\text{conf}}(o,n), & \text{if feasible}(s,t) = 1 \end{cases} \tag{1}$$

$$\text{feasible}(s,t) := \begin{cases} 1, & \text{if } \forall o \in \text{co}(s) : |\Delta i(o,s,t)| \leq 1 \\ 0, & \text{else} . \end{cases} \tag{2}$$

$$\text{co}\left(s^{h}\right) := \left\{ o : o \in \{1, \ldots, o_{max}\} \wedge \left| \text{ci}\left(o, s^{h}\right) \right| > 1 \right\} \tag{3}$$

$$\text{ci}\left(o, s^{h}\right) := \left\{ i : i \in \{1, \ldots, i_{max}\} \wedge s_{i}^{h} = o \right\} \tag{4}$$

$$\Delta i(o,s,t) = \text{ci}(o,s) \setminus \text{ci}(o,t) \tag{5}$$

$$\text{nci}\left(s^{h}\right) := \{1, \ldots, i_{max}\} \setminus \bigcup_{o \in \text{co}\left(s^{h}\right)} \text{ci}\left(o, s^{h}\right) \tag{6}$$

$$p_{st}^{\text{noConf}}(i, n+1) = \begin{cases} 1 - \tilde{q}(i,n), & \text{if } s_i = 0 \wedge t_i = 0 \\ \ell_{it_i}^{l}(n) \cdot \tilde{q}(i,n), & \text{if } s_i = 0 \wedge t_i \neq 0 \\ r(s_i, n) \cdot \text{lps}(i,n), & \text{if } s_i \neq 0 \wedge t_i = 0 \\ r(s_i, n) \cdot \ell_{it_i}^{l}(n) \cdot \text{nfp}(i,n), & \text{if } s_i \neq 0 \neq t_i \wedge s_i \neq t_i \\ r(s_i, n) \cdot \ell_{it_i}^{l}(n) \cdot \text{nfp}(i,n) \\ \quad + (1 - r(s_i, n)) & \text{if } s_i \neq 0 \wedge t_i = s_i \end{cases} \tag{7}$$

$$\text{lps}(i,n) = \begin{cases} \frac{\nu_1^q(i,n) \cdot (1 - \tilde{q}(i,n))}{1 - \nu_0^q(i,n)}, & \text{if } \nu_0^q(i,n) < 1 \\ 0, & \text{if } \nu_0^q(i,n) = 1 \end{cases} \tag{8}$$

$$\text{nfp}(i,n) = \begin{cases} \frac{\nu_1^q(i,n) \cdot \tilde{q}(i,n)}{1 - \nu_0^q(i,n)} + \frac{1 - \nu_0^q(i,n) - \nu_1^q(i,n)}{1 - \nu_0^q(i,n)}, & \text{if } \nu_0^q(i,n) < 1 \\ 0, & \text{if } \nu_0^q(i,n) = 1 \end{cases} \tag{9}$$

$$p_{st}^{\text{conf}}(o, n+1) = \begin{cases} \frac{r(o,n)}{|\text{ci}(o,s)|} \cdot \text{lps}(\Delta i(o,s,t), n), & \text{if } t_{\Delta i(o,s,t)} = 0 \\ \frac{r(o,n)}{|\text{ci}(o,s)|} \cdot \ell_{\Delta i(o,s,t) t_{\Delta i(o,s,t)}}^{l}(n) \\ \quad \cdot \text{nfp}(\Delta i(o,s,t), n), & \text{if } t_{\Delta i(o,s,t)} \neq 0 \\ \frac{r(o,n)}{|\text{ci}(o,s)|} \cdot \sum_{c \in \text{ci}(o,s)} \left(\ell_{co}^{l}(n) \cdot \text{nfp}(c,n) \right) \\ \quad + (1 - r(o,n)) & \text{if } \Delta i(o,s,t) = \emptyset \end{cases} \tag{10}$$

Fig. 2. State transition probabilities $\mathbf{P}^{h}(n)$ for the head-of-line Markov chains

And second, all HOL packets of inputs that are competing for an output o at state s have to be considered together due to their interdependence and are calculated by $p_{st}^{\text{conf}}(o,n)$, defined in (10). The conflict sets for different outputs are calculated independently.

Non-conflicting Inputs. The state transition probability $p_{st}^{\text{noConf}}(i, n+1)$ for input buffer i from state s to state t for time step $n+1$ is determined by (7). Five cases have to be distinguished for $p_{st}^{\text{noConf}}(i, n+1)$:

- First, if the queue of input buffer i in state s is empty and stays empty in state t, then no new packet comes to i during this state transition with probability $(1 - \tilde{q}\,(i,n))$, where $\tilde{q}\,(i,n)$ is the receiving probability of input buffer i in case that buffer space is available. It is determined by the sending probability of the preceding component $r_{\text{Pred}}\,(o,n)$ normalized by the probability that the buffer is not full: $\tilde{q}\,(i,n) = \frac{r_{\text{Pred}}(o,n)}{1 - \nu^{q}_{m_{\max}(i)}(i,n)}$.

  In case the preceding component is a network source, the sending probability r_{Pred} represents the geometric load distribution $r(x)$ of this source (see Sect. 3.1). For switches it is given by the probability that their virtual output buffer is not empty, hence, a packet is sent: $r_{\text{Pred}}\,(o,n) = 1 - \nu^{o}_{\text{Pred},0}(o,n)$.

- Second, if the queue was empty in s and has a HOL packet in t, a packet is received during the state transition multiplied with the probability $\ell^{1}_{it_i}\,(n)$ that a packet that enters the switch from input i is targeted to the output t_i, i.e., the output that is given by the i-th element of t.

- Third, if the queue of i had only one packet left at s, no new packet were coming ($\text{lps}\,(i,n)$) and this last packet was successfully sent during the state transition to t because the output was available ($r\,(s_i,n)$), the new HOL state for i would be $t_i = 0$. The last packet sent probability $\text{lps}\,(i,n)$ (see (8)) is defined as follows: if i is definitely empty ($\nu^{q}_0\,(i,n) = 1$), the probability equals 0; in all other cases, the probability that only one packet is present ($\nu^{q}_1(i,n)$) is multiplied by the probability that no new packet is received ($1 - \tilde{q}\,(i,n)$). The term has to be normalized by the sum of state probabilities of a non-empty buffer, since $\text{lps}(i,n)$ is only defined for state transitions where the buffer is not empty.

- Fourth, if the HOL packet of i changed to a different target during the state transition from state s to t, then the packet with the output s_i was successfully sent ($r\,(s_i,n)$) and a new first packet ($\text{nfp}\,(i,n)$) targeted to t_i is present with the routing probability $\ell^{1}_{it_i}\,(n)$. The new first packet probability ($\text{nfp}\,(i,n)$, see (9)) is defined as follows: in case the input buffer is definitely empty ($\nu^{q}_0\,(i,n) = 1$) the probability equals 0.

  In all other cases, the probability for a new first packet is combined by the following two cases: input buffer i had only one packet stored ($\nu^{q}_1(i,n)$) and a new packet is received ($\tilde{q}\,(i,n)$). Or i had more than one packet, thus, receiving a new packet does not influence the HOL state ($1 - \nu^{q}_0\,(i,n) - \nu^{q}_1\,(i,n)$). Since a new first packet only covers the case where the buffer is not empty, the probability has to be normalized by the term $1 - \nu^{q}_0\,(i,n)$.

- Finally, if the target of the HOL packet in i at state s is equal to the target at t, then two situations have to be distinguished. Similar to the previous case, a packet could have been sent and the new packet is heading to the same output, or the packet was blocked ($1 - r\,(s_i,n)$) due to a filled queue in the destination switch.

Conflicting Inputs. The probability $p^{\text{conf}}_{st}\,(o,n+1)$ for a set of inputs that are competing for output o at state s cannot be calculated independently. Thus, the probability for each conflict set that is identified by the common output o at state s is given by (10). Only one of the inputs can change from state s to t in each conflict set, otherwise, this state transition would not be feasible. Three cases for $p_{st}^{conf}(o, n+1)$ have to be distinguished:

- First, if the only changed input, identified by $\Delta i(o, s, t)$ (see (5)), changed its HOL state to $t_{\Delta i(o,s,t)} = 0$, output o was not blocked ($r(o, n)$) at time step n, its HOL packet won the conflict in this conflict set ($1/|ci(o, s)|$, random conflict resolution) and it was the last packet in this buffer ($lps(\Delta i(o, s, t), n)$).
- Second, if one HOL packet changes its target from output o to $t_{\Delta i(o,s,t)}$ during the state transition from s to t, then this input has won the conflict ($1/|ci(o, s)|$), the output was not blocked ($r(o, n)$), and a new first packet ($nfp(\Delta i(o, s, t), n)$) has the target output $t_{\Delta i(o,s,t)}$ with the routing probability $\ell_{\Delta i(o,s,t)t_{\Delta i(o,s,t)}}^l(n)$.
- Third, if no input changes in the conflict set of output o ($\Delta i(o, s, t) = \emptyset$) from s to t, two things could have happened: output o was blocked ($1 - r(o, n)$), so no HOL packet of this conflict set has moved. Alternatively, o was not blocked ($r(o, n)$), one of the inputs won the conflict ($1/|ci(o, s)|$), and a new packet moved to its HOL position ($nfp(c, n)$) with the same output o as target ($\ell_{co}^l(n)$). Since this could be the case for all inputs of the conflict set, the probabilities have to be added up for each winning input.

State Transition Probabilities for Queue Length DTMCs. The state transition matrix for the queue length of input buffer i at time step n is denoted as $\mathbf{P}^q(i, n)$. The element (j, k) of $p_{jk}^q(i, n)$ gives the probability that the queue length of i changes from j to k at n. The queue length can only increase or decrease by one packet from time step n to $n + 1$. Thus, only the following elements of $\mathbf{P}^q(i, n)$ are non-zero:

$$p_{jk}^q(i, n) \geq 0, \text{ if } 0 \leq |j - k| \leq 1.$$

To calculate all non-zero elements of $p_{jk}^q(i, n+1)$ for the next time step $n+1$, all possible HOL states have to be considered since the decision whether a packet can leave this buffer depends on the other input buffers as well. Equations (11)–(18) in Fig. 3 show the state transition probabilities for the queue length DTMCs.

The state transition probability of whether the buffer is empty at time step n and no new packet is coming at time step $n + 1$ is given by (11).

If the buffer is empty and a new packet is coming, the state transition probability is equal to the receiving probability of this input buffer (see (12)).

All remaining cases for which the state transition probability is not null are given by (13). All states of S^h with their state probability $\nu_s^h(n)$ have to be considered except states where no HOL packet is present at input buffer i ($s \in S^h |_{s_i \neq 0}$). The resulting probability has to be normalized with the sum of all considered state probabilities. Five different situations are distinguished (defined in (14)–(18)):

The probability for a full buffer ($j = m_{max}(i)$) at time step n and a full buffer ($k = m_{max}(i)$) at $n + 1$ is given by (14). Two situations could have happened:

$$p_{00}^q (i, n+1) = 1 - \tilde{q}(i,n) \qquad (11) \qquad\qquad p_{01}^q (i, n+1) = \tilde{q}(i,n) \qquad (12)$$

$$p_{jk|_{j>0}}^q (i, n+1) = \frac{1}{\sum\limits_{s \in S^h|_{s_i \neq 0}} \nu_s^h(n)} \cdot \sum\limits_{s \in S^h|_{s_i \neq 0}} \nu_s^h(n)$$

$$\cdot \begin{cases} \tilde{p}_{m_{\max}(i)m_{\max}(i)}^q (i, s, n), & \text{if } j = k = m_{\max}(i) \\ \tilde{p}_{jj}^q (i, s, n), & \text{if } j = k < m_{\max}(i) \\ \tilde{p}_{j(j-1)}^q (i, s, n), & \text{if } j < m_{\max}(i) \wedge k = j - 1 \\ \tilde{p}_{m_{\max}(i)(m_{\max}(i)-1)}^q (i, s, n), & \text{if } m_{\max}(i) = j = k+1 \\ \tilde{p}_{j(j+1)}^q (i, s, n), & \text{if } k = j+1 \wedge j < m_{\max}(i) \end{cases}$$

$$(13)$$

$$\tilde{p}_{m_{\max}(i)m_{\max}(i)}^q (i, s, n+1) = 1 - r(s_i, n) + r(s_i, n) \cdot \frac{|ci(s_i, s)| - 1}{|ci(s_i, s)|} \qquad (14)$$

$$\tilde{p}_{jj}^q (i, s, n+1) = (1 - r(s_i, n)) \cdot (1 - \tilde{q}(i,n))$$
$$+ r(s_i, n) \cdot \frac{|ci(s_i, s)| - 1}{|ci(s_i, s)|} \cdot (1 - \tilde{q}(i,n)) + r(s_i, n) \cdot \frac{1}{|ci(s_i, s)|} \cdot \tilde{q}(i,n)$$

$$(15)$$

$$\tilde{p}_{j(j-1)}^q (i, s, n+1) = (1 - \tilde{q}(i,n)) \cdot \frac{r(s_i, n)}{|ci(s_i, s)|} \qquad (16)$$

$$\tilde{p}_{m_{\max}(i)(m_{\max}(i)-1)}^q (i, s, n+1) = \frac{r(s_i, n)}{|ci(s_i, s)|} \qquad (17)$$

$$\tilde{p}_{j(j+1)}^q (i, s, n+1) = \tilde{q}(i,n) \cdot \left((1 - r(s_i, n)) + r(s_i, n) \cdot \frac{|ci(s_i, s)| - 1}{|ci(s_i, s)|} \right) \qquad (18)$$

Fig. 3. State transition probabilities $\mathbf{P}^q(n)$ for the queue length Markov chains

the succeeding buffer is not able to receive a packet $(1 - r(s_i, n))$ or it is able to receive $(r(s_i, n))$ but the packet of this buffer lost the conflict with another HOL packet $((|ci(s_i, s)| - 1) / |ci(s_i, s)|)$.

If the queue length does not change and the buffer is neither full nor empty, the probability is determined by (15). The following three situations result in an unchanged queue length: First, the HOL packet is blocked $(1 - r(s_i, n))$ because the succeeding buffer is full and no new packet is received $(1 - \tilde{q}(i,n))$. Second, the target is available but this HOL packet loses the conflict to another HOL packet. If no competition is present in this HOL state, the term becomes 0 because $|ci(s_i, s)|$ would be 1.

The last case is similar to the previous case but now this HOL packet wins the conflict and a new packet is coming to this buffer.

The state transition probability for a decreasing queue length, if the queue is not full, is calculated by (16). The queue length can only decrease if no new packet arrives $(1 - \tilde{q}(i,n))$, the output is not blocked $(r(s_i, n))$, and the HOL packet of this buffer wins a potential conflict $(1/ |ci(s_i, s)|)$.

Equation (17) also describes a decreasing queue length but for the case of a full buffer. Here, the probability that a new packet is received has to be omitted $(1 - \tilde{q}(i, n))$ due to the local backpressure mechanism. No packet is accepted at a full buffer.

Finally, the state transition probability for an increased queue length is determined by (18). This situation arises if a packet is received $(\tilde{q}(i, n))$ and the output is blocked $(1 - r(s_i, n))$, or the output is not blocked but the HOL packet lost a conflict $(((|ci(s_i, s)| - 1) / |ci(s_i, s)|))$.

State Probabilities of the Virtual Output Buffers. As mentioned before, the virtual output buffers are a tool to simplify the calculation of the spatial distribution within the network. Their state is derived from the state probability vector of the HOL Markov chain $\nu^h(n)$ at the switch in question and the receiving probabilities of the succeeding component.

The probability for virtual output buffer o to "hold" a packet from input buffer i at time step n is given by

$$\nu_i^o\Big|_{0 < i \leq i_{max}} (o, n) = r(o, n) \cdot \sum_{s \in S^h \wedge s_i = o} \frac{\nu_s^h(n)}{|ci(o, s)|}.$$

A packet can only enter o if the succeeding component is unblocked $(r(o, n))$. This probability is multiplied with the sum of all HOL states where input i has this output o as target $(s_i = o)$ and the probability for winning a possible conflict with other inputs to the same target $(1/|ci(o, s)|)$.

The probability for an empty virtual output buffer, i.e., no packet will move through this buffer during time step n, can be calculated in the same manner. Either the succeeding component is blocked $(1 - r(o, n))$ or if the succeeding component is available $(r(o, n))$, then the state probability $(\nu_s^h(n))$ of all HOL states where no HOL packet is targeted to output o $(s \in S^h \wedge o \notin s)$ have to be considered:

$$\nu_0^o(o, n) = (1 - r(o, n)) + r(o, n) \cdot \sum_{s \in S^h \wedge o \notin s} \nu_s^h(n).$$

Routing. To support irregular network topologies and non-uniform traffic, spatial traffic distributions and routing have to be regarded. The probability that a packet wants to proceed from source s to destination d of the network is given by the spatial distribution matrix ℓ_{sd}^g. These global routing probabilities are mapped to local ones for each switch. This local matrix $\ell_{io}^l(n)$ denotes the probabilities that a packet at switch input i moves to output o at time step n.

To achieve this, routing tables at each switch are established. With Dijkstra's algorithm, all possible shortest paths from each source to each destination are calculated. Each path is traversed and at every switch along the path, a routing entry is generated. This entry is identified by the source/destination pair as well as the local input/output pair and consists of the number of redundant paths a packet could take and the probability that a packet moves along this specific path (from s to d via i/o at this switch). The probabilities of the routing entries are iteratively set by getting the probability of the related entry from the preceding switches or sources on the routing path and weighted with the probability that a packet leaves the preceding router on this path (given by $\nu^o_{\text{Pred}}(n)$). If a redundant path is available at a switch, the probability for a routing entry is divided by the number of outputs a packet could take (random selection of paths).

The local probability matrices ℓ^l are iteratively determined during the fixed-point iteration by summation of the probabilities of the routing entries for each input/output pair.

3.3 Performance Measures

The steady-state performance of a network can be determined by this model since the Markov chains are ergodic; they are aperiodic and positive recurrent. This property is ensured by having time-independent load and spatial distributions as model inputs.

Network performance is usually determined by the normalized throughput, delay times for packets, and queue lengths. These measures can be determined within the network or at the destinations. The normalized throughput λ_o at output o of a switch is given by the probability that a packet moves through its virtual output buffer: $\lambda_o(n) = 1 - \nu^o_0(o,n)$.

The normalized throughput of input buffer i is then given by the sum of the probabilities that this input sends a packet through a virtual output buffer: $\lambda_i(n) = \sum_{o \in \{1,...,o_{\max}\}} \nu^o_i(o,n)$.

If the switch output o is connected to destination d, the throughput $\lambda_d(n)$ is given by $\lambda_d(n) = \lambda_{\text{Pred}(o)}(n) = 1 - \nu^o_{\text{Pred}(o)}(o,n)$.

The weighted sum of state probabilities of the queue length DTMC determines the mean queue length of input buffer i: $\bar{m}(i,n) = \sum_{j=1}^{m_{\max}(i)} j \cdot \nu^q_j(i,n)$. Little's Law calculates the mean delay of a packet residing in input buffer i within the network: $d(i,n) = \frac{\bar{m}(i,n)}{\lambda_i(n)}$. To get the packet's mean delay at the network destinations, each routing entry also holds, besides its probability, the accumulated mean delay of its path. At the destination, these accumulated delays are weighted with their path probability and then summed.

These steady-state performance measures of the network are calculated by the same iterative method which has been applied to similar models (e.g. [3]): start with an empty network and iterate over the above equations and transition tables until all performance measures reach steady-state. Each time step is divided into smaller steps to calculate the different equations presented above. These calculations are executed concurrently for all network components within the following minor steps:

1. The State transition probabilities for the queue length of the input buffers $\mathbf{P}^q$ and the state probabilities for the HOL DTMCs $\mathbf{P}^h$ are determined for the next time step..

2. The state vectors ν^q and ν^h are calculated for the input buffers and the HOL Markov chain as well as the receiving probabilities of the destinations.
3. During this minor step, the state vectors of the virtual output buffers ν^o are derived and the mean queue lengths $\bar{m}$ are calculated.
4. Based on the results of the former steps, the normalized receiving probability $\tilde{q}$ at input buffers in case space is available, the new routing probabilities for each routing entry, and the mean delays are determined.
5. With the results of the former step, the local routing tables ℓ^l and the updated delays for the routing entries are calculated.
6. Finally, the delays for each router output are computed.

These minor steps are necessary to support the concurrent calculations without evaluating the topology of the network and some equations are dependent to equations in preceding or succeeding network components.

The intermediate results during iteration show the transient behavior of the network. Up to now, the existence of the fixed point has not been proven. However, all of our tests show that it exists. We have tested the model with a large variety of networks with regular and irregular topologies and traffic distributions ranging from 1×2 to 32×32 networks.

4   Results

The DTMC model is implemented in a tool, called CINAn (Component-based Interconnection Network Analysis). It uses caches and precalculates feasible state transitions at setup to reduce computation time dramatically during fixed-point iteration. During the iteration steps, only non-dynamic data structures are used and the calculation of the state transition probabilities is parallelized. CINAn uses the same file format for network descriptions as CINSim. Thus, CINSim's Graphical User Interface (GUI) (see Fig. 4) can be used for model description and the results of the DTMC model can easily be compared to simulation results.

As an example, we evaluate a 8×8 bidirectional MIN with three stages (shown in Fig. 4). This example shows the support of different switch sizes (4×4 in the first two stages and 2×2 in the last stage) and different route length in the network. The input buffers have a queue length of four packets. This example has eight HOL DTMCs with 625 states each, four HOL DTMCs with nine states each and forty queue length DTMCs with 5 states each.

The spatial distribution of each source is set to $\ell^g_{sd} = \frac{d}{36}$ for Destinations $\{1, \ldots, 8\}$, i.e., Destination 2 is targeted with the probability $\frac{2}{36} = \frac{1}{18}$ and Destination 8 with $\frac{8}{36} = \frac{2}{9}$. The offered load is increased at each run. The results of the proposed model are compared with simulation runs and depicted in Fig. 5. The normalized throughput and the mean delay are measured for Destinations 1–4 and Destinations 5–8 separately. The simulation termination criteria was set to a confidence level of 99 % and a precision of 0.1.

The results indicate that the normalized throughput of the proposed model matches very well with the simulation. The mean delay also matches very well until the network reaches saturation. In the part of the network where the load is

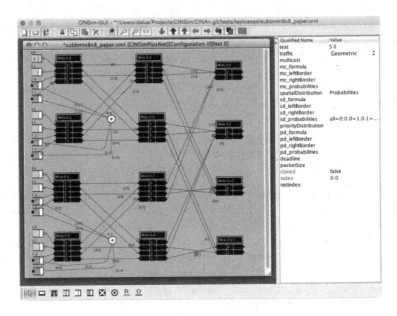

Fig. 4. Screenshot of the GUI of CINSim with a three-stage 8×8 bidirectional MIN

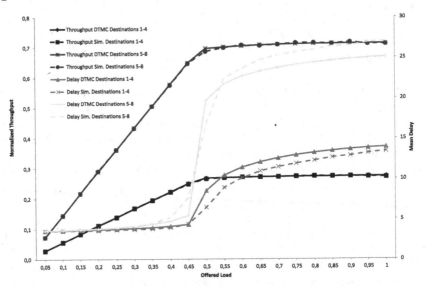

Fig. 5. Steady-state normalized throughput and mean delay of an 8×8 bidirectional MIN shown in Fig. 4 with an asymmetric spatial traffic distribution results obtained by the DTMC model and simulation.

very high (Destinations 5–8), the DTMC model underestimates the mean delay when the network load increases. This is typical for these kind of models and has been reported before (e.g. [3,11]). We expect that these errors are caused by the decomposition of the model.

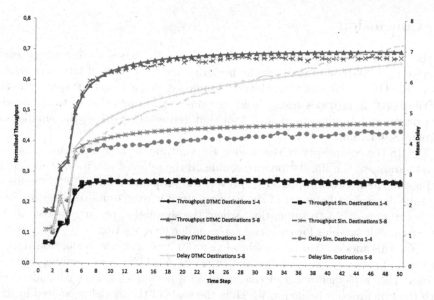

Fig. 6. Normalized throughput and mean delay for time step 0–50 of the 8 × 8 bidirectional MIN with an asymmetric spatial traffic distribution shown in Fig. 4 with an offered load of 0.5

For Destinations 1–4 another effect predominates. In case a conflict happens at a switch, the DTMC model does not consider a possible redundant routing path for packets that have lost the conflict. CINSim, however, tries to find an alternative path for packets that have lost a conflict. Hence, the DTMC model estimates the mean delay higher if a network has redundant paths.

Figure 6 shows the finite horizon performance measures of the same network for an offered load of 0.5 for the first fifty time steps. Again, the throughput is close to the simulation and with some more differences for the delay.

The observed errors of the DTMC model are representative to other scenarios we evaluated, for instance, asymmetric topologies, larger networks, etc.

Table 1 shows the runtime and memory usage of the DTMC model and the simulation for the steady-state (Fig. 5) and the finite horizon (Fig. 6) case with an offered load of 0.5, respectively. The results were obtained on a Linux machine with two Intel Xeon E5 processors with four cores each with a clock frequency of 3.3 GHz.

Table 1. Runtime and memory usage for results of Figs. 5 and 6

	Runtime DTMC	Runtime sim.	Memory DTMC	Memory sim.
Steady-state	13 s	18 s	195 MiB	73 MiB
Time-depended	2 s	82 s	158 MiB	74 MiB

5   Conclusions

In this paper we presented a new analytical model, based on inhomogeneous discrete time Markov chains, for the performance evaluation of interconnection networks. The model supports arbitrary topologies and asymmetric spatial traffic distributions. It provides results that are close to a simulation of the network. The model was implemented in a tool that reuses the GUI of the simulation framework CINSim.

Due to the complexity of the model, the runtime of the fixed-point iteration is comparable to the simulation, depending on the selected precision of the simulation and the network topology. However, the proposed model directly delivers the time-dependent behavior of the network until it reaches its steady-state in a single run in contrast to simulation, where the simulation restarts repeatedly to gather enough samples for each time step under investigation.

With this model, two approaches exists for performance evaluation during the design process of future onboard computers. This increases confidence in the results. The implementation of the SpaceWire network switches coincides well with the synchronous behavior, which is the basis of the model presented in this paper. Future work will compare actual measurements of the SpaceWire network in different configurations with the results of the presented model.

Currently, the DTMC model is extended to support Wormhole routing by adding an additional Markov chain to keep track of the part of the packet currently at the HOL position at a switch. This model will then be used for optimizations of configurations for future spacecraft onboard computers. Furthermore, other routing algorithms can easily be added to the model, e.g., west-first or xy routing, by replacing Dijkstra's algorithm during model setup.

References

1. Amiri-Zarandi, M., Safaei, F., Roozikhar, M.: Performance evaluation of generic multi-stage interconnection networks with blocking and back-pressure mechanism. J. Supercomputing **71**(3), 1038–1066 (2015)
2. Atiquzzaman, M., Akhtar, M.: Performance of buffered multistage interconnection networks in a nonuniform traffic environment. J. Parallel Distrib. Comput. **30**(1), 52–63 (1995)
3. Dias, D., Jump, J.: Analysis and simulation of buffered delta networks. IEEE Trans. Comput. **C-30**(4), 273–282 (1981)
4. Hamid, N., Walters, R.J., Wills, G.B.: An analytical model of multi-core multi-cluster architecture (MCMCA). Open J. Cloud Comput. (OJCC) **2**(1), 1–12 (2015)
5. Kiasari, A., Lu, Z., Jantsch, A.: An analytical latency model for networks-on-chip. IEEE Trans. Very Large Scale Integr. (VLSI) Syst. **21**(1), 113–123 (2013)
6. Lüdtke, D., Westerdorff, K., Stohlmann, K., Börner, A., Maibaum, O., Peng, T., Weps, B., Fey, G., Gerndt, A.: OBC-NG: towards a reconfigurable on-board computing architecture for spacecraft. In: 2014 IEEE Aerospace Conference, pp. 1–13, March 2014
7. Lüdtke, D., Tutsch, D.: The modeling power of CINSim: performance evaluation of interconnection networks. Comput. Netw. **53**(8), 1274–1288 (2009)

8. Moadeli, M., Shahrabi, A., Vanderbauwhede, W., Ould-Khaoua, M.: An analytical performance model for the spidergon NoC. In: 21st International Conference on Advanced Information Networking and Applications, AINA 2007, pp. 1014–1021, May 2007

9. Parkes, S., Armbruster, P.: SpaceWire: a spacecraft onboard network for real-time communications. In: 14th IEEE-NPSS Real Time Conference, pp. 6–10 (2005)

10. Sabbaghi-Nadooshan, R., Patooghy, A.: Analytical performance modeling of de Bruijn inspired mesh-based network-on-chips. Microprocess. Microsyst. **39**(1), 27–36 (2015)

11. Tutsch, D., Hommel, G.: Generating systems of equations for performance evaluation of multistage interconnection networks. J. Parallel Distrib. Comput. (JPDC) **62**(2), 228–240 (2002)

12. Yoon, H., Lee, K.Y., Liu, M.T.: Performance analysis of multibuffered packet-switching networks in multiprocessor systems. IEEE Trans. Comput. **29**(3), 319–327 (1990)

13. Youn, H.Y., Mun, Y.: On multistage interconnection networks with small clock cycles. IEEE Trans. Parallel Distrib. Syst. **6**(1), 86–93 (1995)

14. Zhou, B., Atiquzzaman, M.: Efficient analysis of multistage interconnection networks using finite output-buffered switching elements. Comput. Netw. ISDN Syst. **28**(13), 1809–1829 (1996)

Whittle's Index Policy for Multi-Target Tracking with Jamming and Nondetections

José Niño-Mora$^{(\boxtimes)}$

Department of Statistics, Carlos III University of Madrid, Getafe, Spain
jnimora@alum.mit.edu
http://alum.mit.edu/www/jnimora

Abstract. This paper proposes a tractable priority-index policy based on the Whittle index for multi-target tracking with jamming and nondetections. The policy is to be used by M phased-array radars tracking the positions of $N > M$ targets moving according to independent scalar Gauss–Markov linear dynamics, which allows use of a random Kalman filter for track-error variance updates. The dynamics incorporate random jamming or nondetections decreasing the quality of measurement. The paper exploits the natural problem formulation as a multi-armed restless bandit problem with real-state projects by deploying Whittle's index policy. The issues of indexability (existence of the index) and index evaluation are addressed by deploying a method introduced by the author in earlier work, based on a set of sufficient indexability conditions. Preliminary numerical results are reported providing evidence that such conditions hold and evaluating the index.

Keywords: Multi-target tracking · Jamming · Nondetections · Random Kalman filter · Index policies · Restless bandits · Whittle index · Indexability

1 Introduction

1.1 Motivation and Background

Phased-array radars are able to dynamically steer the radar's beam towards desired directions. The flexibility of such a capability raises the research challenge of optimizing tracking performance, through the design of appropriate *scheduling control policies* for dynamic prioritization of target track updates when more targets than radars are present. In practice, the quality of radar measurements can be deteriorated in an unpredictable fashion by *jamming*, which can be due to several factors, e.g., occlusions, clutter or electronic countermeasures. See, e.g., [1–3]. Such a deterioration of the measurement can reach the extreme of *nondetection* of the target.

This paper extends the approach to multi-target tracking model introduced in [4] by incorporating the possibility of random jamming, for the dynamic tracking of a fixed number N of moving targets using $M < N$ radars. We refer the reader

© Springer International Publishing Switzerland 2016
S. Wittevrongel and T. Phung-Duc (Eds.): ASMTA 2016, LNCS 9845, pp. 210–222, 2016.
DOI: 10.1007/978-3-319-43904-4_15

to [5] for a survey of the literature in the area. For early work on the subject of track-update scheduling, see, e.g., [6–8].

The model we consider here is based on and extends that formulated in [9], in which targets and target track measurements follow scalar, linear Gauss–Markov dynamics, and target *track-error variances* (TEVs) are updated via Kalman filter equations. The objective to optimize in [9] is the sum of the targets track error variances over a finite horizon. The authors propose a *greedy* scheduling policy, which at each time updates a target of largest TEV, thus taking a target's current TEV as its *priority index*. They further claim such a *greedy-index* policy to be optimal in the case of two symmetric targets.

In [10], the authors extend the results in [9] on optimality of the greedy-index scheduling policy for tracking two symmetric targets to the case of more general scalar Gauss–Markov dynamics under the same performance objective. They note that such a problem falls within the framework of the *multiarmed restless bandit problem* (MARBP) introduced by Whittle in [11], although they do not use the indexation approach proposed there.

The MARBP is a powerful modeling framework that concerns the optimal dynamic scheduling of N stochastic projects, M of which must be engaged at each time. Each project (or *bandit*) is modeled as a binary-action (active or passive) *Markov decision process* (MDP). The goal is to find a scheduling policy that maximizes the expected total discounted (ETD) or the long-run expected time-average reward earned over an infinite horizon. In the special classic case where passive projects do not change state and $M = 1$, the optimal policy is given by the *Gittins index* [12] policy: an index $\lambda^{*,n}(x^n)$ is attached to each project n as a function of its state x^n, and then a project of largest index is engaged at each time. In [13], a Gittins index policy is proposed for multi-target tracking in a classic (nonrestless) multiarmed bandit model, where projects represent targets, assuming that a target's motion from one period to the next is negligible. As pointed out in [10], however, such an assumption is typically inadequate, calling for use of the MARBP model where passive projects can change state.

Although *index policies* are generally suboptimal for the MARBP, Whittle introduced in [11] a heuristic index policy for the MARBP, which emerges from a Lagrangian relaxation and decomposition approach that also yields a bound on the optimal problem value. The *Whittle index* raises substantial research challenges such as (i) *indexability* (i.e., existence of the index) needs to be established for the model at hand; and (ii) the index needs to be evaluated in a tractable fashion. We remark that [14] has make considerable progress towards such goals in the special version of the present model without jamming.

The author has developed a methodology for resolving such issues on *restless bandit indexation* in the discrete-state case, in a stream of work starting in [15–17], which is reviewed in [18]. More recently he has announced in [19] extensions to real-state restless bandits, which are the cornerstone of the present paper's approach, and whose theoretical justification is provided in [20].

1.2   Goals and Contributions

This paper extends such a line of work by investigating an MARBP formulation of dynamic tracking of multiple asymmetric targets with scalar linear Gauss–Markov dynamics, which incorporates jamming or nondetection, the main goal being to obtain a tractable index policy that performs well based on restless bandit indexation. Such as model extends that considered in [4,14], which does not consider the possibility of jamming.

The paper deploys the methodology for real-state restless bandit indexation announced in [19] and proven in [4] to address the issues of indexability and evaluation of the Whittle index in an efficient fashion for the model of concern.

1.3   Organization of the Paper

The remainder of the paper is organized as follows. Section 2 describes the multi-target tracking model and its MARBP formulation. Section 3 discusses the restless bandit indexation methodology for real-state restless bandits introduced in [19] as it applies to the design of index policies for multi-target tracking. Section 4 discusses how to deploy such a methodology in the present model to verify indexability and approximately evaluate the Whittle's index. Section 4.1 reports the results of an exploratory numerical study on satisfaction of the sufficient indexability conditions and Whittle index evaluation. Section 5 concludes.

2   Multi-target Tracking and Restless Bandits

2.1   Random Kalman Filter Model of Multi-target Tracking

We consider a system consisting of M identical radars labeled by $m \in \mathsf{M} \triangleq \{1, \ldots, M\}$, which are used to track $N > M$ heterogeneous moving targets labeled by $n \in \mathsf{N} \triangleq \{1, \ldots, N\}$. All radars are synchronized to operate over a common set of discrete time periods $t = 0, 1, \ldots$

At the start of each time period t the system controller selects a subset of M targets to update their track information, by steering towards them as many radar beams to measure their positions.

The targets are assumed to move in one dimension over the real line $\mathbb{R}$ according to independent linear Gauss–Markov dynamics. Thus, letting $\xi_t^n \in \mathbb{R}$ be the (unobservable) *position* of target n at the start of period t, we assume that

$$\xi_t^n = \xi_{t-1}^n + \omega_t^n, \quad t = 0, 1, \ldots, \tag{1}$$

where the ω_t^n are i.i.d. Normal random variables with mean 0 and variance q^n.

If a radar beam is steered towards target n in period t, a noisy measurement

$$y_t^n = \xi_t^n + \nu_t^n, \tag{2}$$

is obtained, where the ν_t^n are i.i.d. random variables, each being a mixture with weights $\rho > 0$ and $1 - \rho > 0$ of two Normal random variables with mean 0 and variances $r_1 < r_2$, respectively.

We thus model the situation in which a high quality measurement with variance r_1 is obtained with probability ρ, and a low quality measurement with a larger variance r_2 is obtained with probability $1 - \rho$. The low quality measurement models the phenomenon of *jamming* or clutter. The extreme case $r_2 = \infty$ can be used to model nondetection of the target.

The decisions on which targets to track at each time period are formulated by binary *actions* $a_t^n \in \{0, 1\}$, where $a_t^n = 1$ if target n is tracked in period t, and $a_t^n = 0$ otherwise.

The *track error variance* (TEV) x_t^n for target n at the beginning of period t evolves according to the following *random Kalman filter* dynamics (see [9] for a model with the standard deterministic Kalman filter dynamics). If a radar beam is steered towards target n in period t, so $a_t^n = 1$, then

$$
x_t^n = \begin{cases}
\dfrac{x_{t-1}^n + q^n}{1 + (x_{t-1}^n + q^n)/r_1} & \text{with probability } \rho \\[4mm]
\dfrac{x_{t-1}^n + q^n}{1 + (x_{t-1}^n + q^n)/r_2} & \text{with probability } 1 - \rho,
\end{cases}
\tag{3}
$$

whereas if it is not steered towards target n ($a_t^n = 0$) then

$$
x_t^n = x_{t-1}^n + q^n.
\tag{4}
$$

We will find it more convenient to take the *state* of target n to be its *scaled TEV* (STEV) $s_t^n \triangleq x_t^n/q^n$, which evolves over the state space of nonnegative reals $\mathsf{S} \triangleq \mathbb{R}_+$. Hence, writing $\alpha_k^n \triangleq q^n/r_k$ for $k = 1, 2$, the state s_t^n evolves according to the following dynamics. If $a_t^n = 1$ then

$$
s_t^n = \begin{cases}
\phi_1^{1,n}(s_{t-1}^n) & \text{with probability } \rho \\[3mm]
\phi_2^{1,n}(s_{t-1}^n) & \text{with probability } 1 - \rho,
\end{cases}
\tag{5}
$$

whereas if $a_t^n = 0$ then

$$
s_t^n = \phi^0(s_{t-1}^n),
\tag{6}
$$

where

$$
\phi^0(s) \triangleq s + 1 \quad \text{and} \quad \phi_k^{1,n}(s) \triangleq \frac{s+1}{\alpha_k^n(s+1)+1}, \quad k = 1, 2.
\tag{7}
$$

Note that the case $\alpha_2^n > 0$ models deterioration of the measurement quality due to jamming or clutter whereas $\alpha_2^n = 0$ models nondetection of the target.

Actions are selected according to a *scheduling policy* π, taken from the class Π of *admissible policies*, which are nonanticipative (based on the history of states and actions) and track exactly M targets per period, so

$$
\sum_{n \in \mathsf{N}} a_t^n = M, \quad t = 0, 1, \ldots
\tag{8}
$$

We take the *track-error cost* in period t to be x_t^n. Hence, the one-period cost incurred on target n when it occupies state s is $C^n(s) \triangleq q^n s_t^n$.

2.2 Multiarmed Restless Bandit Formulation

Consider the following dynamic optimization problem: for a given discount factor $0 < \beta < 1$, find a policy that minimizes the *expected total discounted* (ETD) cost starting from any initial joint state $\mathbf{s} = (s^n)$, i.e., which solves the problem

$$\underset{\pi \in \Pi}{\text{minimize}} \; \mathsf{E}_{\mathbf{s}}^{\pi} \left[\sum_{t=0}^{\infty} \sum_{n \in \mathsf{N}} \beta^t C^n(s_t^n) \right]; \tag{9}$$

for any $\mathbf{s}$, where $\mathsf{E}_{\mathbf{s}}^{\pi}[\cdot]$ denotes expectation under policy π starting from $\mathbf{s}_0 = \mathbf{s}$ and Π is the class of admissible scheduling policies referred to above. We denote by $V^*(\mathbf{s})$ the minimum value of problem (9)'s cost objective.

Problem (9) is a multiarmed restless bandit problem with real-state projects. Such problems are known to be generally intractable. See [21].

2.3 Index Policies

We aim to design and compute a heuristic policy of priority-index type. Such policies attach an index $\lambda^n(s^n)$ to each target n as a function of its state s^n. At time t, the resulting index policy selects M targets to track using $\lambda^n(s_t^n)$ as a priority index for measuring target n, where a larger index value means a higher priority, breaking ties arbitrarily.

3 Restless Bandit Indexation

3.1 Relaxed Problem, Lagrangian Relaxation and Decomposition

As introduced by Whittle in [11], we first construct a *relaxation* of (9), replacing the sample-path constraint (8) that M targets be tracked at each time by the aggregate constraint that the ETD number of tracked targets be $M/(1-\beta)$, i.e.,

$$\mathsf{E}_{\mathbf{s}}^{\pi} \left[\sum_{t=0}^{\infty} \sum_{n \in \mathsf{N}} \beta^t a_t^n \right] = \frac{M}{1 - \beta}. \tag{10}$$

Denoting by $\widehat{\Pi}$ the class of nonanticipative scheduling policies that are allowed to track any number of targets at any time, consider the *relaxed problem*

$$\text{minimize} \; \mathsf{E}_{\mathbf{s}}^{\pi} \left[\sum_{t=0}^{\infty} \sum_{n \in \mathsf{N}} \beta^t C^n(s_t^n) \right].$$

$$\text{subject to}: (10), \pi \in \widehat{\Pi} \tag{11}$$

Since $\Pi \subset \widehat{\Pi}$, the minimum cost objective $V^{\mathrm{R}}(\mathbf{s})$ of problem (11) gives a *lower bound* on the minimum cost $V^*(\mathbf{s})$ of the original problem (9).

To address (11) we deploy a Lagrangian approach, attaching a multiplier $\lambda \in \mathbb{R}$ to the aggregate constraint (10) and dualizing it. The resulting problem

$$\underset{\pi \in \widehat{\Pi}}{\text{minimize}} \, \mathsf{E}_{\mathbf{s}}^{\pi} \left[\sum_{t=0}^{\infty} \sum_{n \in \mathsf{N}} \beta^t \left\{ C^n(s_t^n) + \lambda a_t^n \right\} \right] - \frac{M\lambda}{1-\beta} \tag{12}$$

is a *Lagrangian relaxation* of (11), whose minimum cost objective value $V^{\mathrm{L}}(\mathbf{s}; \lambda)$ gives a lower bound on $V^{\mathrm{R}}(\mathbf{s})$. The *Lagrangian dual problem* is to find an optimal value $\lambda^*(\mathbf{s})$ of λ giving the best such lower bound, which we denote by $V^{\mathrm{D}}(\mathbf{s})$:

$$\underset{\lambda \in \mathbb{R}}{\text{maximize}} \, V^{\mathrm{L}}(\mathbf{s}; \lambda). \tag{13}$$

Note that $V^{\mathrm{L}}(\mathbf{s}; \lambda)$ is concave in λ, which simplifies the solution of (13).

Coming back to problem (12), it *decomposes* into the N subproblems

$$\underset{\pi^n \in \Pi^n}{\text{minimize}} \, \mathsf{E}_{s^n}^{\pi^n} \left[\sum_{t=0}^{\infty} \beta^t \left\{ C^n(s_t^n) + \lambda a_t^n \right\} \right], \tag{14}$$

where Π^n is the class of nonanticipative tracking policies for target n *in isolation*. Note that in (14) multiplier λ represents a *measurement cost*.

3.2 Indexability and Whittle's Index Policy

Consider now target n's subproblem (14) treating the measurement charge λ as a *parameter* ranging over $\mathbb{R}$. We next define a key structural property of such a parametric collection of subproblems, termed *indexability*, which simplifies its solution and hence that of (13). The indexability property of restless bandits was introduced by Whittle in [11].

Definition 1. We say that the parametric collection of subproblems (14), as $\lambda \in \mathbb{R}$, is *indexable* if there exists an *index* function $\lambda^{*,n}: S \to \mathbb{R}$ such that, for any $\lambda \in \mathbb{R}$, it is optimal in (14)—regardless of the initial state—to take action $a_t^n = 1$ when the target is in state $s_t^n = s$ if and only if $\lambda^{*,n}(s) \geqslant \lambda$, and it is optimal to take action $a_t^n = 0$ if and only if $\lambda^{*,n}(s) \leqslant \lambda$. In such a case we say that such $\lambda^{*,n}$ is the *Whittle index* of target n.

If each single-target subproblem (14) were indexable and if a tractable procedure were available to evaluate the Whittle index $\lambda^{*,n}$, then we would have a tractable scheme to solve Lagrangian dual problem (13)—provided the objective of (14) could also be efficiently evaluated—and thus compute the lower bound $V^{\mathrm{D}}(\mathbf{s})$ referred to above. Further, we could then use for multi-target problem (9) the *Whittle index policy*, which uses $\lambda^{*,n}$ as target n's priority index.

3.3 Sufficient Indexability Conditions and Index Evaluation

Yet, indexability needs to be established for the model at hand. For such a purpose, the author introduced in work reviewed in [18] sufficient indexability conditions for discrete-state restless bandits based on satisfaction on *partial conservation laws* (PCLs), along with an index algorithm. The author has further extended the scope of such conditions to real-state restless bandits in results first announced in [19] and proven in [20], as reviewed next. The ensuing discussion focuses on a single-project restless bandit modeling the optimal tracking of a single target, whose label n is henceforth dropped. We thus write, e.g., the target's state and action processes as s_t and a_t, respectively.

We evaluate the performance of an admissible single-target tracking policy $\pi \in \Pi$ along two dimensions: the *cost metric*

$$F(s, \pi) \triangleq \mathsf{E}_s^\pi \left[\sum_{t=0}^{\infty} \beta^t C(s_t) \right],$$

giving the ETD cost under policy π starting from $s_0 = s$, and the *work metric*

$$G(s, \pi) \triangleq \mathsf{E}_s^\pi \left[\sum_{t=0}^{\infty} \beta^t a_t \right],$$

giving the corresponding ETD number of times the target is tracked.

The target's optimal tracking subproblem (14) is thus formulated as

$$\underset{\pi \in \Pi}{\text{minimize}}\, F(s, \pi) + \lambda G(s, \pi). \tag{15}$$

We will refer to (15) as the target's λ-*charge subproblem*. Problem (15) is a real-state MDP, whose optimal cost function we denote by $V^*(s; \lambda)$.

In order to solve (15) it suffices to consider *deterministic stationary policies*, which are naturally represented by their *active (state) sets*, i.e., the set of states where they prescribe the active action (track the target). For an active set $B \subseteq \mathsf{S}$, we will refer to the B-*active policy*.

We focus attention on the family of *threshold policies*. For a given *threshold level* $z \in \overline{\mathbb{R}} \triangleq \mathbb{R} \cup \{-\infty, \infty\}$, the z-*threshold policy* tracks the target when it occupies state s if and only if $s > z$, so its active set is $B(z) \triangleq \{s \in \mathsf{S} : s > z\}$. Note that $B(z) = (z, \infty)$ for $s \geqslant 0$, $B(z) = \mathsf{S} = [0, \infty)$ for $z < 0$, and $B(z) = \emptyset$ for $z = \infty$. We denote by $F(s, z)$ and $G(s, z)$ the corresponding metrics.

For fixed z, the cost metric $F(s, z)$ is characterized as the unique solution to the functional equation

$$F(s, z) = \begin{cases} C(s) + \beta \rho F(\phi_1^1(s), z) + \beta(1 - \rho) F(\phi_2^1(s), z), & s > z \\ C(s) + \beta F(\phi^0(s), z), & s \leqslant z. \end{cases} \tag{16}$$

whereas the work metric $G(s, z)$ is characterized by

$$G(s, z) = \begin{cases} 1 + \beta \rho G(\phi_1^1(s), z) + \beta(1 - \rho) G(\phi_2^1(s), z), & s > z \\ \beta G(\phi^0(s), z), & s \leqslant z, \end{cases} \tag{17}$$

We will use the marginal counterparts of such measures. For a threshold z and an action a, let $\langle a, z \rangle$ denote the policy that takes action a at time $t = 0$ and then adopts the z-threshold policy thereafter. Define the *marginal cost metric*

$$f(s, z) \triangleq \Delta_{a=1}^{a=0} F(s, \langle a, z \rangle), \tag{18}$$

where $\Delta_{a=b_0}^{a=b_1} h(a) \triangleq h(b_1) - h(b_0)$, and the *marginal work metric*

$$g(s, z) \triangleq \Delta_{a=0}^{a=1} G(s, \langle a, z \rangle). \tag{19}$$

If $g(s, z) \neq 0$, define further the *marginal productivity* (MP) metric

$$m(s, z) \triangleq \frac{f(s, z)}{g(s, z)}. \tag{20}$$

We further define the *MP index* by

$$m^*(s) \triangleq m(s, s). \tag{21}$$

The following definition extends to the real-state setting a corresponding definition introduced by the author in [15, 16, 22] for discrete-state restless bandits.

Definition 2. We say that subproblem (15) is *PCL-indexable* (with respect to threshold policies) if the following conditions hold:

- (PCLI1) $g(s, z) > 0$ for every state s and threshold z;
- (PCLI2) m^* is monotone nondecreasing, continuous and bounded below;
- (PCLI3) for each state s, $F(s, \cdot)$, $G(s, \cdot)$ and m^* are related by

$$F(s, z_2) - F(s, z_1) = \int_{(z_1, z_2]} m^*(z) \, G(s, dz), \quad -\infty < z_1 < z_2 < \infty. \tag{22}$$

The next result, which is proven in [20], extends the scope of corresponding results in [15, 16, 22] for discrete-state restless bandits to the real-state setting.

Theorem 1. *If subproblem (15) is PCL-indexable, then it is indexable and the MP index m^* is its Whittle index.*

4   Performance Metrics and MP Index Computation

The computation of performance metrics and of the MP index can be carried out recursively as discussed in [20, Appendix C]. It is shown there how to evaluate, for $k = 0, 1, \ldots$, k-horizon counterparts to the performance metrics and MP index defined above, which are denoted by $F_k(s, z)$, $G_k(s, z)$, $f_k(s, z)$, $g_k(s, z)$, $m_k(s, z)$ and $m_k^*(s)$. Under mild conditions, such finite-horizon metrics converge to $F(s, z)$, $G(s, z)$, $f(s, z)$, $g(s, z)$, $m(s, z)$ and $m^*(s)$ as $k \to \infty$.

In particular, we consider the *k-horizon performance metrics*

$$F_k(s, z) \triangleq \mathsf{E}_s^z \left[\sum_{t=0}^k \beta^t C(s_t) \right] \quad \text{and} \quad G_k(s, z) \triangleq \mathsf{E}_s^z \left[\sum_{t=0}^k \beta^t a_t \right]. \tag{23}$$

The function sequences $\{F_k(\cdot, z)\}_{k=0}^{\infty}$ and $\{G_k(\cdot, z)\}_{k=0}^{\infty}$ are determined by the following *value iteration* recursions: $F_0(s, z) \triangleq C(s)$, $G_0(s, z) \triangleq 1_{\{s > z\}}$ and, for $k = 0, 1, \ldots$,

$$F_{k+1}(s, z) \triangleq \begin{cases} C(s) + \beta \rho F_k(\phi_1^1(s), z) + \beta(1 - \rho)F_k(\phi_2^1(s), z), & s > z \\ C(s) + \beta F_k(\phi^0(s), z), & s \leqslant z. \end{cases} \quad (24)$$

$$G_{k+1}(s, z) \triangleq \begin{cases} 1 + \beta \rho G_k(\phi_1^1(s), z) + \beta(1 - \rho)G_k(\phi_2^1(s), z), & s > z \\ \beta G_k(\phi^0(s), z), & s \leqslant z. \end{cases} \quad (25)$$

Consider further the *k-horizon marginal metrics* $f_k(s, z) \triangleq \Delta_{a=1}^{a=0} F_k(s, \langle a, z \rangle)$ and $g_k(s, z) \triangleq \Delta_{a=0}^{a=1} G_k(s, \langle a, z \rangle)$, which are computed by $f_0(s, z) = 0$, $g_0(s, z) = 1$, and

$$f_{k+1}(s, z) = \beta \left[F_k(\phi^0(s), z) - \rho F_k(\phi_1^1(s), z) - \beta(1 - \rho)F_k(\phi_2^1(s), z) \right], \quad (26)$$

$$g_{k+1}(s, z) = 1 + \beta \rho G_k(\phi_1^1(s), z) + \beta(1 - \rho)G_k(\phi_2^1(s), z) - \beta G_k(\phi^0(s), z). \quad (27)$$

Finally, consider the *k-horizon MP metric* $m_k(s, z) \triangleq f_k(s, z)/g_k(s, z)$ and the *k-horizon MP index* $m_k^*(s) \triangleq m_k(s, s)$, which are defined when $g_k(s, z) \neq 0$ and $g_k(s, s) \neq 0$, respectively.

4.1 Exploring Numerically Satisfaction of PCL-indexability

Attempting to prove satisfaction of the PCL-indexability conditions above for the present model is beyond the scope of this paper. Instead, we report herein the results of an exploratory numerical study for checking satisfaction of conditions (PCLI1–PCLI3) in Definition 2 in several instances.

Regarding condition (PCLI1), i.e., positivity of the marginal work metric $g(s, z)$, we have approximately evaluated and plotted $g_k(s, z)$ for a wide range of model parameters, obtaining that $g_k(s, z) > 0$ in each instance considered.

As a sample result, consider a single-target instance with parameters $q = 1$, $\alpha_1 = 2$, $\alpha_2 = 1.3$, $\rho = 0.7$ and $\beta = 0.8$. Figures 1, 2 and 3 plot the k-horizon marginal work metric $g_k(s, z)$ as a function of the state s for horizon $k = 15$ and threshold values $z = 3, 7, 10$. In each case it is seen that $g_k(s, z)$ is positive, in agreement with (PCLI1). Note also that $g_k(s, z)$ is piecewise constant in s.

Regarding condition (PCLI2), i.e., that the MP index $m^*(s)$ be nondecreasing, continuous and bounded below, we have also evaluated and plotted $m_k^*(s)$ for a wide range of model parameters and horizons $k \leqslant 20$. For any given horizon k, such plots reveal that the function m_k^* is neither continuous nor nondecreasing, having a number of jump discontinuities. However, as k increases the magnitudes of such jumps get smaller, consistently with the conjecture that continuity holds in the limit as $k \to \infty$.

As a sample result, consider a single target model with parameters $\alpha_1 = 2$, $\alpha_2 = 1.3$, $\rho = 0.6$ and $\beta = 0.7$. Figure 4 plots the k-horizon MP index $m_k^*(s)$ as a function of the state s for horizon $k = 15$. Even though we have observed that

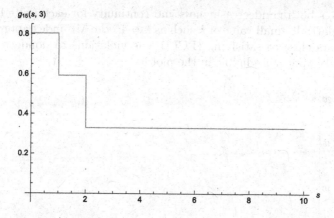

Fig. 1. Marginal work metric $g_k(s, z)$ vs. s for $k = 15$ and $z = 3$.

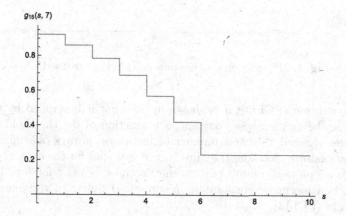

Fig. 2. Marginal work metric $g_k(s, z)$ vs. s for $k = 15$ and $z = 7$.

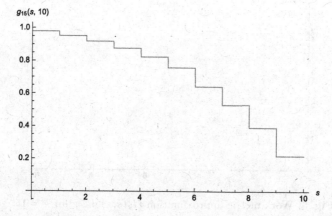

Fig. 3. Marginal work metric $g_k(s, z)$ vs. s for $k = 15$ and $z = 10$.

$m_k^*(s)$ violates both nondecreasingness and continuity for each finite horizon k, even for a relatively small value of k such as $k = 15$ the MP index approximation $m_k^*(s)$ appears close to satisfying (PCLI1), as violations to nondecreasingness and continuity appear negligible in the plot.

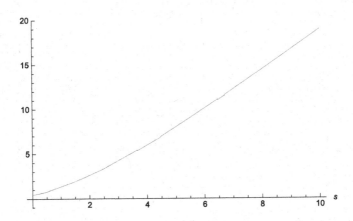

Fig. 4. MP index approximation $m_k^*(s)$ vs. s for $k = 15$.

As for condition (PCLI3), it is shown in [20] that it is implied by the work metric $G(s, z)$ being piecewise constant as a function of the threshold variable. We have also checked this in a number of instances, always obtaining that it holds. As an example, consider the same instance as that for the $g_k(s, z)$ reported above. Figure 5 plots the finite horizon metric $G_k(s, z)$ vs. z for $k = 15$, showing that it is a piecewise constant (nonincreasing) function, consistently with satisfaction of (PCLI3).

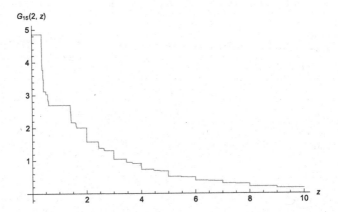

Fig. 5. Work metric approximation $G_k(s, z)$ vs. z for $k = 15$.

5   Concluding Remarks

This paper has introduced a model for multi-target tracking with jamming and misdetections, and has extended the approach in [4] to deploy Whittle's index policy for dynamically allocating M sensors to N targets in such a setting. The issues of establishing the indexability and evaluating the index have been addressed via the PCL-indexability sufficient conditions introduced and proven by the author in earlier work. Although no proof is given that such conditions hold in the present model, preliminary numerical evidence is provided that such is the case. Tasks for future work include testing empirically the performance of the proposed policy, and establishing theoretically satisfaction of the PCL-indexability conditions.

Acknowledgment. This work was partially supported by the Spanish Ministry of Economy and Competitiveness project ECO2015-66593-P.

References

1. Koch, W.: On exploiting 'negative' sensor evidence for target tracking and sensor data fusion. Inf. Fusion **8**, 28–39 (2007)
2. Pao, L., Powers, R.: A comparison of several different approaches for target tracking with clutter. In: Proceedings of 2013 American Control Conference, pp. 3919–3924. IEEE (2003)
3. Hou, J., Rong Li, X., Jing, Z.: Multiple model tracking of manoeuvring targets accounting for standoff jamming information. IET Radar Sonar Navig. **7**, 342–350 (2013)
4. Niño-Mora, J., Villar, S.S.: Multitarget tracking via restless bandit marginal productivity indices and Kalman filter in discrete time. In: Proceedings of 2009 CDC/CCC, Joint 48th IEEE Conference on Decision and Control and 28th Chinese Control Conference, pp. 2905–2910. IEEE, New York (2009)
5. Moran, W., Suvorova, S., Howard, S.: Application of sensor scheduling concepts to radar. In: Hero, A.O., Castañón, D., Cochran, D., Kastella, K. (eds.) Foundations and Applications of Sensor Management, pp. 221–256. Springer, New York (2008)
6. Van Keuk, G., Blackman, S.S.: On phased-array radar tracking and parameter control. IEEE Trans. Aerosp. Electron. Syst. **29**, 186–194 (1993)
7. Strömberg, D.: Scheduling of track updates in phased array radars. In: Proceedings of IEEE 1996 National Radar Conference, Ann Arbor, MI, pp. 214–219. IEEE (1996)
8. Hong, S.M., Jung, Y.H.: Optimal scheduling of track updates in phased array radars. IEEE Trans. Aerosp. Electron. Syst. **34**, 1016–1022 (1998)
9. Howard, S., Suvorova, S., Moran, B.: Optimal policy for scheduling of Gauss-Markov systems. In Svensson, P., Schubert, J. (eds.) Procedings of 7th International Conference on Information Fusion, pp. 888–892. International Society of Information Fusion, Mountain View (2004)
10. La Scala, B.F., Moran, B.: Optimal target tracking with restless bandits. Digital Signal Process. **16**, 479–487 (2006)
11. Whittle, P.: Restless bandits: activity allocation in a changing world. J. Appl. Probab. **25A**, 287–298 (1988)

12. Gittins, J.C.: Bandit processes and dynamic allocation indices. J. Roy. Stat. Soc. Ser. B **41**, 148–177 (1979). With discussion
13. Krishnamurthy, V., Evans, R.J.: Hidden Markov model multiarm bandits: a methodology for beam scheduling in multitarget tracking. IEEE Trans. Signal Process. **49**, 2893–2908 (2001)
14. Dance, C.R., Silander, T.: When are Kalman-filter restless bandits indexable? In: Cortes, C., Lawrence, N.D., Lee, D.D., Sugiyama, M., Garnett, R. (eds.) Advances in Neural Information Processing Systems, vol. 28, pp. 1711–1719. Curran Associates, Inc., Dundee (2015)
15. Niño-Mora, J.: Restless bandits, partial conservation laws and indexability. Adv. Appl. Probab. **33**, 76–98 (2001)
16. Niño-Mora, J.: Dynamic allocation indices for restless projects and queueing admission control: a polyhedral approach. Math. Program. **93**, 361–413 (2002)
17. Niño-Mora, J.: Marginal productivity index policies for scheduling a multiclass delay-/loss-sensitive queue. Queueing Syst. **54**, 281–312 (2006)
18. Niño-Mora, J.: Dynamic priority allocation via restless bandit marginal productivity indices. TOP **15**, 161–198 (2007)
19. Niño-Mora, J.: An index policy for dynamic fading-channel allocation to heterogeneous mobile users with partial observations. In: Proceedings of NGI 2008, 4th Euro-NGI Conference on Next Generation Internet Networks, pp. 231–238. IEEE, New York (2008)
20. Niño-Mora, J.:A verification theorem for indexability of discrete time real state-discounted restless bandits (2015). arXiv:1512.04403v1 [math.OC]
21. Papadimitriou, C.H., Tsitsiklis, J.N.: The complexity of optimal queuing network control. Math. Oper. Res. **24**, 293–305 (1999)
22. Niño-Mora, J.: Restless bandit marginal productivity indices, diminishing returns and optimal control of make-to-order/make-to-stock M/G/1 queues. Math. Oper. Res. **31**, 50–84 (2006)

Modelling Unfairness in IEEE 802.11g Networks with Variable Frame Length

Choman Othman Abdullah[1,2(✉)] and Nigel Thomas[2]

[1] School of Science Education, University of Sulaimani, Sulaymaniyah, Iraq
choman.abdullah@univsul.edu.iq
[2] School of Computing Science, Newcastle University, Newcastle upon Tyne, UK
{c.o.a.abdullah,nigel.thomas}@ncl.ac.uk

Abstract. In this paper we consider variations in performance between different communicating pairs of nodes within a restricted network topology. This scenario highlights potential unfairness in network access, leading to one or more pair of communicating nodes being adversely penalised, potentially meaning that high bandwidth applications could not be supported. In particular we explore the effect that variable frame lengths can have on fairness, which suggests that reducing relative frame length variance at affected nodes might be one way to alleviate some of the effect of unfairness in network access.

Keywords: WLAN · IEEE 802.11g · Performance modelling · PEPA · Fairness

1 Introduction

Wireless network access has been adopted across the world as the network medium of choice due primarily to ease of installation, ease of access from a wide range of devices and flexibility of access for roaming users. Amongst the range of access protocols available, the IEEE 802.11 family of protocols has become the standard for wireless networks [1]. The different protocols (a/b/g) all have a similar structure, but different operating ranges (power, data rate, frame length etc.) [9]. These protocols are controlled with the two main standards: Medium Access Control (MAC) and the PHY layer. Access control is managed by the Distributed Coordination Function (DCF) and the Point Coordination Function (PCF) which support collision free and time restricted services.

Understanding the performance of wireless systems is clearly crucial in making appropriate choices for the provision of infrastructure and services. Clearly we need to know at least the expected network throughput and latency in order to know whether the network is able to support a given level of service. Fairness is concerned with the forced variability of throughput and latency at different nodes leading to different parts of the network attaining different levels of performance. Fairness in 802.11g has been assessed by studying the backoff and contention window mechanisms [11]. Here poor fairness arises as unsuccessful nodes are obliged to remain unsuccessful in term of channel access, while the standard

© Springer International Publishing Switzerland 2016
S. Wittevrongel and T. Phung-Duc (Eds.): ASMTA 2016, LNCS 9845, pp. 223–238, 2016.
DOI: 10.1007/978-3-319-43904-4_16

backoff protocol allows successful nodes are able to access the medium successfully for long periods. In our previous works we considered models of unequal network access in 802.11b and g [2,3], based on an original model by Kloul and Valois [10]. From this we observed that fairness is affected by both transmission rate and frame length. In our modelled scenario short frames transmitted faster promoted a greater opportunity sharing of access, even under a pathologically unfair network topology. In practice it is not possible to simply set an arbitrarily short frame length and fast transmission rate as these factors also dictate the transmission range; in CSMA/CA neighbouring nodes need to be able to 'sense' a transmission in order to minimise and detect interference. For this reason wireless protocols generally provide only a small set of possible transmission rates with fixed, or at least minimum, frame lengths, allowing the network provider to choose an option which best fits its operating environment. In this paper we seek to relax these conditions to explore the effect of frame length variability on the fairness of network access. The model we propose and explore has many of the features of IEEE 802.11g, including the same average frame lengths. However, by introducing greater variability to the frame lengths we allow frames to be shorter than the prescribed IEEE 802.11g frame length, which would not be permitted in practice. Notwithstanding this practical limitation, the results provide greater insight into the fairness of wireless systems with highly variable frame lengths, including frame bursting provision in IEEE 802.11n.

This paper extends the model presented in [3] to study a number of deployment scenarios in IEEE 802.11g with variable frame lengths modelling using the stochastic process algebra PEPA [8]. The paper is organized as follows. Section 2 describes the model that we used in PEPA for each scenario and the parameters are presented in Sect. 3. The results and figures are given in Sect. 4. Section 5 explores the contribution of this work with some related work on the performance of IEEE 802.11 and in particular modelling with PEPA. Finally, conclusion and future works are provided in Sect. 5.

2   The Model

2.1   Basic Access Mechanism

The Basic Access (BA) method is widely used in 802.11 up to 802.11g [4]. It cooperates in either the Point Coordination Function (PCF needs a central control object) or the Distributed Coordination Function (DCF based on CSMA/CA). The DCF mechanism specifies two techniques for data transmission, which are the basic access method and two way handshake mechanism, in our study we focused solely on the basic access method. In BA, shown in Fig. 1, a WLAN node listens to the channel to access it, when the medium is free to use with no congestion, then it can make its transmission. On successful receipt, the receiving node will transmit an acknowledgement (ACK). However, if two nodes attempt to transmit simultaneously, then collision occurs resulting in an unsuccessful transmission and an initiation of a back-off algorithm. An unsuccessful transmitting node waits for a random time (back-off) in the range [0, CW], where contention

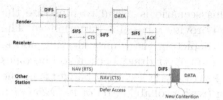

(a) RTC-CTS and Data-ACK scheme.

Attribute	Values of 802.11g
CW *min* and CW *max*	15(pure) and 1023
Slot time	$20\mu s$, $9\mu s$
SIFS	$10\mu s$
DIFS	$50\mu s$, $28\mu s$
EFIS	$364\mu s$

(b) Attribute values of 802.11g.

Fig. 1. Basic access method with 802.11g attributions

window CW is based on the number of transmission failures. The initial value of CW is 15 for 802.11g; it is doubled after every unsuccessful transmission, until it reaches to the maximum number (1023) and CW returns to the initial value after each ACK received (see [6,9] for more detail). If the channel is not free to use, the node monitors the channel until it becomes idle. However, the node will not attempt to transmit immediately (as this approach clearly cause a collision with any other waiting nodes), but instead continues to listen for a further backoff period until it is satisfied that the channel is idle.

2.2 Scenarios Modelled with PEPA

We now consider a model of pairs of transmitting nodes competing to use the transmission channel, as illustrated in Fig. 2. We only consider cases where the demand for access is very high, in order to determine the maximum channel utilisation and throughput that can be achieved. The basic model (the one pair scenario) is used to derive a baseline throughput when there is no contention. The other two models (two and three pair scenarios) are used to explore how competition for access affects throughput and utilisation. If the system is fair then all nodes should experience the same throughput and utilisation (when all nodes have the same demand). However, the three pair scenario is pathologically unfair due to its rigid topology; the inner pair will be out-competed by their neighbours which can transmit simultaneously, whereas the inner pair must wait until neither outer pair is transmitting. We seek to explore how variable frame lengths affect the fairness in each scenario, using two transmission rates, one for "normal" short frames and one for "occasional" long frames.

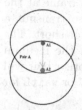
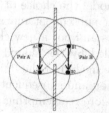
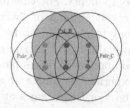

(a) Scenario 1.   (b) Scenario 2.   (c) Scenario 3.

Fig. 2. (One pair, Two pairs and Three pairs) scenarios.

One Pair Scenario (Scenario 1): This scenario is useful to illustrate the behaviour of the transmitting pairs and to provide a baseline performance. The model consists of two components; *Pair*, depicting the communicating nodes, and Med_F, depicting the transmission medium. `Pair` draws backoff and becomes `Pair0`, `Pair0` starts to count *DIFS* to `Pair1`. `Pair1` counts `backoff` in the same `Pair1` or it ends `backoff` to `Pair2a` (with probability α) or `Pair2b` (with probability $1-\alpha$). `Pair2a` depicts transmission of short frames, whereas `Pair2b` specifies transmission of long frames ($\mu data_1 > \mu data_2$). `Pair3` counts the *SIFS* period, then an *ACK* is received in `Pair4`.

$$Pair \overset{def}{=} (draw\_backoff, r).Pair0$$
$$Pair0 \overset{def}{=} (count\_difs, \mu difs).Pair1$$
$$Pair1 \overset{def}{=} (count\_backoff, p\mu bck).Pair1 + (end\_backoff, \alpha q\mu bck).Pair2a$$
$$+(end\_backoff, (1 - \alpha)q\mu bck).Pair2b$$
$$Pair2a \overset{def}{=} (transmit, \mu data_1).Pair3$$
$$Pair2b \overset{def}{=} (transmit, \mu data_2).Pair3$$
$$Pair3 \overset{def}{=} (count\_sifs, \mu sifs).Pair4$$
$$Pair4 \overset{def}{=} (ack, \mu ack).Pair$$

$$Med\_F \overset{def}{=} (transmit, \top).Med\_F1 + (count\_difs, \top).Med\_F$$
$$+(count\_backoff, \top).Med\_F + (end\_backoff, \top).Med\_F$$
$$Med\_F1 \overset{def}{=} (transmit, \top).Med\_F2 + (ackA, \top).Med\_F$$
$$+(count\_difs, \top).Med\_F1 + (count\_backoff, \top).Med\_F1$$
$$+(end\_backoff, \top).Med\_F1$$
$$Med\_F2 \overset{def}{=} (ack, \top).Med\_F1 + (count\_difs, \top).Med\_F2$$
$$+(count\_backoff, \top).Med\_F2 + (end\_backoff, \top).Med\_F2$$
$$Scenario1 \overset{def}{=} Pair \underset{\mathcal{K}}{\bowtie} Med\_F$$

Where $\mathcal{K} = \{transmit, ack, count\_difs, count\_backoff, end\_backoff\}$.

Two Pairs Scenario (*Scenario 2*): Here we have two asymmetric pairs interacting with a shared medium (see Fig. 2 (b)). If, one node in a pair attempts to transmit, its partner node waits to receive an *ACK*. `Pair_A` behaves as in the previous model, having long and short frames, whereas `Pair_B` has only one frame length. Unlike the previous case we also need to consider contention and subsequent waiting for access, which adds additional behaviours to both model components. For this reason we model the choice of long or short frame at the very beginning of `Pair_A`, so that subsequent repeat attempts to transmit a long frame will also be long frames and not a new choice of long or short. The availability to transmit is controlled by the shared actions with the medium component. Frames blocked by the medium being busy with the other pair will experience a `queue` or `queueB` action and subsequent `wait` (`waitS` or `waitL` for short or long frames at `Pair_A`) before reattempting to transmit.

$$Pair\_A \overset{def}{=} (draw\_backoff, \alpha * r).Pair\_A0S + (draw\_backoff, (1 - \alpha) * r).Pair\_A0L$$
$$Pair\_A0S \overset{def}{=} (count\_difs, \mu difs).Pair\_A1S + (queue, \top).Pair\_A5S$$

$Pair\_A0L \stackrel{def}{=} (count\_difs, \mu difs).Pair\_A1L + (queue, \top).Pair\_A5L$

$Pair\_A1S \stackrel{def}{=} (count\_backoff, p\mu bck).Pair\_A1S + (end\_backoff, q\mu bck).Pair\_A2S$
$\qquad + (queue, \top).Pair\_A5S$

$Pair\_A1L \stackrel{def}{=} (count\_backoff, p\mu bck).Pair\_A1L + (end\_backoff, q\mu bck).Pair\_A2L$
$\qquad + (queue, \top).Pair\_A5L$

$Pair\_A2S \stackrel{def}{=} (transmit, \mu data1).Pair\_A3S + (queue, \top).Pair\_A5S$

$Pair\_A2L \stackrel{def}{=} (transmit, \mu data2).Pair\_A3L + (queue, \top).Pair\_A5L$

$Pair\_A3 \stackrel{def}{=} (count\_sifs, \mu sifs).Pair\_A6$

$Pair\_A4S \stackrel{def}{=} (count\_difs, \mu difs).Pair\_A1S + (count\_eifs, \mu eifs).Pair\_A1S$
$\qquad + (queue, \top).Pair\_A5S$

$Pair\_A4L \stackrel{def}{=} (count\_difs, \mu difs).Pair\_A1L + (count\_eifs, \mu eifs).Pair\_A1L$
$\qquad + (queue, \top).Pair\_A5L$

$Pair\_A5S \stackrel{def}{=} (waitS, \mu data).Pair\_A4S$

$Pair\_A5L \stackrel{def}{=} (waitL, \mu data).Pair\_A4L$

$Pair\_A6 \stackrel{def}{=} (ack, \mu ack).Pair\_A$

$Pair\_B \stackrel{def}{=} (draw\_backoff, r).Pair\_B0$

$Pair\_B0 \stackrel{def}{=} (count\_difsB, \mu difs).Pair\_B1 + (queueB, \top).Pair\_B5$

$Pair\_B1 \stackrel{def}{=} (count\_backoffB, p\mu bck).Pair\_B1 + (end\_backoffB, q\mu bck).Pair\_B2$
$\qquad + (queueB, \top).Pair\_B5$

$Pair\_B2 \stackrel{def}{=} (transmitB, \mu data).Pair\_B3 + (queueB, \top).Pair\_B5$

$Pair\_B3 \stackrel{def}{=} (count\_sifs, \mu sifs).Pair\_B6$

$Pair\_B4 \stackrel{def}{=} (count\_difsB, \mu difs).Pair\_B1 + (count\_eifsB, \mu eifs).Pair\_B1$
$\qquad + (queueB, \top).Pair\_B5$

$Pair\_B5 \stackrel{def}{=} (wait, \mu data).Pair\_B4$

$Pair\_B6 \stackrel{def}{=} (ackB, \mu ack).Pair\_B$

$Med\_F \stackrel{def}{=} (transmit, \top).Med\_F2 + (transmitB, \top).Med\_F1$
$\qquad + (count\_difs, \top).Med\_F + (count\_backoff, \top).Med\_F$
$\qquad + (end\_backoff, \top).Med\_F + (count\_eifs, \top).Med\_F$
$\qquad + (count\_difsB, \top).Med\_F + (count\_backoffB, \top).Med\_F$
$\qquad + (end\_backoffB, \top).Med\_F + (count\_eifsB, \top).Med\_F$

$Med\_F1 \stackrel{def}{=} (ackB, \top).Med\_F + (queue, \lambda oc).Med\_F1$

$Med\_F2 \stackrel{def}{=} (transmit, \top).Med\_F3 + (ack, \top).Med\_F + (queueB, \lambda oc).Med\_F2$
$\qquad + (count\_difs, \top).Med\_F2 + (count\_backoff, \top).Med\_F2$
$\qquad + (end\_backoff, \top).Med\_F2 + (count\_eifs, \top).Med\_F2$

$Med\_F3 \stackrel{def}{=} (ack, \top).Med\_F2 + (queueB, \lambda oc).Med\_F3 + (count\_difs, \top).Med\_F3$
$\qquad + (count\_backoff, \top).Med\_F3 + (end\_backoff, \top).Med\_F3$
$\qquad + (count\_eifs, \top).Med\_F3$

$$Scenario2 \stackrel{def}{=} ((Pair\_A \underset{\mathcal{K}}{\bowtie} Med\_F) \underset{\mathcal{L}}{\bowtie} Pair\_B$$

Where $\mathcal{K} = \{transmit, ack, queue, count\_difs, count\_backoff, end\_backoff,$
$count\_eifs\}$. And $\mathcal{L} = \{transmitB, ackB, queueB, count\_difsB, count\_$
$backoffB, end\_backoffB, count\_eifsB\}$.

Three Pairs Scenario (*Scenario 3*): This final scenario has two symmetric outer pairs (Pair_A and Pair_C), one inner pair (Pair_B) and a shared medium (Med_F) (see Fig. 2 (c)). The outer pairs cannot hear one another and so may transmit independently. However both outer pairs are within the interference range of the inner pair, hence the inner pair can only transmit when the medium is quiescent. In our model the outer pairs have both long and short frames (modelled as Pair_A in the previous scenario) whereas the inner pair has only one frame type (modelled as Pair_B in the previous scenario). The model therefore only differs from the previous scenario in having two instances of Pair_A (the second renamed for clarity) and having a modified cooperation set.

$$Scenario3 \overset{def}{=} ((Pair\_A \| Pair\_C) \underset{\mathcal{K}}{\bowtie} Med\_F) \underset{\mathcal{L}}{\bowtie} Pair\_B$$

where $\mathcal{K}$ = {$transmit, ack, queue, count\_difs, count\_backoff, end\_backoff,$ $count\_eifs$}. And $\mathcal{L}$ = {$transmitB, ackB, queueB, count\_difsB, count\_$ $backoffB, end\_backoffB, count\_eifsB$}.

3   Parameters

IEEE 802.11 has a very specific inter-frame spacing, which coordinates access to the medium to transmit frames. If any pair wants to transmit and if it senses the channel is idle then it transmits with the probability of 'p'. For convenience, each pair in this paper has count back-off and end back-off actions with ($p \times \mu bck$) and ($q \times \mu bck$) rates respectively; we assume the values of **p** and **q** (q = 1-p) are equal to 0.5. According to the definition of 802.11g and PHY standards, the possible data rate per stream are (6, 9, 12, 18, 24, 36, 48, and 54) Mbits/s [6,9]. In this paper we considered 6, 12, 36 and 54 Mbits/s as a sample of data rates. These rates have been applied with each of the frame payload size (700, 900, 1000, 1200, 1400 and 1500) bytes. The frames per time unit for arrival and departure rate are λoc = 100000 and μ = 200000 respectively. In this model (μack) shows as a rate of ACK, where μack = Channel throughput $\div$ (Ack length=1 byte). If μack = 1644.75 for 1 Mbits/s then for 6 Mbits/s it is 9868.5 (6 × 1644.75).

Inter-Frame Space (IFS): At the beginning, the length of the IFS is dependent on the previous frame type, if noise occurs, the (IFS) is used. Possibly, if transmission of a particular frame ends and before another one starts the IFS applies a delay for the channel to stay clear. It is an essential idle period of time needed to ensure that other nodes may access the channel. The aim of the IFS is to supply a waiting time for each frame transmission in a specific node, to allows the transmitted signal to reach another node (essential for listening) [5,6,12].

Short Inter-Frame Space (SIFS): SIFS is the shortest IFS for highest priority transmissions used with DCF, used to process a received frame. In 802.11b/g/n SIFS is 10 μs.

DCF Inter-Frame Space (DIFS): DIFS is a medium priority waiting time used to monitor the medium for a longer period than SIFS. If the channel is idle, the node waits for the *DIFS* to determined that the channel is not being used, then it waits for another (*backoff*).

DIFS = SIFS + (2× (Slot time = 20 μs in 802.11b/g/n)).

Extended Inter-Frame Space (EIFS): When the node can detect a signal, the transmission node uses EIFS, instead of *DIFS*, (used with erroneous frame transmission). It is the longest of *IFS* but has the lowest priority after DIFS. In *DCF*

EIFS = SIFS + DIFS + transmission time(Ack-lowest basic rate).

Contention Window (CW): In CSMA/CA, if a node wants to transmit any frame, it senses whether the channel whether is free or not. If it is free then the node transmits, if not the node waits for a random backoff, selected by node from a Contention Window (*CW*), until it becomes free. The node waits to minimise any collision once it experiences an idle channel for an appropriate *IFS*, otherwise many waiting nodes might transmit simultaneously. The node needs less time to wait if there is a shorter backoff period, so transmission will be faster, unless there is a collision. Backoff is chosen in [0, *CW*]. CW=CW*min* for all nodes if a node successfully transmits a packet and then receives an *ACK*. Otherwise, the node draws another (*backoff*) and the *CW* increases exponentially, until it reaches CW*max*. Finally, when the backoff reaches 0, the node starts to transmit and the CW resets to CW*min* when the packet is received.

CW*min* = 15, CW*max* = 1023. CW*min* augmented by 2n−1 on each retry.

Backoff Time = (Random () mod (CW + 1)) × Slot Time.

If BackoffTimer = b, where b is a random integer, also CW*min* ≤ b ≤ CW*max*. The mean of CW is calculated by: μ bck = 10^6 ÷ (Mean of CW × Time Slot).

The mean of μbck=7.5 and Time slot=20μs. The receiver sends an *ACK* if it gets a packet successfully, it is a precaution action to notify when collisions occur.

μData and Variance: The value of $\mu data$ can be obtained by (Data rate × (10^6÷8))÷ Packet payload size. The pair with a single frame length uses $\mu data$ for the frame portion in the transmit and wait actions (see the depictions of Pair_B). For the pairs with two frame lengths we obtained two different values, $\mu data1$ for the transmission of the short frames (proportion α) and $\mu data2$ for the long frames (proportion $1 - \alpha$). We assume that the average frame length is the same for the two size pair as the one size pair, hence,

$$\frac{\alpha}{\mu data1} + \frac{1-\alpha}{\mu data2} = \frac{1}{\mu data}$$

If we assume that $\mu data1 = a\mu data2$, where $a > 1$, then

$$\mu data2 = \frac{\mu data(\alpha + a(1 - \alpha))}{a}$$

In PEPA all actions are negative exponentially distributed. Hence the case where there is a single frame length, the transmit action duration is simply a negative exponentially distributed random variable. However, when we model the choice of two frame lengths this delay becomes a hyper-exponential. In the following experiments we fix $a = 100$ and vary the proportion of short frames, α, in order to change the variance of the frame transmission.

4 Results and Figures

The models specified above give rise to a continuous time Markov chain [8] which can be solved to give the steady state performance.

4.1 Results of One Pair Scenario: Scenario 1

Figure 3 shows the average utilisation and throughput for the one pair scenario for different average frame lengths and transmission rates. In this scenario there is no competition and so altering the proportion of long and short frames makes no difference if the average frame length remains the same. There is a small amount of variation with frame length; the utilisation increases slightly and the throughput decreases slightly as the payload increases.

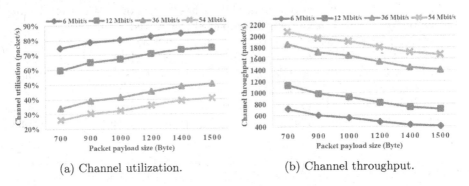

(a) Channel utilization.   (b) Channel throughput.

Fig. 3. Channel utilization and throughput for one pair (Scenario 1)

4.2 Results of Two Pairs Scenario: Scenario 2

In this case the frame length variance is greater at Pair A than at Pair B. From Figs. 4 and 5 we can see that the medium utilisation is greater by Pair B than Pair A, but this effect is less when the proportion of long frames is reduced. When $\alpha = 0.89$ Pair B gains around a 15 % utilisation advantage over Pair A, whereas when $\alpha = 0.99$ (fewer long frames) this advantage is around 8 %. This effect is fairly consistent regardless of transmission rate. Figures 6 and 7 show the corresponding throughput results. In Fig. 6 we show the throughput for both pairs when $\alpha = 0.89$. It is clear that Pair B has significantly better performance than Pair A under these conditions. However, if $\alpha = 0.99$ there is only a slight difference between the throughput at each pair. In each case

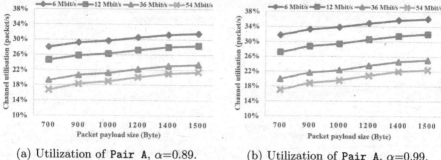

(a) Utilization of Pair A, α=0.89.   (b) Utilization of Pair A, α=0.99.

Fig. 4. Utilization for Pair A, $\alpha = 0.89$ and $\alpha = 0.99$ in (Scenario 2)

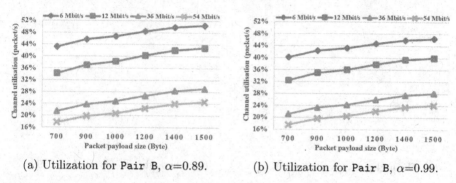

(a) Utilization for Pair B, α=0.89.   (b) Utilization for Pair B, α=0.99.

Fig. 5. Utilization for Pair B, $\alpha = 0.89$ and $\alpha = 0.99$ in (Scenario 2)

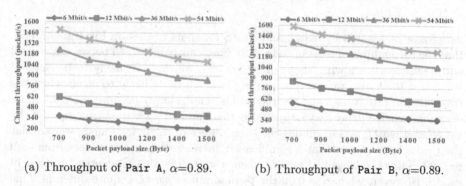

(a) Throughput of Pair A, α=0.89.   (b) Throughput of Pair B, α=0.89.

Fig. 6. Throughput for Pair A, $\alpha = 0.89$ and Pair B $\alpha = 0.89$ in (Scenario 2)

the general trends of utilisation and throughput are consistent with the non-competitive case in Scenario 1. However, it is clear that variance in frame length is having a major impact on the share of resources available to each pair. In order to better understand what is causing this unbalance in performance, we also studied the throughput of the *wait* action. In Pair A waiting to transmit a

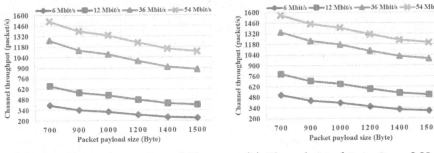

(a) Throughput of **Pair A**, α=0.99.   (b) Throughput of **Pair B**, α=0.99.

Fig. 7. Throughput for **Pair A**, $\alpha = 0.99$ and **Pair B** $\alpha = 0.99$ in (Scenario 2)

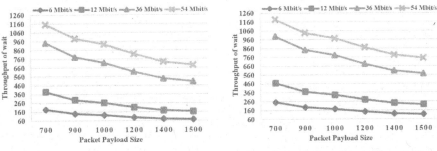

(a) Throughput of wait if α=0.89.   (b) Throughput of wait if α=0.99.

Fig. 8. Throughput of wait for **Pair B**, if $\alpha = 0.89$ and $\alpha = 0.99$ in (Scenario 2)

long frame is denoted by $waitL$ and $waitS$ for short frames. In Fig. 8 we observe that the throughput of $wait$ in **Pair B** is significantly increased between $\alpha = 0.89$ and 0.99, clearly showing that transmission is much more likely to be delayed when $\alpha = 0.99$. also we see that waiting is much more likely to occur when the transmission rate is high and the payload is small, simply because there are more occasions when a delay may happen. Figures 9 and 10 show the throughput of the $waitS$ and $waitL$ actions under the same values of α. As for Pair B, we see that the throughput of $waitS$ at A significantly increases when α increases from 0.89 to 0.99. The throughput of $waitL$ when $\alpha = 0.89$ is almost identical to that of $waitS$. However, when $\alpha = 0.99$ the throughput of $waitL$ is quite different. One aspect of this is that there are far fewer long frames when $\alpha = 0.99$. However we also observe that when the payload is small then the throughput of $waitL$ is less for high transmission rates than for lower transmission rates. The cumulative throughput of wait actions($waitS$ and $waitL$) at Pair A far exceeds that at B when $\alpha = 0.89$. This corresponds to the lower performance of **Pair A** shown in Figs. 4, 5, 6 and 7. However, when $\alpha = 0.99$ the cumulative throughput of wait actions at Pair A is only slightly higher than for B, leading to the much closer performance noted earlier. It is clear that the behaviour of the wait actions,

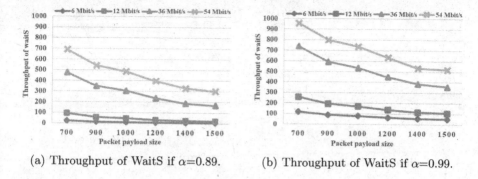

(a) Throughput of WaitS if α=0.89.   (b) Throughput of WaitS if α=0.99.

Fig. 9. Throughput of WaitS for `Pair A`, if $\alpha = 0.89$ and $\alpha = 0.99$ in (Scenario 2)

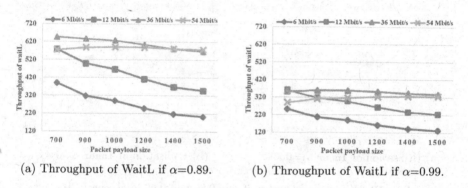

(a) Throughput of WaitL if α=0.89.   (b) Throughput of WaitL if α=0.99.

Fig. 10. Throughput of WaitL for `Pair A`, if $\alpha = 0.89$ and $\alpha = 0.99$ in (Scenario 2)

particularly *waitL* has a significant impact on the fairness exhibited by this scenario.

4.3   Results of Three Pairs Scenario (Scenario 3)

We have observed that the higher variance of the hyper-exponential distribution can have a significant negative impact on performance in competitive situations. We now seek to exploit this observation in the three pair scenario which has been previously seen to be pathologically unfair [3]. By causing the outer pairs to have a higher variance we aim to reduce their topological advantage over the inner pair. Figure 11 shows the combined utilisation of the outer pairs when $\alpha = 0.89$ and 0.99. We see that when the transmission rate is low there is little variation with payload size, but greater variation as transmission rate increases. We also observe that the utilisation by the outer pairs increases by around 8 % as α increases from 0.89 to 0.99. Figure 12 shows the corresponding utilisation by the inner pair. As expected, when $\alpha = 0.99$ then in all cases the outer pairs significantly outperform the inner pair. However, when $\alpha = 0.89$ and the transmission rate is 6 Mbit/s then the inner pair actually has a greater share of the

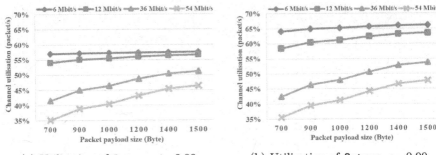

(a) Utilisation of Outers, α=0.89.   (b) Utilisation of Outers, α=0.99.

Fig. 11. Utilisation of Outers if $\alpha = 0.89$ and 0.99 in (Scenario 3)

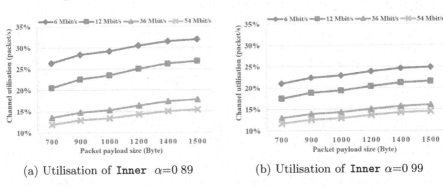

(a) Utilisation of Inner α=0 89   (b) Utilisation of Inner α=0 99

Fig. 12. Utilisation of Inner, if $\alpha = 0.89$ and 0.99 in (Scenario 3)

medium than each of the outer pairs, except when the payload is very small. In all other cases the outer pairs still outperform the inner pairs, although the unfairness is clearly reduced compared with $\alpha = 0.99$. The corresponding throughput results are shown in Figs. 13 and 14. We see that when $\alpha = 0.89$ and the transmission rate is 6 Mbit/s then the small advantage in utilisation when the payload is larger, leads to a significant advantage in throughput. This reversal of the pathological unfairness shows that modifying the variance can have a profound effect on the overall performance. But, this effect is limited in most cases and particularly at higher transmission rates, where the topological advantage still holds sway. As in the previous scenario we now consider the throughput of the various wait actions in order to better understand the observed behaviour. Figure 15 shows the throughput of the *wait* action at the inner pair. The throughput of *wait* at high transmission rates is hardly affected by α. However the slower transmission rates show some differences between $\alpha = 0.89$ and $\alpha = 0.99$. It is especially interesting to observe that the throughput of *wait* is very low when the transmission rate is 6 Mbit/s and α is 0.89. This shows that very few transmissions are being queued. Figures 16 and 17 show the corresponding throughputs for *waitS* and *waitL* respectively, at the outer pairs.

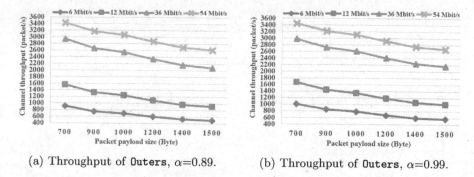

(a) Throughput of Outers, α=0.89.   (b) Throughput of Outers, α=0.99.

Fig. 13. Throughput of Outers, if $\alpha = 0.89$ and 0.99 in (Scenario 3)

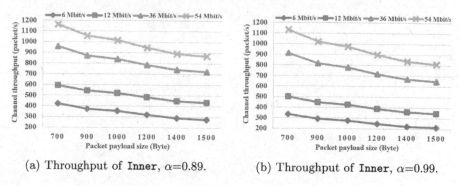

(a) Throughput of Inner, α=0.89.   (b) Throughput of Inner, α=0.99.

Fig. 14. Throughput of Inner, if $\alpha = 0.89$ and 0.99 in (Scenario 3)

The throughput of $waitS$ (Fig. 17) increases significantly as α increases from 0.89 to 0.99. Moreover we see that the throughput of $waitS$ is very low when the transmission rate is 6 Mbit/s. The throughput of $waitL$ (Fig. 16) is substantially higher. This is not surprising given that long frames are much more likely to be delayed under competition. Again we see that when $\alpha = 0.99$ the throughput of $waitL$ is much less than when $\alpha = 0.89$, in part due to the much lower proportion of long frames. We also observe the different profiles of the throughput of $waitL$ with different transmission rates. As in Scenario 2 we see that there is very little variation with payload when the transmission rate is high, but a decreasing profile when the transmission rate is low. This difference in behaviour is due to the interaction between the different inter-frame spaces and different frame transmission durations. The longer frames have less impact when the payload is large and transmission is slower, as all frames then take a significant length of time to transmit compared with the accumulated inter-frame spaces. However, if the transmission rate is faster or the payload is less, then the effect of variance is clearly greater.

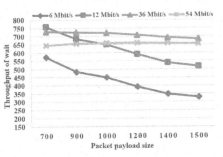
(a) Throughput of Wait, α=0.89.

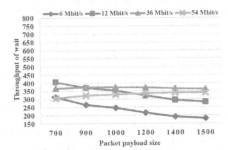
(b) Throughput of Wait, α=0.99.

Fig. 15. Throughput of Wait, if α = 0.89 and 0.99 in (Scenario 3)

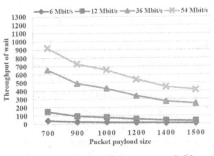
(a) Throughput of WaitL, α=0.89.

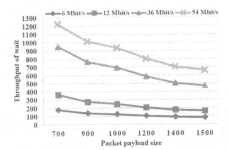
(b) Throughput of WaitL, α=0.99.

Fig. 16. Throughput of WaitL, if α = 0.89 and 0.99 in (Scenario 3)

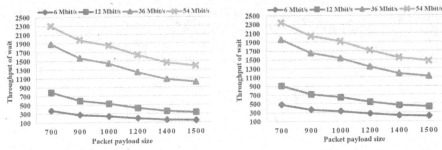
(a) Throughput of WaitS, α=0.89.   (b) Throughput of WaitS, α=0.99.

Fig. 17. Throughput of WaitS, if α = 0.89 and 0.99 in (Scenario 3)

5   Conclusion

In this paper, we analysed the performance modelling of fairness properties of the IEEE 802.11g using PEPA under heavy load. Thus we aim to derive the maximum total throughput and utilisation, but under such conditions we also demonstrate the maximum imbalance in the behaviour of the different pairs. We have introduced a hyper-exponential transmission of frames in order to study the effect of increased variance in frame length distribution. In the case where there is no competition for the medium it is clear that increased variance has no impact on the average utilisation and throughput. However, when there is competition, nodes with a higher variance experience a weaker performance as disproportionately long frames are more likely to be delayed.

The three pairs scenario demonstrates a topologically unfair situation which has been previously shown to massively hinder the performance of the inner pair [2,3,10]. By introducing greater variance in the outer pairs not only reduced overall unfairness, but under certain conditions actually gave a slight advantage to the inner pair. These results show that controlling variance in transmission duration, as well as average duration, can have a significant impact on relative performance. A node which is severely impacted by topological unfairness might therefore attempt to decrease variance and mean by limiting frame sizes in order to increase performance.

This work forms part of an ongoing study into fairness and unfairness in wireless networks. The obvious next step is to consider recent versions of 802.11, the most obvious candidate being 802.11n. This protocol includes a number of measures aimed at reducing the use of inter-frame spacing and consequently increasing efficiency and performance. One of those features most relevant to this work is that of so-called frame bursting, where several frames are sent by the same node in quick succession, negating the need to wait between frames to listen for other senders. Under the extreme topological conditions considered in the scenarios in this paper this may lead to nodes being treated unfairly, or conversely might allow deprived nodes a chance to send a backlog of frames. A further feature of potential interest when considering fairness is the co-existence of 802.11n with legacy systems operating 802.g, see [7] for example.

References

1. IEEE Draft Standard for IT-Telecommunications and Information Exchange Between Systems Local and Metropolitan Area Networks-Specific Requirements Part 11: WLAN Medium Access control (MAC) and Physical Layer (PHY) Specifications, pp. 1–3701, February 2015
2. Abdullah, C.O., Thomas, N.: Formal performance modelling and analysis of IEEE 802.11 wireless LAN protocols. In: UK Performance Engineering Workshop (2015)
3. Abdullah, C.O., Thomas, N.: Performance modelling of IEEE 802.11g wireless LAN protocols. In: 9th EAI International Conference on Performance Evaluation Methodologies and Tools (valuetools) (2015)
4. Alekhya, T.N.V.L., Mounika, B., Jyothi, E., Bhandari, B.N.: A waiting-time based backoff algorithm in the IEEE 802.11 based wireless networks. In: 2012 National Conference on Communications (NCC), pp. 1–5 (2012)
5. Chou, C., Shin, K.G., Shankar, S.N.: Inter-Frame Space (IFS) based service differentiation for IEEE 802.11 wireless LANs. In: Proceedings of the 58th IEEE Vehicular Technology Conference, vol. 3, pp. 1412–1416. IEEE (2003)

6. Duda, A.: Understanding the performance of 802.11 networks. In: Proceedings of the 19th International Symposium on Personal, Indoor and Mobile Radio Communications, vol. 8, pp. 1–6 (2008)
7. Galloway, M.: Performance measurements of coexisting IEEE 802.11 g/n networks. In: Proceedings of the 49th Annual Southeast Regional Conference, pp. 173–178. ACM (2011)
8. Hillston, J.: A Compositional Approach to Performance Modelling. Cambridge University Press, Cambridge (2008)
9. Khanduri, R., Rattan, S., Uniyal, A.: Understanding the features of IEEE 802. 11g in high data rate wireless LANs. IJCA **64**(8), 1–5 (2013)
10. Kloul, L., Valois, F.: Investigating unfairness scenarios in MANET using 802.11b. In: Proceedings of the 2nd ACM International Workshop on Performance Evaluation of Wireless Ad Hoc, Sensor, and Ubiquitous Networks (2005)
11. Kuptsov, D., Nechaev, B., Lukyanenko, A., Gurtov, A.: How penalty leads to improvement: a measurement study of wireless backoff in IEEE 802.11 networks. Comput. Netw. **75**, 37–57 (2014)
12. Xi, S., Kun, X., Jian, W., Jintong, L.: Performance analysis of medium access control protocol for IEEE 802.11g-over-fiber networks. China Commun. **10**, 81–92 (2013)

Optimal Data Collection in Hybrid Energy-Harvesting Sensor Networks

Kishor Patil[1]([✉]), Koen De Turck[2], and Dieter Fiems[1]

[1] Department of Telecommunications and Information Processing,
Ghent University, Ghent, Belgium
Patil.kishor@ugent.be
[2] Central Supélec, Laboratoire des Signaux et Systèmes, Gif-sur-Yvette, France

Abstract. In hybrid energy harvesting sensor networks, there is a trade-off between the cost of data collection by a wireless sink and the timeliness of the collected data. The trade-off further depends on the energy harvesting capability of the sensor nodes as sensors cannot transmit data if they do not have sufficient energy. In this paper, we propose an analytic model for assessing the value of the information that a sensor node brings to decision making. We account for the timeliness of data by discounting the value of the information at the sensor over time and adopt the energy-chunk approach (i.e. discretise the energy level) to track energy harvesting and expenditure over time. Finally, by numerical experiments, we study the optimal data collection rate for the sensor node at hand.

Keywords: Age of information · Sensor networks · Energy harvesting · Markov process

1 Introduction

Wireless sensor networks (WSNs) are one of the key constituents of the Internet of Things (IoT) [1,2], and have attracted considerable research interest over the past couple of years. Sensor networks collect and monitor spatially distributed data like temperature, humidity, movement and noise [3,4], extract information from the collected data and deliver relevant information to the user. WSNs have a variety of applications including military, environmental, home and healthcare. Applications of WSNs are surveyed in [5], whereas [6] focuses on applications of WSNs in the context of the IoT.

Combining sensor networks with data analytics enable fast data-to-decision applications that act in real time on the collected data such that the value the information brings to the decision not only depends on the quality but also on the timeliness of the information. Therefore, analogous to Quality of Service which measures the performance of a data communications network, the term "Quality of Information" (QoI) has been introduced to evaluate performance of sensor networks [7–9].

© Springer International Publishing Switzerland 2016
S. Wittevrongel and T. Phung-Duc (Eds.): ASMTA 2016, LNCS 9845, pp. 239–252, 2016.
DOI: 10.1007/978-3-319-43904-4_17

The present paper investigates the QoI in energy harvesting sensor networks by calculating the value of information transmitted by energy-harvesting sensors. Energy harvesting sensor nodes mitigate their dependence on batteries by harvesting energy from their environment [10]. More precisely, energy harvesting sensor nodes (EHSN) use ambient sources of energy like solar, wind or heat, convert the energy into electricity which can then be used for sensing or transmitting data. Most often, the amount of energy harvested is not constant over time. Hence, energy harvesting is an additional source of uncertainty that a performance evaluation should account for. The sensor nodes under consideration operate energy neutral: all energy for sensing and transmissions is harvested, a small on-board battery providing for temporary energy storage. Summarising, the sensor nodes at hand can only transmit when the mobile sink is in range and the sensor has sufficient energy for transmitting its data.

We focus on optimal data collection, adopting the hybrid WSN of Zhou et al. [11] which consists of static sensors responsible for sensing environmental variables, and mobile sensors called IoT mobile sinks that move to designated sink locations where they gather data sensed by static sensors. Mobile sinks were introduced to overcome the hot-spot effect in sensor networks [12]. Both static and mobile sink nodes (or base terminals) collect data from sensor nodes and sometimes act as gateways to other users by processing and sending relevant information. If all sensor data is relayed by the sensor nodes to a (static) sink node, nodes closer to the static sink are more heavily loaded as they need to relay more packets to the static sink in comparison with nodes further away. As a result, they consume more energy and may die at early stage, or will frequently run out of energy if they can harvest energy. Mobile sinks overcome this problem by moving the sink around. See e.g. [13] for a discussion on design issues and challenges in existing distributed protocols for mobile sinks. Although mobility increases the network lifetime by balanced utilisation of power [14], it also introduces new challenges as delay in packet delivery should be sufficiently small [15].

The remainder of this paper is organised as follows. In the next section, we introduce a discrete-time Markov model for studying the optimal data collection probability in energy-harvesting hybrid WSNs that are unaware of the value of their information. Section 3 is then concerned with refining the model: while the sensor node is still unaware of the value of the information, it is aware whether or not there is any value. We then illustrate our approach by numerical examples in Sect. 4, prior to drawing conclusions in Sect. 5.

2   Mathematical Model

We consider an energy harvesting sensor node. The node is equipped with a battery for storing harvested energy chunks and on-board memory to store sensed information.

We assume that time is discrete, i.e., time is divided into fixed length intervals or slots, and denote the value of the sensed data and the amount of harvested energy during slot n by S_n and H_n, respectively. The sequence $\{H_n, n \in \mathbb{N}\}$ constitutes a sequence of independent and identically distributed random variables, taking values in $\mathbb{N}$. Let h_k denote the probability that k energy chunks arrive in a slot. For further use, we always assume that the sensor node can harvest energy, that is, we assume $h_0 < 1$. Moreover, we introduce the following notation for the tail distribution function of H_n,

$$\bar{H}_k = \sum_{m=k}^{\infty} h_m = 1 - \sum_{m=0}^{k-1} h_m.$$

For the sequence $\{S_n, n \in \mathbb{N}\}$ of the value of harvested information, no independence assumptions are required. We only assume that the mean value does not depend on time; let $\bar{S} = \mathsf{E}[S_n]$ for $n \in \mathbb{N}$.

The battery level at the beginning of slot n is denoted by B_n. We assume that transmitting data and sensing data requires $N > 0$ and $M > 0$ energy chunks, respectively, and that at most C_B energy chunks can be stored in the battery. Let T_n be the indicator that there is a transmission during slot n. As the sensor node harvests H_n chunks during slot n, we have,

$$B_{n+1} = \min(B_n - M1_{\{B_n \geq M\}} - NT_n + H_n, C_B).$$

Let V_n be the value of the sensed data in the sensor node at the beginning of the nth slot. We assume that the value of the sensed data is additive and discounted over time. Discounting is introduced to account for the timeliness of data, older data being less valued than recent data. Let $0 \leq \alpha \leq 1$ denote the discount factor. The values of the data at two consecutive slot boundaries then relate as,

$$V_{n+1} = \alpha V_n (1 - T_n) + S_n 1_{\{B_n \geq M\}}.$$

The recursion above implies that any data sensed during slot n cannot be transmitted during slot n. In addition, any chunks of energy harvested during slot n cannot be used to sense and/or transmit data in slot n.

It now remains to express the indicator T_n that there is a transmission in terms of B_n. To this end, let P_n be the indicator that the data is collected from the sensor during slot n, that is, P_n is 1 if data is collected at time n. Let $p = \mathsf{P}[P_n = 1] = \mathsf{E}[P_n]$ be the collection probability, $0 < p \leq 1$. We assume that the sensor node cannot evaluate the value of its information, and therefore transmits if sufficient energy is available. We therefore have,

$$T_n = \begin{cases} 1 & \text{for } P_n = 1 \text{ and } B_n \geq M + N, \\ 0 & \text{otherwise.} \end{cases}$$

The set of recursions above now allows for determining the value of the information collected by the mobile sink. To this end, let $v_k = \mathsf{E}[V_n 1\{B_n = k\}]$ be the mean value of the information at the sensor node for battery level k, and let $b_k = \mathsf{P}[B_n = k]$ be the probability of having battery level k. In view of the equations for V_n, B_n and T_n and by conditioning on the battery level and the availability of a transmission opportunity in the preceding slot, we find that the mean value of the information at battery level k adheres to the following set of equations,

$$v_k = \alpha \sum_{\ell=0}^{M-1} v_\ell h_{k-\ell} + \alpha \sum_{\ell=M}^{C_B} v_\ell h_{k-\ell+M} - \alpha p \sum_{\ell=M+N}^{C_B} v_\ell h_{k-\ell+M}$$

$$+ \bar{S}(b_k - \sum_{\ell=0}^{M-1} b_\ell h_{k-\ell}), \quad (1)$$

for $k = 0, 1, \ldots, C_B - 1$, whereas for $k = C_B$ we have,

$$v_{C_B} = \alpha \sum_{\ell=0}^{M-1} v_\ell \bar{H}_{C_B-\ell} + \alpha \sum_{\ell=M}^{C_B} v_\ell \bar{H}_{C_B-\ell+M} - \alpha p \sum_{\ell=M+N}^{C_B} v_\ell \bar{H}_{C_B-\ell+M}$$

$$+ \bar{S}(b_{C_B} - \sum_{\ell=0}^{M-1} b_\ell \bar{H}_{C_B-\ell}). \quad (2)$$

To find the battery level probabilities, we again condition on the battery level and the availability of a transmission opportunity in the preceding slot. We have the following equations for the battery level probabilities,

$$b_k = \sum_{\ell=0}^{M-1} b_\ell h_{k-\ell} + \sum_{l=M}^{C_B} b_\ell h_{k-\ell+M} + p \sum_{l=M+N}^{C_B} b_\ell (h_{k-\ell+M+N} - h_{k-\ell+M}), \quad (3)$$

for $k = 0, 1, \ldots, C_B - 1$.

The system of Eq. (3) along with the normalisation condition

$$\sum_{k=0}^{C_B} b_k = 1,$$

allows for solving for the probabilities b_k. We can then solve the system of Eq. (1) for the conditional mean values v_k.

Finally, we express the mean value of the sensed data per time slot that is actually collected in terms of the v_k's as follows,

$$\bar{V} = p \sum_{k=M+N}^{C_B} v_k,$$

as there are only transmissions if the battery level exceeds the threshold and there is a transmission opportunity.

Remark 1. If C_B is small, both systems of equations (for b_k and v_k) are easily solved. For larger C_B, we can exploit structural properties of the systems of equations. As $h_k = 0$ for $k < 0$, we see that Eq. (3) expresses b_k in terms of the probabilities b_ℓ for $\ell \le k + M + N$. This implies that the Markov chain B_k is a $G/M/1$-type Markov chain. Hence, solution methods for a $G/M/1$-type Markov chain (e.g. [16]) can be applied to solve the system of equations of the b_k's. Analogously, we see that Eq. (1) expresses v_k in terms of v_ℓ, for $\ell < k + M$.

Remark 2. The assumption $h_0 > 0$ and Eq. (1) imply that $\lim_{p \to 0} v_k < \infty$ for $k < C_B$ and any value of α, as well as for $k = C_B$ and $\alpha < 1$. For $\alpha = 1$, we have $0 < \lim_{p \to 0} p v_{C_B} < \infty$. This shows that for $\alpha < 1$ we have $\lim_{p \to 0} \bar{V} = 0$, whereas for $\alpha = 1$ we have $\lim_{p \to 0} \bar{V} = \lim_{p \to 0} p v_{C_B} > 0$.

Remark 3. We assumed that the battery can be modelled by a queueing-like system with "arrivals" and "departures" of chunks of energy. Such battery models were considered several times in literature, see e.g. [17,18]. While being modelled as a queueing system, one prefers that the battery/queue is full in contrast to most other queueing systems. That is, the preferred operation differs significantly.

3   A Refinement

While it is reasonable to assume that the sensor node cannot value its data, it is not always reasonable to assume that the sensor node is unaware of the presence of data. We therefore now refine the model such that the sensor node does not transmit when there is no data to transmit. To this end, we introduce the indicator A_n which is 1 if the node has data to transmit and 0 otherwise. Of course we have $A_n = 1_{\{V_n > 0\}}$, but it is more convenient to track A_n separately, see below. In addition, we assume that the sequence $\{S_n, n \in \mathbb{N}\}$ constitutes a sequence of independent and identically distributed random variables with common mean $\bar{S} = E[S_n]$ and with non-zero probability mass $s_0 = P[S_n = 0]$ when there is no data sensed. Following similar arguments as in the preceding section, we find that the system variables V_n, A_n and B_n adhere the following set of recursive equations,

$$V_{n+1} = \alpha V_n(1 - T_n) + S_n 1_{\{B_n \ge M\}},$$
$$A_{n+1} = 1_{\{S_n > 0\}} 1_{\{B_n \ge M\}} + (1 - 1_{\{S_n > 0\}} 1_{\{B_n \ge M\}}) A_n(1 - T_n),$$
$$B_{n+1} = \min(B_n - M 1_{\{B_n \ge M\}} - N T_n + H_n, C_B),$$

where T_n denotes the indicator that there is a transmission,

$$T_n = 1_{\{B_n \ge M+N\}} A_n P_n.$$

That is, there is a transmission when (i) there is sufficient energy, (ii) there is something to send and (iii) there is a transmission opportunity.

In contrast to the preceding section, the sequence $\{B_n, n \in \mathbb{N}\}$ is not a Markov chain as the evolution of B_n depends on the presence or absence of information. Therefore we focus on the sequence $\{(A_n, B_n), n \in \mathbb{N}\}$ which is a Markov chain. Let $\widetilde{b}_k = \mathsf{P}[B_n = k, A_n = 0]$ be the probability that there are k chunks of energy in the battery at slot boundaries and there is no information at the sensor node. By conditioning on the values of P_n, H_n and A_n, we find,

$$\widetilde{b}_k = \sum_{\ell=0}^{M-1} \widetilde{b}_\ell h_{k-\ell} + s_0 \sum_{\ell=M}^{C_B} \widetilde{b}_\ell h_{k-\ell+M} + p\, s_0 \sum_{\ell=M+N}^{C_B} \widetilde{b}_\ell h_{k-\ell+M+N}, \qquad (4)$$

for $k = 0, 1, \ldots, C_B - 1$, and,

$$\widetilde{b}_{C_B} = \sum_{\ell=0}^{M-1} \widetilde{b}_\ell \bar{H}_{C_B-\ell} + s_0 \sum_{\ell=M}^{C_B} \widetilde{b}_\ell \bar{H}_{C_B-\ell+M} + p\, s_0 \sum_{\ell=M+N}^{C_B} \widetilde{b}_\ell \bar{H}_{C_B-\ell+M+N}.$$

Here, $\widehat{b}_k = \mathsf{P}[B_n = k, A_n = 1]$ is the probability to have k chunks and information at a slot boundary. Again, by conditioning on the values of P_n, H_n and A_n, we have,

$$\widehat{b}_k = \sum_{\ell=0}^{M-1} \widehat{b}_\ell h_{k-\ell} + \sum_{\ell=M}^{C_B} (\widehat{b}_\ell + \widetilde{b}_\ell(1 - s_0)) h_{k-\ell+M}$$

$$+ p \sum_{\ell=M+N}^{C_B} \widehat{b}_\ell((1 - s_0) h_{k-\ell+M+N} - h_{k-\ell+M}), \qquad (5)$$

for $n = 0, 1, \ldots, C_B - 1$. Complementing the set of equations above with the normalisation condition,

$$\sum_{n=0}^{C_B} \widehat{b}_n + \widetilde{b}_n = 1,$$

allows for determining the probabilities $\widetilde{b}_n$ and $\widehat{b}_n$. For further use, we also define the probability b_n to have n chunks of energy at the node, irrespective of whether there is information at the node,

$$b_n = \widetilde{b}_n + \widehat{b}_n.$$

We now focus on the mean value of the information at the sensor node. Let $v_n = \mathsf{E}[V_k 1_{\{B_k=n\}}]$ be the mean value of information when there are n chunks of energy available. Notice that by the definition of A_k we have that $A_k = 0$ implies $V_k = 0$. Hence, there is no need to focus on the expectation of V_k for $A_k = 0$ and $A_k = 1$ as $v_n = \mathsf{E}[V_k 1_{\{B_k=n\}}] = \mathsf{E}[V_k 1_{\{B_k=n, A_k=1\}}]$ and $\mathsf{E}[V_k 1_{\{B_k=n, A_k=0\}}] = 0$. By conditioning on the presence of transmission opportunities, the availability of data and the amount of harvested energy, we find,

$$v_k = \sum_{\ell=0}^{M-1} v_\ell \alpha h_{k-\ell} + \sum_{\ell=M+N}^{C_B} \widehat{b_\ell} p \bar{S} h_{k-\ell+M+N} + \sum_{\ell=M}^{C_B} (\alpha v_\ell + \bar{S} b_\ell) h_{k-\ell+M}$$

$$- \sum_{\ell=M+N}^{C_B} (\alpha v_\ell + \widehat{\bar{S} b_\ell}) p h_{k-\ell+M},$$

for $k = 0, \ldots, C_B - 1$, while for $k = C_B$ we have,

$$v_{C_B} = \sum_{\ell=0}^{M-1} v_\ell \alpha \bar{H}_{C_B-\ell} + \sum_{\ell=M+N}^{C_B} \widehat{b_\ell} p \bar{S} \bar{H}_{C_B-\ell+M+N} + \sum_{\ell=M}^{C_B} (\alpha v_\ell + \bar{S} b_\ell) \bar{H}_{C_B-\ell+M}$$

$$- \sum_{\ell=M+N}^{C_B} (\alpha v_\ell + \widehat{\bar{S} b_\ell}) p \bar{H}_{C_B-\ell+M}.$$

Finally, as we collect information when there is energy and information at the sensor at a transmission opportunity, the mean value of the information collected at a slot boundary equals,

$$\bar{V} = p \sum_{\ell=M+N}^{C_B} v_\ell.$$

Remark 4. We note that the model of this section does not reduce to the model of Sect. 2 when the sensor always picks up information, that is, for $s_0 = 0$. Even for $s_0 = 0$, it is still possible that there is no new value of information during a time slot as there may be no sensing due to a lack of energy. If we additionally assume that sensing does not take energy ($M = 0$), both models do correspond. Indeed, Eq. (4) then implies $\widetilde{b}_k = 0$, whereas the sets of Eqs. (3) and (5) are the same.

4   Optimal Data Collection and Numerical Results

We now investigate the optimal data collection policy for the sensor node at hand. We assume that there is a cost c associated to data collection such that the average value after collection equals,

$$\bar{V}_p = -cp + \bar{V}.$$

We first illustrate the analysis of the initial model, introduced in Sect. 2 by some numerical examples. We then complement these with some numerical results for the refinements which were discussed in Sect. 3. In either case, we particularly focus on the optimal collection probability p.

4.1   Information-Agnostic Transmissions

We first investigate how the battery capacity and discount factor affect the mean value of information. To this end, we consider the initial model assuming Poisson energy harvesting — the energy harvesting distribution is Poisson with mean λ — and energy discretisation such that $M = 1$ and $N = 4$ chunks of energy are required for sensing and transmitting, respectively.

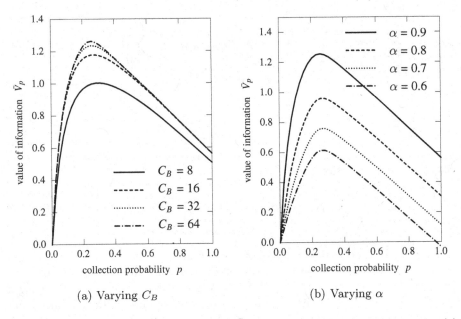

(a) Varying C_B (b) Varying α

Fig. 1. The mean value of the information $\bar{V}_p$ versus the polling probability p for (a) different values of the battery capacity C_B and (b) different values of the decay rate α as indicated.

Figures 1(a) and (b) depict the value of information $\bar{V}_p$ in terms of the data collection rate p. We assume that the cost of collection is half the mean value of information collected in a slot: $c = 1$ and $\bar{S} = 2$. Figure 1 fixes the discount factor to $\alpha = 0.9$ and shows $\bar{V}_p$ for various values of the battery capacity C_B as indicated. In contrast, Fig. 1(b) fixes the battery capacity to $C_B = 32$ and shows $\bar{V}_p$ for various values of the discount factor. For both figures, the mean number of chunks of harvest energy equals $\lambda = 2$.

It can be seen from both figures that the value of information $\bar{V}_p$ first increases for increasing values of the collection probability and then decreases again. This observation can be explained by noting that for higher values of p the chance of having insufficient energy increases as more energy is consumed for transmissions. For high p, it is quite likely that there is sufficient energy to transmit such that the possible gain of frequent data collection cannot compensate the collection cost. Further, Fig. 1(a) shows that it is beneficial to increase the battery size.

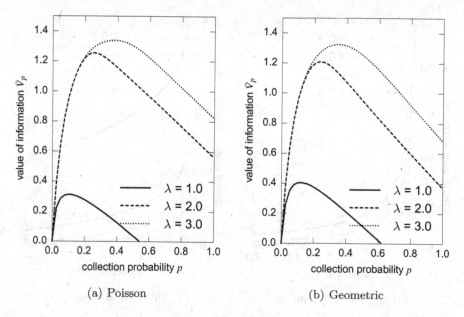

(a) Poisson   (b) Geometric

Fig. 2. The mean value of the information $\bar{V}_p$ versus the polling probability p for a Poisson (a) and geometric (b) energy harvesting distribution, for different values of the mean amount of harvested energy λ as indicated.

Having a battery with more capacity facilitates compensating periods with little energy harvesting. However, the marginal gain obtained by increasing the battery capacity quickly disappears. Increasing the discounting factor is equally beneficial as can be seen from Fig. 1(b). A higher discounting factor implies that the value of information decays more slowly such that more information is available during collection.

We now focus on the effects of the distribution of the harvested energy. To this end, Figs. 2(a) and (b) depict the mean value of information versus the collection probability p for Poisson distributed (Fig. 2(a)) and geometrically distributed (Fig. 2(b)) energy harvesting. Different values for the mean number λ of harvested chunks in a slot are assumed as indicated. As for the preceding figures, $M = 1$ and $N = 4$ chunks of energy are required for sensing and transmitting, respectively. Moreover, the discounting factor is equal to $\alpha = 0.9$, the mean value of sensed information $\bar{S} = 2$ is twice the collection cost $c = 1$, and the battery can store up to $C_B = 32$ chunks of energy.

Comparing Figs. 2(a) and (b) reveals that the distribution of the harvested energy considerable affects the value of information $\bar{V}_p$ as well as the optimal collection probability p. Increasing the harvesting capability of the sensor node (increasing λ) is initially beneficial, but the marginal gain from a further increase quickly disappears. Indeed, if there is already sufficient energy, one cannot expect that a further increase of the harvesting capability considerably improves the performance of the sensor node.

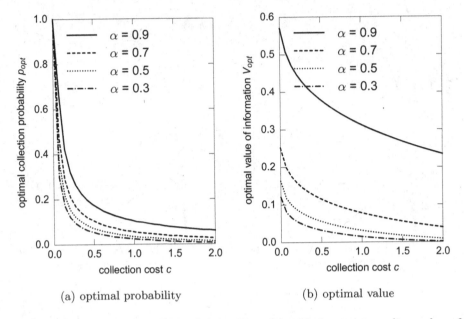

(a) optimal probability   (b) optimal value

Fig. 3. The optimal collection probability p_{opt} (a) and the corresponding value of information V_{opt} (b) for different values of α as indicated.

Finally, we consider the effect of the collection cost on the optimal collection probability and the associated optimal value of information. Figure 3(a) shows the optimal collection probability p versus the collection cost c for different values of the discount factor α as depicted. Figure 3(b) depicts the value of information corresponding to this optimal probability versus the collection cost c. Apart from the discount factor and the collection cost, the parameters are chosen as in Figs. 1(a) and (b): the mean value of sensed information equals $\bar{S} = 2$, the energy harvesting distribution is a Poisson distribution with mean 1, $M = 1$ and $N = 4$ chunks of energy are required for sensing and transmissions and the battery can store up to $C_B = 32$ of these chunks.

The optimal collection probability quickly decreases for increasing collection costs. As the collection cost increases, any gain of collecting quickly drops due to the cost of collecting. Further, if α is higher, it is more beneficial to collect (see Fig. 1(b)) such that the optimal collection probability is higher as well. In addition, the value of information $\bar{V}_p$ at the optimal collection probability is higher for increasing values of α such that it is not only beneficial to collect more, but the net gain of collecting more is higher as well.

4.2   Information-Aware Transmissions

To evaluate the use of the refined model, we now focus on how the absence of information probability s_0 affects the value of information. To make its influence clear, Figs. 4(a) and (b) depict the value of information versus the collection probability, for high and low s_0 respectively, and for various values of the discount factor α as depicted. To allow for a comparison with the model of Sect. 2 and the results of Sect. 4.1, we largely adopt the parameters of the latter: the mean value of sensed information equals $\bar{S} = 2$ which is twice the cost $c = 1$ of collecting. The energy harvesting distribution is a Poisson distribution with mean 2. In addition, $M = 1$ and $N = 4$ chunks of energy are required for sensing and transmissions and the battery can store up to $C_B = 32$ of these chunks.

As $\bar{S}$ is fixed, a high s_0 not only means that most slots there is no information, but also means that there is considerable information in the slots with information. That is, the sensing is a bursty process. In contrast, small s_0 means that many slots carry a small amount of information. It is not surprising that these considerable differences in information arrival patterns translate into different collected values of information. This is indeed confirmed by comparing Figs. 4(a) and (b). The figures show that burstiness is beneficial. This can be explained by noting that no energy is lost on sending a limited amount of information. Indeed, for $s_0 = 0.9$ the chance that there is no information is considerable, whereas the value of the information is considerable whenever there is something to send. Further comparison reveals that the value of information is considerably larger for bursty sensing compared to non-bursty sensing. Moreover, the optimal collection probability is only sensitive to changes in the discount factor for bursty sensing.

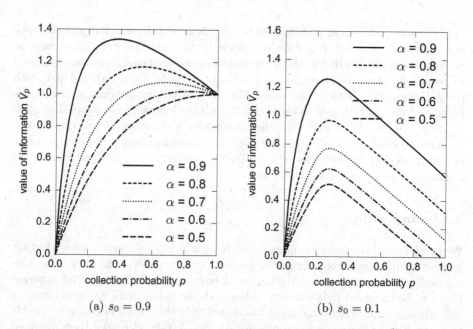

(a) $s_0 = 0.9$ (b) $s_0 = 0.1$

Fig. 4. The mean value of information versus the collection probability for (a) $s_0 = 0.9$ and (b) $s_0 = 0.1$ and different values of α as indicated.

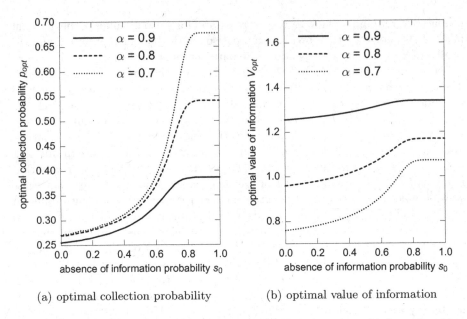

(a) optimal collection probability   (b) optimal value of information

Fig. 5. The optimal collection probability p_{opt} (a) and the value of information V_{opt} (b) versus absence of information probability s_0 for different values of α as indicated.

This observation is also confirmed by Figs. 5(a) and (b) which depict the optimal collection probability and the corresponding value of information versus the probability s_0, respectively. Different values of the discount factor are assumed as depicted. The same parameters are assumed as in Figs. 4(a) and (b), with the exception of s_0 which now varies. We see that the optimal collection probability and the corresponding value of information increases for increasing s_0 as explained before. Moreover, the difference between the optimal collection probabilities for different α is largest for high s_0. Somewhat surprising and opposite to the collection probabilities, the difference between the corresponding values of information is largest for small s_0.

5   Conclusions

We investigated the value of information in hybrid wireless sensor networks that harvest their energy from their environment. For energy-neutral harvesting sensor nodes, we proposed two Markov models to assess the value of information that can be collected by a wireless sink and study the optimal collection rate by this wireless sink. The initial model assumed that the sensor nodes were unable to assess the presence or value of information. A refined model then assumed that sensor nodes were able to assess the presence of information but not the value of information. For both models, numerical examples revealed the complex interplay between battery dynamics and the value of information.

The methodology proposed can be extended in multiple directions. First, the time between collecting currently being geometrically distributed, deterministic or generally distributed intercollection times can be considered with some additional effort. In addition, as it can be expected that energy harvesting is bursty, Markovian energy harvesting can be considered as well. In this case, it can be expected that battery dynamics will have an even more profound impact on the value of information.

Acknowledgements. This research was partially funded by the Interuniversity Attraction Poles Programme initiated by the Belgian Science Policy Office.

References

1. Atzori, L., Iera, A., Morabito, G.: The internet of things: a survey. Comput. Netw. **54**(15), 2787–2805 (2010)
2. Gubbi, J., Buyya, R., Marusic, S., Palaniswami, M.: Internet of Things (IoT): a vision, architectural elements, and future directions. Future Gener. Comput. Syst. **29**, 1645–1660 (2013)
3. Akyildiz, I.F., Su, W., Sankarasubramaniam, Y., Cayirci, E.: Wireless sensor networks: a survey. Comput. Netw. **38**(4), 393–422 (2002)
4. Akyildiz, I.F., Vuran, M.C.: Wireless Sensor Networks. Wiley, New York (2010)
5. Ammari, H.M., Gomes, N., Jacques, M., Maxim, B., Yoon, D.: A survey of sensor network applications and architectural components. Ad Hoc Sens. Wirel. Netw. **25**(1–2), 1–44 (2015)
6. Sundmaeker, H., Guillemin, P., Friess, P., Woelfflé, S. (eds).: Vision and Challenges for Realising the Internet of Things (2010)
7. Sachidananda, V., Khelil, A., Suri, N.: Quality of information in wireless sensor networks: a survey. In: Proceedings of the 15th International Conference on Information Quality (ICIQ 2010), pp. 193–207 (2010)
8. Bisdikian, C., Kaplan, L.M., Srivastava, M.B.: On the quality and value of information in sensor networks. ACM Trans. Sens. Netw. **9**(4), 48 (2013)
9. Ngai, E.C.-H., Gunningberg, P.: Quality-of-information-aware data collection for mobile sensor networks. Pervasive Mob. Comput. **11**, 203–215 (2014)
10. Ahmed, I., Butt, M.M., Psomas, C., Mohamed, A., Krikidis, I., Guizani, M.: Survey on energy harvesting wireless communications: challenges and opportunities for radio resource allocation. Comput. Netw. **88**, 234–248 (2015)
11. Zhou, Z., Du, C., Shu, L., Hancke, G., Niu, J., Ning, H.: An energy-balanced heuristic for mobile sink scheduling in hybrid WSNs. IEEE Trans. Industr. Inf. **12**(1), 28–40 (2016)
12. Bi, Y., Niu, J., Sun, L., Huangfu, W., Sun, Y.: Moving schemes for mobile sinks in wireless sensor networks. In: IEEE International Performance Computing, and Communications Conference, pp. 101–108 (2007)
13. Tunca, C., Isik, S., Donmez, M.Y., Ersoy, C.: Distributed mobile sink routing for wireless sensor networks: a survey. IEEE Commun. Surv. Tutorials **16**(2), 877–897 (2014)
14. Nazir, B., Hasbullah, H.: Mobile sink based routing protocol (MSRP) for prolonging network lifetime in clustered wireless sensor network. In: Proceedings of the 2010 International Conference on Computer Applications and Industrial Electronics, pp. 624–629 (2010)

15. Gu, Y., Ji, Y., Li, J., Zhao, B.: ESWC: efficient scheduling for the mobile sink in wireless sensor networks with delay constraint. IEEE Trans. Parallel Distrib. Syst. **24**(7), 1310–1320 (2013)
16. Neuts, M.: Matrix-Geometric Solutions in Stochastic Models: An Algorithmic Approach. Dover, New York (1981)
17. Seyedi, A., Sikdar, B.: Performance modelling of transmission schedulers capable of energy harvesting. In: Proceedings of the ICC, Cape Town (2010)
18. Seyedi, A., Sikdar, B.: Energy efficient transmission strategies for body sensor networks with energy harvesting. IEEE Trans. Commun. **58**(7), 2116–2126 (2010)

A Law of Large Numbers
for M/M/c/Delayoff-Setup Queues
with Nonstationary Arrivals

Jamol Pender[1] and Tuan Phung-Duc[2(✉)]

[1] School of Operations Research and Information Engineering,
Cornell University, Ithaca, USA
jjp274@cornell.edu
[2] Division of Policy and Planning Sciences, University of Tsukuba,
Tsukuba, Japan
tuan@sk.tsukuba.ac.jp

Abstract. Cloud computing is a new paradigm where a company makes money by selling computing resources including both software and hardware. The core part or infrastructure of cloud computing is the data center where a large number of servers are available for processing incoming data traffic. These servers not only consume a large amount of energy to process data, but also need a large amount of energy to keep cool. Therefore, a reduction of a few percent of the power consumption means saving a substantial amount of money for the company as well as reduce our impact on the environment. As it currently stands, an idle server still consumes about 60 % of its peak energy usage. Thus, a natural suggestion to reduce energy consumption is to turn off servers which are not processing data. However, turning off servers can affect the customer experience. Customers trying to access computing power will experience delays if their data cannot be processed quickly enough. Moreover, servers require setup times in order to move from the off state to the on state. In the setup phase, servers consume energy, but cannot process data. Therefore, there exists a trade-off between power consumption and delay performance. In [7,9], the authors analyze this tradeoff using an $M/M/c$ queue with setup time for which they present a decomposition property by solving difference equations. In this paper, we complement recent stationary analysis of these types of models by studying the sample path behavior of the queueing model. In this regard, we prove a weak law of large numbers or fluid limit theorem for the queue length and server processes as the number of arrivals and number of servers tends to infinity. This methodology allows us to consider the impact of nonstationary arrivals and abandonment, which have not been considered in the literature so far.

Keywords: Setup time · Abandonment · Power-saving

© Springer International Publishing Switzerland 2016
S. Wittevrongel and T. Phung-Duc (Eds.): ASMTA 2016, LNCS 9845, pp. 253–268, 2016.
DOI: 10.1007/978-3-319-43904-4_18

1   Introduction

1.1   Motivation

The core part of cloud computing is the data center where a large number of servers are available to serve the demand generated by the arrival of data traffic. These servers consume a large amount of energy, which translates into a large cost for many cloud computing companies. It is reported that data centers worldwide consume as much as about 20–30 GW of electricity [11]. However, a large part of this energy is consumed by idle servers which do not process any jobs. In fact, it is reported that an idle server still consumes about 60 % of its peak energy usage when processing jobs [3]. Thus, an important issue for the management of these data centers is to minimize the power consumption while maintaining a high quality of service for their customers. A simple way to minimize the power consumption in data center is to turn off idle servers. However, servers that are off eventually need to be turned on in order to process waiting jobs, which causes more delays. In fact, servers require some setup time in order to gain the ability to start processing jobs. Moreover, during this setup time, servers also consume a substantial amount of energy but cannot process waiting jobs. Thus, there exists a trade-off between saving power and the quality of service provided by the company. This motivates our study of multiserver queues with setup times.

In practice, the amount of requests that arrive at a data center varies time to time. It is natural that traffic in daytime is different from that in nighttime. The amount of traffic is also different on weekdays and weekends. This motivates us to consider time-non-homogeneous arrival processes. Furthermore, today data centers are partially operated by renewable energy such that wind energy or solar energy [11]. These energy sources depend on the weather and often vary on time. Thus, the number of available servers is time-dependent. This calls for the need of studying a queueing system with time-dependent number of servers. As mentioned above since setup time not only incurs in extra waiting time it may also incur in increasing energy consumption because a server consumes a large amount of energy during setup. Therefore, it is not a good strategy to turn off a server immediately upon idle. In our model, we allow an idle time before shutdown. A job arriving during the idle time is served immediately while if there is no arriving customer during the idle time, the server is switched off. Requests to data centers have time limiting nature and thus they will abandon after some waiting time. This may cause by impatient user or by the timeout of a web browser. We incorporate customer abandonment in our model. We allow all the setup rate and abandonment rate to be time-dependent. To the best of our knowledge, this paper is the first to consider a time-dependent queueing model for power-saving data centers.

1.2   Literature Review

Artalejo et al. [2] present a thorough analysis for multiserver queues with setup times where the authors consider the case in which at most one server can be in the setup mode at a time. This policy is later referred to as staggered setup in the literature [9]. Artalejo et al. [2] show an analytical solution by solving the set of balance equations for the joint stationary distribution of the number of active servers and that of jobs in the system using a difference equation approach. The solution of the staggered setup model is significantly simplified by Gandhi et al. [9] who also present a decomposition property for the queue length and the waiting time.

Recently, motivated by applications in data centers, multiserver queues with setup times have been extensively investigated in the literature. In particular, Gandhi et al. [9] present a stationary analysis for multiserver queues with setup times. They obtain some closed form approximations for the ON-OFF policy where any number of servers can be in the setup mode. As is pointed out in Gandhi et al. [9], from an analytical point of view the most challenging model is the ON-OFF policy where the number of servers in setup mode is not limited. Recently, Gandhi et al. [7,8] analyze the M/M/c/Setup model with ON-OFF policy using a recursive renewal reward approach. Gandhi et al. [7,8] claim that the model is difficult to be solved using conventional methods such as generating function or matrix analytic methods. As a result, the recursive renewal reward approach is presented as a new mathematical tool to resolve the problem. Phung-Duc [28] analyzes the same model via generating function and matrix analytic methods. It should be noted that in all the work above, arrival, service and setup processes are time-homogeneous and abandonment of customers is not taken into account.

However, as is mentioned above, in reality, traffic to data center has time-inhomogeneous nature because it is generated by human users whose activities clearly depend on time. Furthermore, nowadays, many data centers partially operate using renewable resources such as wind or solar energies [1,10]. As a result, the number of available servers also depends on time. On the other hand, ON-OFF control of servers may also incur in extra delays which cause abandonment of customers. Therefore, there is a need to develop and analyze a model taking into account all of these factors and that is the aim of the current paper.

1.3 Main Contributions of Paper

In this work, we make the following contributions to the literature on queueing theory:

1. We develop a new queueing model that incorporates a stochastic number of servers with the Delay-off feature, abandonment of jobs, and nonstationary arrival times of jobs.
2. We propose a heuristic mean field limit for the queue length and non-idle server processes.
3. We prove that the mean field heuristic is asymptotically true when the arrival rate and number of servers tend to infinity.

1.4   Organization of Paper

The rest of this paper is organized as follows. Section 2 presents the model in detail while Sect. 3 is devoted to the analysis where we present a mean field approximation and fluid limit. Section 4 shows some numerical examples showing insights into the performance of the system. Concluding remarks are presented in Sect. 5.

1.5   Notation

The paper will use the following notation:

- $\lambda(t)$ is the external arrival rate of jobs to the data center at time t
- $\mu(t)$ is the service rate of all of the servers at time t
- $\theta(t)$ is the abandonment rate of jobs at time t
- $\beta(t)$ is the rate at which needed servers transition from the OFF state to the ACTIVE (BUSY) state at time t
- $\gamma(t)$ is the rate at which unneeded servers transition from the IDLE state to the OFF state at time t
- C_{max} is the bound on the number of servers in the data center facility
- $x \wedge y = \min(x, y)$
- $(x - y)^+ = \max(0, x - y)$

2 $M_t/M/c/\text{Delayoff-Setup}+M$ Queueing Model

We consider $M_t/M/C_{max}(t)/\text{Setup}+M$ queueing systems with ON-OFF policy. Jobs arrive at the system according to a time-dependent Poisson process with rate $\lambda(t)$. In this system, after a service completion, if there is a waiting job, the server pickups this job to process immediately. Otherwise, the server stays IDLE for a while and then is switched off. We assume that the switch-off time is instantaneous. The service rate of a server is $\mu(t)$. The rate at which the server changes to OFF state is $\gamma(t)$. However, if there is some waiting customer, an OFF server is switched to the ON state with rate $\beta(t)$. We call $\beta(t)$ the setup rate. Because jobs have time-limiting nature, we assume that each waiting job abandons with rate $\theta(t)$. For this system, let $Q(t)$ denote the number of jobs in the system at time t and $C(t)$ denote the total number of BUSY and IDLE servers at time t. In our system, a server can take one of the following states: OFF, IDLE (not serving a job), BUSY (serving a job), SETUP. In the OFF state, the server does not consume energy but also does not process a job. In the IDLE state, the server consumes energy but does not process any job. In this the IDLE state, the server can process an arriving job immediately. If a job arrives at the system and there are not idle servers, the job is queued and an OFF server is activated and that server changes to the SETUP state. After the setup time, the server processes the waiting job.

Under our setting, the number of servers in setup at time t is given by $S(t) = ((Q(t) - C(t))^+ \wedge (C_{max}(t) - C(t))$ and the number of IDLE servers at time t is given by $(C(t) - Q(t))^+$, respectively.

We will model our version of the setup queue model with abandonment with a two dimensional Markov process. In fact, it is possible to derive a sample path representation of the queueing model via the work of [17] or [18] that is given by the following stochastic integral equation

$$Q(t) = Q(0) + \Pi_1 \left(\int_0^t \lambda(s)ds \right) - \Pi_2 \left(\int_0^t \mu(s) \cdot (Q(s) \wedge C(s))ds \right)$$

$$- \Pi_3 \left(\int_0^t \theta(s) \cdot (Q(s) - C(s))^+ ds \right) \tag{1}$$

$$C(t) = C(0) + \Pi_4 \left(\int_0^t \beta(s) \cdot S(s)ds \right)$$

$$- \Pi_5 \left(\int_0^t \gamma(s) \cdot (C(s) - Q(s))^+ ds \right). \tag{2}$$

The first stochastic process $Q(t)$ is for the queue length and the second stochastic process $C(t)$ is to keep track of the total number of busy servers and idle servers. With this construction, we need to define the Poisson processes Π_i that are used. For the first Poisson process Π_1, it counts the number of arrivals of jobs to be processed at the data center during the interval $(0, t]$. For Π_2, we have that it counts the number of service completions from the data center in the interval $(0, t]$. Similarly, Π_3, we have that it counts the number of jobs that have abandoned or timed out from the data center in the interval $(0, t]$. For the Poisson process Π_4 we have that it counts the number of servers that have been turned on when there is sufficient number of jobs that need to be processed. Lastly, Π_5 represents the number of servers that have been turned off because the idle times expire and jobs do not arrive.

With our stochastic model representation for a data center, there are several important observations to make under certain parameter settings. When we let the delay-off parameter $\gamma = 0$, we construct a situation where none of the servers can be turned off. Thus, the number of servers will increase until it reaches its maximum and when the maximum is reached, the queue will behave as a nonstationary multiserver or $M_t/M/C_{max}$ queue. When $\gamma = \infty$ server is turned off immediately when they are considered to be idle. In this case, the number of servers mimics the number of jobs in the system as the number of jobs decreases. Moreover, in this setting, the least amount of energy is used since servers are immediately turned off when they become idle. However, turning a server off immediately can cause unnecessary delays for future jobs.

Since the joint process $(Q(t), C(t))$ is clearly Markovian with time-dependent infinitesimal generator A_t defined on continuous and bounded functions $h : \mathbb{R} \times \mathbb{R} \to \mathbb{R}$ which has the following representation

$$A_t h(x, y) \equiv \lim_{\Delta \to 0} \frac{E[g(t, \Delta)|(Q(t), C(t)) = (x, y)] - h(x, y)}{\Delta}$$

$$= \sum_{c \in C} r_c(x, y, t) \cdot [h(x + \delta_{(x,c)}, y + \delta_{(y,c)}) - h(x, y)],$$

where $g(t, \Delta) = h(Q(t + \Delta), C(t + \Delta))$. It thus follows by Dynkin's formula, see for example Lemma 17.21 of [15], that for $t \in \mathbb{R}_+$

$$E[h(Q(t), C(t))] = h(x_0, y_0) + \int_0^t E[\mathcal{A}_s h(Q(s), C(s))] ds$$

Using the Dynkin's formula for Markov processes and the Poisson process representation of the stochastic queueing model, we can subsequently derive the functional Kolmogorov forward equations for the two dimensional Markov process as

$$\frac{d}{dt} E[h(Q(t), C(t))] \equiv \overset{\bullet}{E}[h(Q(t), C(t))]$$

$$\equiv \overset{\bullet}{E}[h(Q(t), C(t)) \mid Q(0) = Q_0, C(0) = C_0]$$
$$= E[\lambda(t) \cdot (h(Q + 1, C) - h(Q, C))]$$
$$+ E[\mu(t) \cdot (Q \wedge C) \cdot (h(Q - 1, C) - h(Q, C))]$$
$$+ E[\theta(t) \cdot (Q - C)^+ \cdot (h(Q - 1, C) - h(Q, C))]$$
$$+ E[\beta(t) \cdot (S \cdot (h(Q, C + 1) - h(Q, C))]$$
$$+ E[\gamma(t) \cdot (C - Q)^+ \cdot (h(Q, C - 1) - h(Q, C))],$$

where we omit (t) of $Q(t), C(t)$ and $S(t)$ in the right hand side for simplicity. When we let $h(x, y) := x$ or $h(x, y) := y$, we have the following equations for the mean queue length and mean number of non-idle servers

$$\overset{\bullet}{E}[Q(t)] = \lambda(t) - \mu(t) \cdot E[(Q \wedge C)] - \theta(t) \cdot E[(Q - C)^+]$$
$$\overset{\bullet}{E}[C(t)] = \beta(t) \cdot E[((Q - C)^+ \wedge (C_{max} - C)] - \gamma(t) \cdot E[(C - Q)^+].$$

Equations for second-order moments can be obtained by choosing $h(x, y) := (x \cdot y, x^2, y^2)$. In fact, monomial functions of any order can be used to obtain equations for moments of arbitrary orders by letting $h(x, y) := x^i \cdot y^j$. However, if the rate functions, which define the time changed Poisson processes, are non-linear (as is usually the case and is the case here), the term $E[\mathcal{A}_s h(Q(s), C(s))]$ will involve expectations of non-linear functions of the stochastic processes and will thus need to be simplified by applying some form of moment-closure approximation. One type of moment closure technique is the mean field approximation.

2.1   Mean Field Approximation

Using the functional Kolmogorov forward equations as outlined in [5, 12, 19, 20], we have the following system of differential equations for the mean queue length and the mean number of non-idle servers

$$\overset{\bullet}{E}[Q(t)] = \lambda(t) - \mu(t) \cdot E[(Q \wedge C)] - \theta(t) \cdot E[(Q - C)^+]$$
$$\overset{\bullet}{E}[C(t)] = \beta(t) \cdot E[((Q - C)^+ \wedge (C_{max} - C)] - \gamma(t) \cdot E[(C - Q)^+]$$

Now if we use a mean field approximation i.e.

$$E[f(X)] = f(E[X]) \tag{3}$$

we have that

$$\dot{E}[Q(t)] \approx \lambda(t) - \mu(t) \cdot (E[Q] \wedge E[C]) - \theta(t) \cdot (E[Q] - E[C])^+$$

$$\dot{E}[C(t)] \approx \beta(t) \cdot ((E[Q] - E[C])^+ \wedge (C_{max} - E[C]) - \gamma(t) \cdot (E[C] - E[Q])^+$$

Unlike the exact equations for the mean queue length and the mean number of non-idle servers, the system of equations for the mean field approximation is an autonomous dynamical system and can be solved numerically quite easily. However, this approximation is a heuristic and is not rigorous. We will show in the sequel that the mean field approximation can be made rigorous by an appropriate scaling limit of our queueing model.

3   A Weak Law of Large Numbers Limit

In order to prove a fluid limit for the queue length process and the number of servers, we need to scale our system appropriately. We define $Q^\eta(t)$ and $C^\eta(t)$ as the following stochastic processes in terms of time changed Poisson processes.

$$Q^\eta(t) = Q^\eta(0) + \Pi_1\left(\int_0^t \eta\lambda(s)ds\right) - \Pi_2\left(\eta\int_0^t \mu(s) \cdot (\bar{Q}^\eta(s) \wedge \bar{C}^\eta(s))ds\right)$$

$$-\Pi_3\left(\int_0^t \eta\theta(s) \cdot (\bar{Q}^\eta(s) - \bar{C}^\eta(s))^+ds\right)$$

$$C^\eta(t) = C^\eta(0) - \Pi_5\left(\int_0^t \eta\,\gamma(s) \cdot (\bar{C}^\eta(s) - \bar{Q}^\eta(s))^+ds\right) + \Pi_4\left(\eta\int_0^t \beta(s) \cdot \bar{S}^\eta(s)ds\right)$$

where

$$\bar{Q}^\eta(t) = \frac{1}{\eta}Q^\eta(t), \quad \bar{C}^\eta(t) = \frac{1}{\eta}C^\eta(t), \quad \bar{S}^\eta(t) = \frac{1}{\eta}S^\eta(t).$$

Let $\mathcal{D}\left([0,\infty), \mathbb{R}^2\right)$ be the space of right continuous functions with left limits in $\mathbb{R}^2$ having the domain $[0,\infty)$. We give the space $\mathcal{D}\left([0,\infty), \mathbb{R}^2\right)$ the standard Skorokhod J_1 topology. Suppose $\{X^\eta\}_{\eta=1}^\infty$ is a sequence of stochastic processes, then $X^\eta \Rightarrow x$ means that X^η converges weakly to the stochastic process x.

Definition 1. *If there exists a limit in distribution for the scaled processes* $\{\bar{Q}^\eta\}_{\eta=1}^\infty$ *and* $\{\bar{C}^\eta\}_{\eta=1}^\infty$ *i.e.* $\bar{Q}^\eta(t) \Rightarrow q(t)$ *and* $\bar{C}^\eta(t) \Rightarrow c(t)$, *then* $(q(t), c(t))$ *is called the fluid limit for the original stochastic model.*

Proposition 1. *The sequence of scaled stochastic processes* $(\bar{Q}^\eta, \bar{C}^\eta)$ *are relatively compact and all weak limits are almost surely continuous.*

Proof. In order to show that $(\bar{Q}^\eta, \bar{C}^\eta)$ is relatively compact with continuous limits, it is sufficient by Theorem 10.2 of [6] to show that the stochastic processes satisfy the following two conditions.

1. *Compact Containment:* for any $T \geq 0$, $\epsilon > 0$, there exists a compact set $\Gamma_T \subset \mathbb{R}^2$ such that

$$\lim_{\eta \to \infty} \mathbb{P}\left((\bar{Q}^\eta, \bar{C}^\eta) \in \Gamma_T, t \in [0, T] \right) \to 1, \tag{4}$$

2. *Oscillation Bound:* for any $\epsilon > 0$, and $T \geq 0$ there exists a $\delta > 0$ such that

$$\limsup_{\eta \to \infty} \mathbb{P}\left(\omega\left((\bar{Q}^\eta, \bar{C}^\eta), \delta, T \right) \geq \epsilon \right) \leq \epsilon, \tag{5}$$

where

$$\omega(x, \delta, T) := \sup_{s, t \in [0,T], |s-t| < \delta} \max_j |x_j(s) - x_j(t)|, \tag{6}$$

The proof of compact containment can be shown easily since there are no initial customers in the queue. Even if there were initial customers in the system, we can still bound the initial customers by a constant. In the case where there are no initial customers in the system, we can bound the queue length process by the arrival process. By defining the following quantity

$$\bar{\lambda} = \sup_{t \in [0,T]} \lambda(t) \tag{7}$$

it is trivial to show using the Law of Large numbers for Poisson processes that

$$\Gamma_T = \left\{ (q, c) \,\middle|\, q + c \leq q(0) + \bar{\lambda} \cdot T + C_{max} \right\} \tag{8}$$

that the compact containment condition holds. Now it remains to prove the oscillation bound for the queueing process. First we bound the difference of the queue length process

$$Q^\eta(t) - Q^\eta(u) \leq \Pi_1 \left(\eta \cdot \int_u^t \lambda(s) ds \right) + \Pi_2 \left(\int_u^t \mu(s) \cdot (Q^\eta(s) \wedge C^\eta(s)) ds \right)$$
$$+ \Pi_3 \left(\int_u^t \theta(s) \cdot (Q^\eta(s) - C^\eta(s))^+ ds \right)$$

Now we bound the difference of the process that keeps track of the number of servers that are not idling.

$$C^\eta(t) - C^\eta(u) \leq \Pi_4 \left(\int_u^t \beta(s) \cdot S^\eta(s) ds \right) + \Pi_5 \left(\int_u^t \gamma(s) \cdot (C^\eta(s) - Q^\eta(s))^+ ds \right)$$

From the compact containment property, we know that there exists a finite constant K^* such that

$$\mathbb{P}\left(\bar{Q}^\eta(s) + \bar{C}^\eta(s) \leq K^*, s \in [0, T] \right) \to 1, \text{ as } \eta \to \infty. \tag{9}$$

Thus on the event $\Omega_\eta = \{ \bar{Q}^\eta(s) + \bar{C}^\eta(s) \leq K^*, s \in [0, T] \}$, then we have the subsequent inequalities for the rate functions for all $u, t \in [0, T]$ where $|t - u| \leq \delta$.

$$\int_u^t \lambda(s)ds \leq \bar{\lambda} \cdot \delta =: c_1(\delta)$$

$$\int_u^t \mu(s) \cdot (\bar{Q}^\eta(s) \wedge \bar{C}^\eta(s))ds \leq (\mu K^*) \cdot \delta =: c_2(\delta)$$

$$\int_u^t \theta(s) \cdot (\bar{Q}^\eta(s) - \bar{C}^\eta(s))^+ \leq (\theta K^*) \cdot \delta =: c_3(\delta)$$

$$\int_u^t \beta(s) \cdot ((\bar{Q}^\eta(s) - \bar{C}^\eta(s))^+ \wedge (\bar{C}^\eta_{max}(s) - \bar{C}^\eta(s))ds \leq (\beta K^*) \cdot \delta =: c_4(\delta)$$

$$\int_u^t \gamma(s) \cdot (\bar{C}^\eta(s) - \bar{Q}^\eta(s))^+ ds \leq (\gamma K^*) \cdot \delta =: c_5(\delta)$$

Now by using the above inequalities, the Law of Large numbers for Poisson processes, and the continuity of the moduli of continuity function, the oscillation bound holds with

$$\delta = \frac{\epsilon}{\bar{\lambda} + (\mu + \theta + \beta + \gamma) \cdot K^*}. \tag{10}$$

Now the proof is complete.

Theorem 1. *If we are given determinsitic values $(q(0), c(0))$ and we assume that $(\bar{Q}^\eta(0), \bar{C}^\eta(0)) \Rightarrow (q(0), c(0))$ as $\eta \to \infty$, then the fluid limit*

$$\lim_{\eta \to \infty} \frac{1}{\eta} Q^\eta(t) \Rightarrow q(t) \quad and \quad \lim_{\eta \to \infty} \frac{1}{\eta} C^\eta(t) \Rightarrow c(t)$$

of the original stochastic queueing model is the unique solution to the following system of ordinary differential equations

$$\frac{d}{dt} q(t) = \lambda(t) - \mu(\tilde{t}) \cdot (q(t) \wedge c(t)) - \theta(t) \cdot (q(t) - c(t))^+$$
$$\frac{d}{dt} c(t) = \beta(t) \cdot (q(t) - c(t))^+ \wedge (c_{max}(t) - c(t)) - \gamma(t) \cdot (c(t) - q(t))^+. \tag{11}$$

Proof. Now that we know that the queueing and server non-idle processes are relatively compact, we can now use this result to prove the fluid limit theorem. Since $(\bar{Q}^\eta(\cdot), \bar{C}^\eta(\cdot))$ is relatively compact, we know that that given any subsequence $(\bar{Q}^{\eta_m}(\cdot), \bar{C}^{\eta_m}(\cdot))$ we can construct another subsequence $(\bar{Q}^{\eta_{m_l}}(\cdot), \bar{C}^{\eta_{m_l}}(\cdot))$ that converges weakly in $\mathbb{D}([0, \infty), \mathbb{R}^2)$, to a continuous process $(q^*(\cdot), c^*(\cdot))$. Thus, we know that $v^*(\cdot)$ is at least one limit of the original stochastic process sequence $(\bar{Q}^\eta(\cdot), \bar{C}^\eta(\cdot))$. Therefore, if we can prove that $(q^*(\cdot), c^*(\cdot))$ satisfies the fluid limit Eq. (11) and the fluid limit equations have a unique solution, then by the arbitrariness of the limit $v^*(\cdot)$, there exists unique fluid limit that is given by the equations of (11). From the representation of $(\bar{Q}^\eta(t), \bar{C}^\eta(t))$ we have that

$$\bar{Q}^\eta(t) = \bar{Q}^\eta(0) + M_Q^\eta (\bar{Q}^\eta(t), \bar{C}^\eta(t)) + \int_0^t A_Q^\eta (\bar{Q}^\eta(u), \bar{C}^\eta(u)) \, du$$

$$\bar{C}^\eta(t) = \bar{C}^\eta(0) + M_C^\eta (\bar{Q}^\eta(t), \bar{C}^\eta(t)) + \int_0^t A_C^\eta (\bar{Q}^\eta(u), \bar{C}^\eta(u)) \, du$$

where

$$M_Q^\eta \left(\bar{Q}^\eta(t), \bar{C}^\eta(t) \right) =$$

$$\left(\frac{1}{\eta} \cdot \Pi_1 \left(\eta \cdot \int_0^t \lambda(s)ds \right) - \int_0^t \lambda(s)ds \right)$$

$$- \frac{1}{\eta} \cdot \Pi_2 \left(\int_0^t \mu(s) \cdot (Q^\eta(s) \wedge C^\eta(s))ds \right) + \int_0^t \mu(s) \cdot (\bar{Q}^\eta(s) \wedge \bar{C}^\eta(s))ds$$

$$- \frac{1}{\eta} \cdot \Pi_3 \left(\int_0^t \theta(s) \cdot (Q^\eta(s) - C^\eta(s))^+ ds \right) + \int_0^t \theta(s) \cdot (\bar{Q}^\eta(s) - \bar{C}^\eta(s))^+ ds$$

$$M_C^\eta \left(\bar{Q}^\eta(t), \bar{C}^\eta(t) \right) =$$

$$\frac{1}{\eta} \cdot \Pi_4 \left(\int_0^t \beta(s) \cdot S^\eta(s)ds \right) - \int_0^t \beta(s) \cdot S^\eta(s)ds$$

$$- \frac{1}{\eta} \cdot \Pi_5 \left(\int_0^t \gamma(s) \cdot (C^\eta(s) - Q^\eta(s))^+ ds \right) + \int_0^t \gamma(s) \cdot (C^\eta(s) - Q^\eta(s))^+ ds$$

and

$$\int_0^t A_Q^\eta \left(\bar{Q}^\eta(u), \bar{C}^\eta(u) \right) du = \int_0^t \lambda(u)du - \int_0^t \mu(u) \cdot (\bar{Q}^\eta(u) \wedge \bar{C}^\eta(u))du$$

$$- \int_0^t \theta(u) \cdot (\bar{Q}^\eta(u) - \bar{C}^\eta(u))^+ du$$

$$\int_0^t A_C^\eta \left(\bar{Q}^\eta(u), \bar{C}^\eta(u) \right) du = \int_0^t \beta(u) \cdot \bar{S}^\eta(u)du - \int_0^t \gamma(u) \cdot (\bar{C}^\eta(u) - \bar{Q}^\eta(u))^+ du$$

Since we know that $\bar{\mathbf{V}}^{\eta m}(\cdot) = (\bar{Q}^{\eta m}(\cdot), \bar{C}^{\eta m}(\cdot)) \stackrel{d}{\Rightarrow} v^*(\cdot) = (q^*(\cdot), c^*(\cdot))$ *and that* $v^*(\cdot)$ *is continuous, then we have that*

$$\bar{Q}^{\eta m}(\cdot) - \bar{Q}^{\eta m}(0) - \int_0^\cdot A_Q(\bar{Q}^{\eta m})ds \stackrel{d}{\Rightarrow} q^*(\cdot) - q(0) - \int_0^\cdot A_Q(q^*(s))ds$$

$$\bar{C}^{\eta m}(\cdot) - \bar{C}^{\eta m}(0) - \int_0^\cdot A_C(\bar{C}^{\eta m})ds \stackrel{d}{\Rightarrow} c^*(\cdot) - c(0) - \int_0^\cdot A_C(c^*(s))ds.$$

Thus, if we can show that

$$\lim_{m \to \infty} \mathbf{M}^{\eta m}(\cdot) \equiv \lim_{m \to \infty} (M_Q^{\eta m}(\cdot), M_C^{\eta m}(\cdot)) = 0, \tag{12}$$

then we have that all of the limits satisfy the fluid limit Eq. (11) and since the functional $A(\cdot)$ *is Lipschitz continuous, the fluid equations have a unique solution. This implies that all of the fluid limits are the same and are all equal to the solution of the fluid limit Eq. (11). Now it remains to prove that*

$$\lim_{m \to \infty} \mathbf{M}^{\eta m}(\cdot) = 0. \tag{13}$$

Using the law of large numbers for Poisson processes, we know that

$$\lim_{\eta \to \infty} \mathbf{Y}(\eta \cdot)/\eta - \cdot \overset{d}{\Rightarrow} 0 \quad \text{in } \mathcal{D}([0, \infty), \mathbb{R}). \tag{14}$$

Moreover, since we have that $\bar{\mathbf{V}}^{\eta m}(\cdot) \overset{d}{\Rightarrow} v^*(\cdot)$ as $m \to \infty$ and we know that the limit $v^*(\cdot)$ is continuous, then we have that

$$\lim_{\eta \to \infty} \int_0^{\cdot} \mu(s) \cdot (\bar{Q}^{\eta}(s) \wedge \bar{C}^{\eta}(s)) ds \overset{d}{\Rightarrow} \int_0^{\cdot} \mu(s) \cdot (q^*(s) \wedge c^*(s)) ds$$

$$\lim_{\eta \to \infty} \int_0^{\cdot} \theta(s) \cdot (\bar{Q}^{\eta}(s) - \bar{C}^{\eta}(s))^+ ds \overset{d}{\Rightarrow} \int_0^{\cdot} \theta(s) \cdot (q^*(s) - c^*(s))^+ ds$$

$$\lim_{\eta \to \infty} \int_0^{\cdot} \beta(s) \cdot \bar{S}^{\eta}(s) ds \overset{d}{\Rightarrow} \int_0^{\cdot} \beta(s) \cdot (q^*(s) - c^*(s))^+ \wedge (c_{max}(s) - c^*(s))) ds$$

$$\lim_{\eta \to \infty} \int_0^{\cdot} \gamma(s) \cdot (\bar{C}^{\eta}(s) - \bar{Q}^{\eta}(s))^+ ds \overset{d}{\Rightarrow} \int_0^{\cdot} \gamma(s) \cdot (c^*(s) - q^*(s))^+ ds.$$

Now by the random time change Theorem of [4], we have that

$$\lim_{\eta \to \infty} Y_{il}^S \left(\eta m \int_0^{\cdot} f(s, \bar{\mathbf{V}}^{\eta m}(s)) ds \right)/\eta - \int_0^{\cdot} f(s, \bar{\mathbf{V}}^{\eta m}(s)) \overset{d}{\Rightarrow} 0$$

and this completes the proof for the fluid limit since the other terms of $\mathbf{M}^{\eta}(\cdot)$ can also be shown to converge to 0.

4   Performance Measures and Numerics

In this section, we compare our limit theorems with a discrete event simulation of the delay-off queueing process. We show that the fluid limit is quite accurate at approximating the mean dynamics of the queueing process.

4.1   Mean Queue Length and Mean Non-Idle Servers

Our first comparison between simulation and our fluid limits is given on the left of Fig. 1. In this example, the turn off rate is of moderate size meaning that it is not too high or too low. On the left of Fig. 1, we see that the simulated mean queue length is well approximated by the fluid limit and we also see similar accuracy for the mean number of idle servers. Our second comparison is given on the right of Fig. 1 and in this example the turn off rate is high. This situation is closest to when the servers are immediately shut off when they become idle. Once again on the right of Fig. 1, we see that the simulated mean queue length and the mean number of idle servers are well approximated by the mean field approximation or the fluid limit.

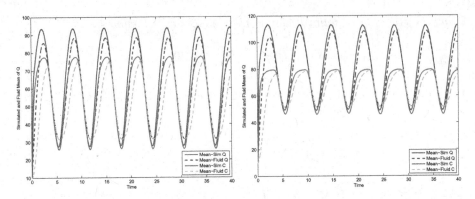

Fig. 1. $\lambda(t) = 80 + 20 \cdot \sin t$, $\mu = 1$, $\theta = 1$, $\beta = 1$, $\gamma = 1$ (*Left*), $\gamma = 1000$ (*Right*), $C_{max} = 80$. (Mean Queue Length and Mean Non-Idle Servers).

4.2   Energy Consumption

In addition to understanding how well our limit theorem approximates the actual stochastic system, it is important to analyze the power consumption of the system in a variety of parameter settings.

The mean energy consumption in the nonstationary setting is now given by

$$E[ActEne(t)] = \int_0^t E[C(u)] \times c_1(u)du,$$

where $c_1(t)$ is the energy cost for an active or idle server at time t. For simplicity we may consider the simple case where $c_1(t) = c_1$.

Furthermore, let $S(t)$ denote the number of servers in setup at time t and let $c_2(t)$ denote the energy cost for a server in setup mode at time t, then the energy consumption by servers in setup mode is given by

$$E[SetEne(t)] = \int_0^t E[S(u)] \times c_2(u)du.$$

By considering a simple case where $c_2(t) = c_2$ and since also in practice, it is empirically seen that $c_2 = c_1$. Thus, in the numerical examples, we consider the case $c_1 = c_2 = 1$. The overall energy consumption in the time interval $[0, t]$ is given by

$$E[TotalEne(t)] = E[ActEne(t)] + E[SetEne(t)].$$

We would like to minimize the above total energy consumption. On the other hand, we also would like to minimize the mean waiting cost which is calculated based on the queue length, i.e., $\int_0^t Q(u)du$. Thus, we need to consider a cost function which is a combination of the power consumption and the waiting cost.

In Figs. 2 and 3, we plot a convex combination of the power used and the queue length or delay of the system integrated over time. These plots represent the trade-off between delays experienced by customers and the power cost of the

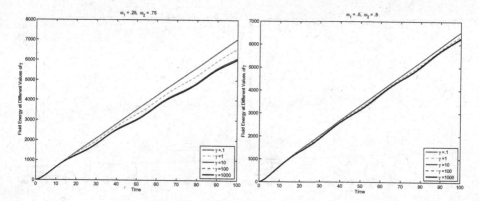

Fig. 2. $\lambda(t) = 60 + 20 \cdot \sin t$, $\mu = 1$, $\theta = 1$, $\beta = 1$, $C_{max} = 100$. Fluid Energy as γ varies and $\omega_1 = .25, \omega_2 = .75$ (Left). Fluid Energy as γ varies and $\omega_1 = .5, \omega_2 = .5$ (Right).

data center. On the left of Fig. 2, we weight the delay by $\omega_1 = .25$ and weight the power by $\omega_2 = .75$. We see that as we increase γ the total power and delay cost decreases. This is partially because we are weighting the power as more costly in this example. On the right of Fig. 2, we weight the delay by $\omega_1 = .5$ and weight the power by $\omega_2 = .5$. We see that as we increase γ the total power and delay cost decreases, but only slightly since the weighting is equal. However, we see that the power is a bit more influential on the cost, but very slight. On the left of Fig. 3, we weight the delay by $\omega_1 = .75$ and weight the power by $\omega_2 = .25$. We see that as we increase γ the total power and delay cost increases. This is partially because we are weighting the delay as more costly in this example.

On the right of Fig. 3, we plot the power consumption as we vary the parameter γ. We see that we increase γ, the power consumption goes down, especially

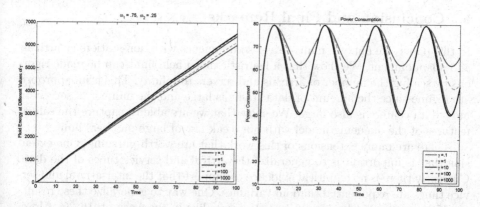

Fig. 3. $\lambda(t) = 60 + 20 \cdot \sin t$, $\mu = 1$, $\theta = 1$, $\beta = 1$, $C_{max} = 100$. Fluid Energy as γ varies and $\omega_1 = .75, \omega_2 = .25$. (Left) $\lambda(t) = 60 + 20 \cdot \sin t$, $\mu = 1$, $\theta = 1$, $\beta = 1$, $C_{max} = 100$. Power Consumption as γ varies. (Right)

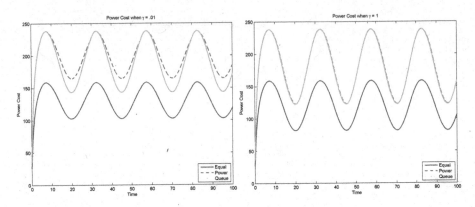

Fig. 4. $\lambda(t) = 60 + 20 \cdot \sin t$, $\mu = 1$, $\theta = 1$, $\beta = 1$, $C_{max} = 100$. Power Cost when $\gamma = .01$ (Left) Power Cost when $\gamma = 1$ (Right)

when the queue moves from the overloaded to underloaded regime. However, in the overloaded setting, the parameter γ does not do much in terms of saving power consumption.

In Fig. 4, we plot the trade-off between delays experienced by customers (in terms of $Q(t)$) and the power cost of the data center over time. In these figures, we have three different scenarios. The first corresponds to 'Equal' where the power and queue length are weighted equally. The second corresponds to power where the power cost is multiplied by a factor of 2. Lastly, the third scenario corresponds to the queue length being multiplied by 2. On the left of Fig. 4, we see that power is very important and therefore the plot that weights power more is higher than the plot that weights delay more. However, as γ gets larger on the right of Fig. 4, this difference between the two plots disappears and is negligible.

5   Conclusion and Final Remarks

In this paper, we analyze multi-server setup queues with non-stationary arrivals and abandonment. We show that a heuristic mean field limit can be made rigorous by scaling the number of arrivals and servers to infinity. This is an appropriate regime since the amount of data traffic is large and the number of servers in most data centers is also large. We show that we are able to capture the salient features of the queueing model with our weak law of large numbers limit.

There are many extensions of this work that are worth pursuing. One extension that is important is to generalize the arrival and service times of the data. Currently there is no empirical evidence to support that the inter-arrival and service times are exponential random variables. One way to generalize these results would be to use Markovian Arrival Processes like in the work of [16, 26]. Moreover, refining the approximations using orthogonal polynomial methods like in the work of [22–25, 27] is also an important area of study. We hope to consider these generalizations in future work. Moreover, we would like to incorporate the energy impact of using renewable energy such as wind and solar. This would involve additional stochastic models for understanding the mix and cost of the energy being provided to the data center. In the case when wind and solar energy are used, it may be cost effective to keep the servers on even though servers are not needed since the energy used is cheaper. We plan to pursue this extension as well. Lastly, we are interested in optimal control methods for these delay-off systems. In this context, we can use the work of [13,14,21] to find optimal turn off or on policies for our delay-off model. We also plan to complete this work in a follow up paper.

Acknowledgments. Tuan Phung-Duc was supported in part by JSPS KAKENHI Grant Number 2673001. The authors would like to thank the four referees for their constructive comments which improve the presentation of the paper.

References

1. Akoush, S., Sohan, R., Rice, A., Moore, A.W., Hopper, A.: Free lunch: exploiting renewable energy for computing. In: Proceedings of HotOS, p. 17 (2011)
2. Artalejo, J.R., Economou, A., Lopez-Herrero, M.J.: Analysis of a multiserver queue with setup times. Queueing Syst. **51**(1–2), 53–76 (2005)
3. Barroso, L.A., Hölzle, U.: The case for energy-proportional computing. Computer **12**, 33–37 (2007)
4. Billingsley, P.: Convergence of Probability Measures, vol. 493. Wiley, London (2009)
5. Engblom, S., Pender, J.: Approximations for the moments of nonstationary and state dependent birth-death queues. Submitted to Stochastic Systems (2014)
6. Ethier, S.N., Kurtz, T.G.: Markov Processes: Characterization and Convergence, vol. 282. Wiley, London (2009)
7. Gandhi, A., Doroudi, S., Harchol-Balter, M., Scheller-Wolf, A.: Exact analysis of the m/m/k/setup class of Markov chains via recursive renewal reward. SIGMETRICS Perform. Eval. Rev. **41**(1), 153–166 (2013)
8. Gandhi, A., Doroudi, S., Harchol-Balter, M., Scheller-Wolf, A.: Exact analysis of the m/m/k/setup class of Markov chains via recursive renewal reward. Queueing Syst. **77**(2), 177–209 (2014)
9. Gandhi, A., Harchol-Balter, M., Adan, I.: Server farms with setup costs. Perform. Eval. **67**(11), 1123–1138 (2010)
10. Gao, V., Zeng, Z., Liu, X., Kumar, P.R.: The answer is blowing in the wind: analysis of powering internet datacenters with wind energy. In: IEEE 2013 Proceedings of INFOCOM, pp. 520–524. IEEE (2013)
11. Goiri, Í., Haque, M.E., Le, K., Beauchea, R., Nguyen, T.D., Guitart, J., Torres, J., Bianchini, R.: Matching renewable energy supply and demand in green datacenters. Ad Hoc Netw. **25**, 520–534 (2015)
12. Grier, N., Massey, W.A., McKoy, T., Whitt, W.: The time-dependent erlang loss model with retrials. Telecommun. Syst. **7**(1–3), 253–265 (1997)
13. Hampshire, R.C., Massey, W.A.: Variational optimization for call center staffing. In: Proceedings of the 2005 Conference on Diversity in Computing, pp. 4–6. ACM (2005)

14. Hampshire, R.C., Massey, W.A.: Dynamic optimization with applications to dynamic rate queues. TUTORIALS in Operations Research, INFORMS Society, pp. 210–247 (2010)
15. Kallenberg, O.: Foundations of Modern Probability. Springer Science & Business Media, New York (2006)
16. Ko, Y.M., Pender, J.: Strong approximations for time varying infinite-server queues with non-renewal arrival and service processes
17. Kurtz, T.G.: Strong approximation theorems for density dependent Markov chains. Stoch. Process. Appl. 6(3), 223–240 (1978)
18. Mandelbaum, A., Massey, W.A., Reiman, M.I.: Strong approximations for Markovian service networks. Queueing Syst. 30(1–2), 149–201 (1998)
19. Massey, W., Pender, J.: Skewness variance approximation for dynamic rate multi-server queues with abandonment. Perform. Eval. Rev. 39, 74 (2011)
20. Massey, W., Pender, J.: Gaussian skewness approximation for dynamic rate multi-server queues with abandonment. Queueing Syst. 75(2), 243–277 (2013)
21. Niyirora, J., Pender, J.: Optimal staffing of clinical revenue centers in health care organizations (2015, under review)
22. Pender, J.: Gram Charlier expansion for time varying multiserver queues with abandonment. SIAM J. Appl. Math. 74(4), 1238–1265 (2014)
23. Pender, J.: A Poisson-Charlier approximation for nonstationary queues. Oper. Res. Lett. 42(4), 293–298 (2014)
24. Pender, J.: An analysis of nonstationary coupled queues. Telecommun. Syst. 61, 823–838 (2016)
25. Pender, J.: Nonstationary loss queues via cumulant moment approximations. Probab. Eng. Inf. Sci. 29(01), 27–49 (2015)
26. Pender, J., Ko, Y.M.: Approximations for the queue length distributions of time-varying many-server queues
27. Pender, J.: Laguerre polynomial expansions for time varying multiserver queueswith abandonment. Working paper (2014). http://www.columbia.edu/~jp3404/LSA.html
28. Phung-Duc, T.: Exact solutions for m/m/c/setup queues. Telecommunication Systems (2016). doi:10.1007/s11235-016-0177-z

Energy-Aware Data Centers with *s*-Staggered Setup and Abandonment

Tuan Phung-Duc[1](✉) and Ken'ichi Kawanishi[2]

[1] Division of Policy and Planning Sciences, University of Tsukuba, Tsukuba, Japan
tuan@sk.tsukuba.ac.jp
[2] Division of Electronics and Informatics, Gunma University, Kiryu, Japan
kawanisi@cs.gunma-u.ac.jp

Abstract. Data centers consume a large amount of energy which has a big impact on the operational cost and the environment. Saving a few percent of power consumption has a considerable impact on the running cost of data centers and on saving our environment. A simple and natural policy for saving power consumption in data centers is to turn off idle servers since these servers still consume about 60 % of their peak consumption. However, servers should be turned on again when the work load increases. Power consumption is needed to turn on an off server so that it can process waiting jobs. Therefore, limiting the number of servers in setup mode is proposed in the literature to reduce the power consumption by setup servers. We study the *s*-Staggered setup policy where at most *s* servers can be in setup mode at a time. While the *s*-Staggered setup policy may save power consumption it may also incur some extra waiting time and then the abandonment of jobs. In this paper, under the finite capacity setting, we present a simple exact analysis for the multiserver queueing model with *s*-Staggered policy and abandonment based on which we carry out the performance evaluation of power-saving data centers.

1 Introduction

1.1 Motivation and Contribution

The core part of cloud computing is data center where a huge number of servers are available. These servers consume a large amount of energy [2]. Thus, the key issue for the management of these server farms is to minimize the power consumption while keeping an acceptable service level for users. It is reported that under the current technology an idle server still consumes about 60 % of its peak processing jobs [2]. Therefore, a natural suggestion to save power consumption is to turn off idle servers. However, off servers need some setup time to be active during which they consume energy but cannot process jobs.

Since the server in setup consumes a considerable amount of energy, the number of servers in setup mode may be limited in order to save energy. Furthermore, a server instantaneously consumes a large amount of energy when it

© Springer International Publishing Switzerland 2016
S. Wittevrongel and T. Phung-Duc (Eds.): ASMTA 2016, LNCS 9845, pp. 269–283, 2016.
DOI: 10.1007/978-3-319-43904-4_19

is switched on. Thus, limiting the number of servers in setup mode also mitigates this instantaneous impact. These are the motivations for the study of the s-Staggered policy where the number of setup servers in a time is limited to s [3].

This policy is proposed and approximately analyzed by Gandhi et al. [3] for an infinite server model. However, to the best of our knowledge, an exact analysis of this policy in a multiserver server model has not been performed except for the case $s = 1$ in a infinite buffer model [1,3,10]. While the s-Staggered policy can reduce the energy consumption it may incur extra waiting time leading to the abandonment of jobs. This is the motivation for us to analyze a multiserver queueing model with s-Staggered setup policy, abandonment and finite buffer based on which we investigate several trade-offs in power-saving data centers. We develop an efficient numerical scheme allowing to calculate the stationary distribution of large-scale systems.

1.2 Related Work

Recently, Gandhi et al. [3] analyze multiserver queues with setup times. They obtain some closed form approximations for the ON-OFF policy. In the ON-OFF policy, upon completion of a service the server is switched off immediately if there is no waiting job in the buffer, otherwise, the server picks up a waiting job to process. Upon the arrival of a job, if there are some OFF servers, one of them is started up and the job is placed at the buffer. Suppose that there are two waiting jobs waiting for two servers in setup mode. In this case, if a server completes a service, the server picks up one waiting job and thus, one setup server is turned off immediately. Under the ON-OFF policy, a server has one of the three states: ACTIVE (serving a job), SETUP or OFF. There is not an idle server (i.e. consuming energy without serving a job). As is pointed out in Gandhi et al. [3], from an analytical point of view the most challenging model is the ON-OFF policy where the number of servers in setup mode is not limited. Gandhi et al. [4] analyze the M/M/c/Setup model with ON-OFF policy using a recursive renewal reward approach. Phung-Duc [14] obtains exact solutions for the same model via generating functions and via matrix analytic methods.

Although, the infinite buffer model has been extensively investigated, less attention is paid to the finite buffer model. Phung-Duc [11] presents a simple recursion for the stationary distribution of the M/M/c/K/Setup without abandonment. The computational complexity of the scheme is significantly reduced in comparison with that of direct methods [8,9]. As a result, models with several hundreds of servers are easily analyzed. This allows us to explore new insights into the performance of large scale systems. Recently, in a closely related paper [5], Kuehn et al. suggest a recursive scheme for finite buffer model with threshold control. However, the stability of the numerical scheme is not proved and the abandonment of jobs is not taken into account. In contrast to [5], we suggest here a new recursive scheme whose numerical stability is guaranteed. In all the work above, the abandonment of jobs is ignored. In our previous work, we considered a special case with abandonment where the number of servers in setup mode is not limited [12], i.e. the ON-OFF policy ($s = c$). It means that the number of servers in setup is equal to the minimum of number of waiting jobs and that of non-active servers.

Some other related works are as follows. Mitrani [6,7] considers models for server farms with setup costs. The author analyzes the models where a group of reserve servers are shutdown simultaneously if the number of jobs in the system is smaller than some lower threshold and are powered up simultaneously when it exceeds some upper threshold. Because of this simultaneous shutdown and setup, the underlying Markov chain in [7] has a birth and death structure which allows closed form solutions. The author investigates the optimal lower and upper thresholds for the system. The same author [6] extends their analysis to the case where each job has an exponentially distributed random timer exceeding which the job leaves the system. Schwartz et al. [15] consider a similar model to that in [6]. A single server model with state-dependent setup time is investigated in [13].

1.3 Organization of the Paper

The rest of this paper is organized as follows. Section 2 presents the model in details while Sect. 3 is devoted to derivation of a recursion for the joint stationary distribution. Section 4 presents some numerical examples showing insights into the performance of the system. Concluding remarks are presented in Sect. 5.

2 Model

We consider a queueing system with c servers and a capacity of K, i.e., the maximum of K ($\geq c$) jobs can be accommodated in the system. Jobs arrive at the system according to a Poisson process with rate λ. In this system, a server is turned off immediately if it has no job to do. Upon arrival of a job, the job is placed in the buffer and an OFF server (if exists) is turned on. However, the server needs some setup time to be active so as to process waiting jobs. We assume that the setup time follows the exponential distribution with mean $1/\alpha$. Thus, basically, our model is the same as the ON-OFF policy (as explained in Sect. 1.2). The new feature is that the maximum number of servers in setup mode at a time is s. The case $s = 1$ is referred to as the staggered setup in [3] while the case $s = c$ is the ON-OFF policy. Let j denote the number of jobs in the system and i denote the number of active servers. The number of servers in setup process is $\min(j - i, c - i, s)$. We call this policy the s-Staggered policy.

Under these assumptions, the number of active servers is smaller than or equal to the number of jobs in the system. Therefore, in this model a server is in either ACTIVE, OFF or SETUP. We assume that the service time of jobs follows the exponential distribution with mean $1/\mu$. Furthermore, we assume that jobs are of impatient nature and they abandon receiving service if the waiting time exceeds some threshold. In particular, we assume that a customer abandons the system after some exponentially distributed time with mean $1/\theta$. Thus, if there are j jobs in the system and i active servers, the abandonment rate is $(j - i)\theta$.

We assume that waiting jobs are served according to a first-come-first-served (FCFS) manner. We call this model an M/M/c/K/Setup queue with abandonment.

3 Analysis

In this section, we present a recursive scheme to calculate the joint stationary distribution. Let $C(t)$ and $N(t)$ denote the number of active servers and the number of jobs in the system, respectively. It is easy to see that $\{X(t) = (C(t), N(t)); t \geq 0\}$ forms a Markov chain on the state space:

$$\mathcal{S} = \{(i,j); 0 \leq i \leq c, j = i, i+1, \ldots, K-1, K\}.$$

See Fig. 1 for transition among states for the case $c = 3, s = 2$ and $K = 5$.

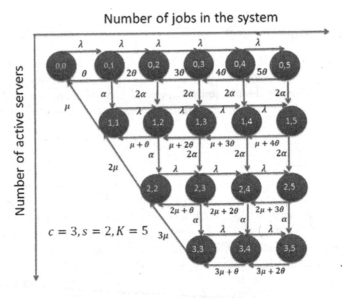

Fig. 1. Transition among states ($c = 3, s = 2, K = 5$).

Let $\pi_{i,j} = \lim_{t \to \infty} P(C(t) = i, N(t) = j)$ $((i,j) \in \mathcal{S})$ denote the joint stationary distribution of $\{X(t)\}$. In this section, we derive a recursion for calculating the joint stationary distribution $\pi_{i,j}$ $((i,j) \in \mathcal{S})$. Lemmas 1, 2, 4 and 6 allow for expressing all the probability $\pi_{i,j}$ $((i,j) \in \mathcal{S})$ in terms of $\pi_{0,0}$. Theorems 1, 3, 5 and 7 show the positivity and bounds of involved quantities that guarantee the stability of our recursive scheme.

Lemma 1. We have $\pi_{0,j} = b_j^{(0)} \pi_{0,j-1}$ where

$$b_j^{(0)} = \frac{\lambda}{\lambda + \min(j, c, s)\alpha + j\theta - (j+1)\theta b_{j+1}^{(0)}}, \tag{1}$$

for $j = 1, 2, \ldots, K - 1$ and

$$b_K^{(0)} = \frac{\lambda}{K\theta + s\alpha}.$$

Proof. Lemma 1 is easily proved using mathematical induction based on the following balance equations for the states $(0, j)$ $(j = 1, 2, \ldots, K)$.

$$(\lambda + \min(j, c, s)\alpha + j\theta)\pi_{0,j} = \lambda\pi_{0,j-1} + (j+1)\theta\pi_{0,j+1}, \quad j = 1, 2, \ldots, K-1,$$
$$(K\theta + s\alpha)\pi_{0,K} = \lambda\pi_{0,K-1},$$

where we have used $s = \min(K, c, s)$.

Theorem 1. *We have the following bound.*

$$0 < b_j^{(0)} \le \frac{\lambda}{j\theta + \min(j, c, s)\alpha}, \quad j = 1, 2, \ldots, K.$$

Proof. It is clear that Theorem 1 is true for $j = K$. Assuming that Theorem 1 is true for $j + 1$, i.e.,

$$0 < b_{j+1}^{(0)} \le \frac{\lambda}{(j+1)\theta + \min(j+1, c, s)\alpha} < \frac{\lambda}{(j+1)\theta}.$$

This inequality and the recursion (1) imply that Theorem 1 is also true for j.

Furthermore, it should be noted that $\pi_{1,1}$ is calculated using the local balance equation in and out the set $\{(0, j); j = 0, 1, \ldots, K\}$ as follows.

$$\mu\pi_{1,1} = \sum_{j=1}^{K} \min(j, c, s)\alpha\pi_{0,j}.$$

Remark 1. We have expressed $\pi_{0,j}$ $(j = 1, 2, \ldots, K)$ and $\pi_{1,1}$ in terms of $\pi_{0,0}$.

Next, we consider the case $i = 1$.

Lemma 2. *We have*

$$\pi_{1,j} = a_j^{(1)} + b_j^{(1)}\pi_{1,j-1}, \quad j = 2, 3, \ldots, K-1, K,$$

where

$$a_j^{(1)} = \frac{(\mu + j\theta)a_{j+1}^{(1)} + \min(j, c, s)\alpha\pi_{0,j}}{\mu + \lambda + \min(j-1, c-1, s)\alpha + (j-1)\theta - (\mu + j\theta)b_{j+1}^{(1)}}, \quad (2)$$

$$b_j^{(1)} = \frac{\lambda}{\mu + \lambda + \min(j-1, c-1, s)\alpha + (j-1)\theta - (\mu + j\theta)b_{j+1}^{(1)}}, \quad (3)$$

for $j = K - 1, K - 2, \ldots, 2$ and

$$a_K^{(1)} = \frac{\min(c, s)\alpha\pi_{0,K}}{\mu + \min(c-1, s)\alpha + (K-1)\theta}, \quad b_K^{(1)} = \frac{\lambda}{\mu + \min(c-1, s)\alpha + (K-1)\theta}.$$

Proof. We prove using mathematical induction. Balance equations are given as follows.

$$(\lambda + \mu + \min(j-1, c-1, s)\alpha + (j-1)\theta)\pi_{1,j} = \lambda\pi_{1,j-1} + (\mu + j\theta)\pi_{1,j+1}$$
$$+ \min(j, c, s)\alpha\pi_{0,j},$$
$$2 \le j \le K-1, \qquad (4)$$
$$(\mu + \min(K-1, c-1, s)\alpha + (K-1)\theta)\pi_{1,K} = \lambda\pi_{1,K-1} + \min(c, s)\alpha\pi_{0,K}. \qquad (5)$$

It follows from (5) that

$$\pi_{1,K} = a_K^{(1)} + b_K^{(1)}\pi_{1,K-1},$$

leading to the fact that Lemma 2 is true for $j = K$. Assuming that Lemma 2 is true for $j+1$, i.e., $\pi_{1,j+1} = a_{j+1}^{(1)} + b_{j+1}^{(1)}\pi_{1,j}$. It then follows from (4) that Lemma 2 is also true for j, i.e., $\pi_{1,j} = a_j^{(1)} + b_j^{(1)}\pi_{1,j-1}$.

Theorem 3. *We have the following bound.*

$$a_j^{(1)} \ge 0, \quad 0 \le b_j^{(1)} \le \frac{\lambda}{\mu + (j-1)\theta + \min(j-1, c-1, s)\alpha},$$

for $j = 2, 3, \ldots, K-1, K$.

Proof. We use mathematical induction. It is easy to see that the theorem is true for $j = K$. Assuming that the theorem is true for $j+1$, i.e.,

$$a_{j+1}^{(1)} \ge 0, \quad 0 \le b_{j+1}^{(1)} \le \frac{\lambda}{\mu + j\theta + \min(j, c-1, s)\alpha},$$

for $j = 1, 2, \ldots, K-1$. Thus, we have $\mu b_{j+1}^{(1)} < \lambda$. From this inequality, (2) and (3), we find that

$$b_j^{(1)} \le \frac{\lambda}{\mu + (j-1)\theta + \min(j-1, c-1, s)\alpha},$$

and $a_j^{(1)} \ge 0$.

It should be noted that $\pi_{2,2}$ can be calculated using the local balance between the flows in and out the set of states $\{(i,j); i = 0, 1, j = i, i+1, \ldots, K\}$ as follows.

$$2\mu\pi_{2,2} = \sum_{j=2}^{K} \min(j-1, c-1, s)\alpha\pi_{1,j}.$$

Remark 2. We have expressed $\pi_{1,j}$ ($j = 1, 2, \ldots, K$) and $\pi_{2,2}$ in terms of $\pi_{0,0}$.

We consider the general case where $2 \le i \le c-1$. Similar to the case $i = 1$, we can prove the following result by mathematical induction.

Lemma 4. *We have*

$$\pi_{i,j} = a_j^{(i)} + b_j^{(i)}\pi_{i,j-1}, \qquad j = i+1, i+2, \ldots, K-1, K,$$

where

$$a_j^{(i)} = \frac{(i\mu + (j+1-i)\theta)a_{j+1}^{(i)} + \min(j-i+1, c-i+1, s)\alpha\pi_{i-1,j}}{s_j^{(i)} - (i\mu + (j+1-i)\theta)b_{j+1}^{(i)}}, \qquad (6)$$

$$b_j^{(i)} = \frac{\lambda}{s_{j.}^{(i)} - (i\mu + (j+1-i)\theta)b_{j+1}^{(i)}}, \qquad (7)$$

where $s_j^{(i)} = \lambda + \min(j-i, c-i, s)\alpha + i\mu + (j-i)\theta$ *and*

$$a_K^{(i)} = \frac{\min(c-i+1, s)\alpha\pi_{i-1,K}}{\min(c-i, s)\alpha + i\mu + (K-i)\theta}, \qquad b_K^{(i)} = \frac{\lambda}{\min(c-i, s)\alpha + i\mu + (K-i)\theta}.$$

Proof. The balance equation for state (i, K) is given as follows.

$$(\min(c-i, s)\alpha + i\mu + (K-i)\theta)\pi_{i,K} = \lambda\pi_{i,K-1} + \min(c-i+1, s)\alpha\pi_{i-1,K},$$

leading to the fact that Lemma 4 is true for $j = K$. Assuming that

$$\pi_{i,j+1} = a_{j+1}^{(i)} + b_{j+1}^{(i)}\pi_{i,j}, \qquad j = i+1, i+2, \ldots, K-1.$$

It then follows from

$$(\lambda + \min(j-i, c-i, s)\alpha + i\mu + (j-i)\theta)\pi_{i,j}$$
$$= \lambda\pi_{i,j-1} + (i\mu + (j+1-i)\theta)\pi_{i,j+1} + \min(j-i+1, c-i+1, s)\alpha\pi_{i-1,j},$$

for $j = K-1, K-2, \ldots, i+1$, that

$$\pi_{i,j} = a_j^{(i)} + b_j^{(i)}\pi_{i,j-1}.$$

Theorem 5. *We have the following bound.*

$$a_j^{(i)} > 0, \quad 0 < b_j^{(i)} \le \frac{\lambda}{i\mu + (j-i)\theta + \min(j-i, c-i, s)\alpha},$$

for $j = i+1, i+2, \ldots, K-1, K$ *and* $i = 1, 2, \ldots, c-1$.

Proof. We also prove using mathematical induction. It is clear that Theorem 5 is true for $j = K$. Assuming that Theorem 5 is true for $j + 1$, i.e.,

$$a_{j+1}^{(i)} > 0, \quad 0 < b_{j+1}^{(i)} \le \frac{\lambda}{i\mu + (j+1-i)\theta + \min(j+1-i, c-i, s)\alpha},$$

for $j = i+1, i+2, \ldots, K-1, i = 1, 2, \ldots, c-1$. It follows from the second inequality that $i\mu b_{j+1}^{(i)} < \lambda$. This together with formulae (6) and (7) yield the desired result.

It should be noted that $\pi_{i+1,i+1}$ is calculated using the following local balance equation in and out the set of states:

$$\{(k,j); k = 0, 1, \ldots, i; j = k, k+1, \ldots, K\}$$

as follows.

$$(i+1)\mu\pi_{i+1,i+1} = \sum_{j=i+1}^{K} \min(j-i, c-i, s)\alpha\pi_{i,j}.$$

Remark 3. We have expressed $\pi_{i,j}$ $(i = 0, 1, \ldots, c-1, j = i, i+1, \ldots, K)$ and $\pi_{i+1,i+1}$ in terms of $\pi_{0,0}$.

Lemma 6. *We have*

$$\pi_{c,j} = a_j^{(c)} + b_j^{(c)}\pi_{c,j-1}, \qquad j = c+1, c+2, \ldots, K,$$

where

$$a_j^{(c)} = \frac{(c\mu + (j+1-c)\theta)a_{j+1}^{(c)} + \alpha\pi_{c-1,j}}{\lambda + c\mu + (j-c)\theta - (c\mu + (j+1-c)\theta)b_{j+1}^{(c)}}, \tag{8}$$
$$j = K-1, K-2, \ldots, c+1,$$

$$b_j^{(c)} = \frac{\lambda}{\lambda + c\mu + (j-c)\theta - (c\mu + (j+1-c)\theta)b_{j+1}^{(c)}}, \tag{9}$$
$$j = K-1, K-2, \ldots, c+1,$$

and

$$a_K^{(c)} = \frac{\alpha\pi_{c-1,K}}{c\mu + (K-c)\theta}, \qquad b_K^{(c)} = \frac{\lambda}{c\mu + (K-c)\theta}.$$

Proof. The global balance equation at state (c, K) is given by

$$(c\mu + (K-c)\theta)\pi_{c,K} = \alpha\pi_{c-1,K} + \lambda\pi_{c,K-1},$$

leading to

$$\pi_{c,K} = a_K^{(c)} + b_K^{(c)}\pi_{c,K-1}.$$

Assuming that $\pi_{c,j+1} = a_{j+1}^{(c)} + b_{j+1}^{(c)}\pi_{c,j}$, it follows from the global balance equation at state (c, j),

$$(\lambda + c\mu + (j-c)\theta)\pi_{c,j} = \lambda\pi_{c,j-1} + (c\mu + (j+1-c)\theta)\pi_{c,j+1} + \alpha\pi_{c-1,j},$$
$$j = c+1, c+2, \ldots, K-1,$$

that $\pi_{c,j} = a_j^{(c)} + b_j^{(c)}\pi_{c,j-1}$ for $j = c+1, c+2, \ldots, K$.

Theorem 7. *We have the following bound.*

$$a_j^{(c)} > 0, \qquad 0 < b_j^{(c)} \le \frac{\lambda}{c\mu + (j-c)\theta}, \qquad j = c+1, c+2, \ldots, K.$$

Proof. We also prove using mathematical induction. It is clear that Theorem 7 is true for $j = K$. Assuming that Theorem 7 is true for $j + 1$, i.e.,

$$a_{j+1}^{(c)} > 0, \qquad 0 \leq b_{j+1}^{(c)} \leq \frac{\lambda}{c\mu + (j + 1 - c)\theta},$$
$$j = c + 1, c + 2, \ldots, K - 1.$$

It follows from the second inequality that $(c\mu + (j + 1 - c)\theta)b_{j+1}^{(c)} < \lambda$. This together with formulae (8) and (9) yield the desired result.

We have expressed all the probability $\pi_{i,j}$ $((i, j) \in \mathcal{S})$ in terms of $\pi_{0,0}$ which is uniquely determined by the normalizing condition.

$$\sum_{(i,j)\in\mathcal{S}} \pi_{i,j} = 1.$$

Remark 4. We see that the computational complexity order for $\{\pi_{i,j}; (i, j) \in \mathcal{S}\}$ is $O(cK)$. A direct method for solving the set of balance equations requires the complexity of $O(c^3 K^3)$ while a level-dependent QBD approach [8] needs the computational complexity of $O(Kc^3)$.

4   Performance Evaluation

4.1   Performance Measures

Let P_B denote the blocking probability. We have

$$P_B = \sum_{i=0}^{c} \pi_{i,K}.$$

Let $\mathbb{E}[N]$ denote the mean number of jobs in the system, i.e.

$$\mathbb{E}[N] = \sum_{i=0}^{c} \sum_{j=i}^{K} \pi_{i,j} \times j.$$

The mean response time $\mathbb{E}[T]$ is given by

$$\mathbb{E}[T] = \frac{\mathbb{E}[N]}{\lambda(1 - P_B)}.$$

Let π_i denote the stationary probability that there are i active servers, i.e., $\pi_i = \sum_{j=i}^{K} \pi_{i,j}$. Let $\mathbb{E}[A]$ and $\mathbb{E}[S]$ denote the mean number of active servers and that in setup mode, respectively. We have

$$\mathbb{E}[A] = \sum_{i=1}^{c} i\pi_i, \qquad \mathbb{E}[S] = \sum_{i=0}^{c} \sum_{j=i}^{K} \min(j - i, c - i, s)\pi_{i,j}.$$

The power for the model with setup time is given by

$$Cost_{on-off} = C_a \mathbb{E}[A] + C_s \mathbb{E}[S], \tag{10}$$

where C_a and C_s are the cost per a unit time for an active server and a server in setup mode, respectively.

For comparison, we also find the power for the corresponding ON-IDLE model, i.e., $M/M/c/K$ with abandonment and without setup times. Letting p_i $(i = 0, 1, \ldots, K - 1, K)$ denote the stationary probability that there are i jobs in the system, we have

$$p_i = \frac{\lambda^i}{\prod_{j=1}^{i} \mu_j} p_0, \quad i = 1, 2, \ldots, K,$$

where $\mu_j = \min(j, c)\mu + \max(0, j - c)\theta$ and p_0 is determined by the normalization condition $\sum_{i=0}^{K} p_i = 1$. Let $\mathbb{E}[\widehat{A}]$ denote the mean number of active servers, we have

$$\mathbb{E}[\widehat{A}] = \sum_{i=0}^{K} \min(i, c)p_i.$$

The mean number of idle servers is given by $c - \mathbb{E}[\widehat{A}]$. Thus, for this model, the power is given by

$$Cost_{on-idle} = C_a \mathbb{E}[\widehat{A}] + (c - \mathbb{E}[\widehat{A}])C_i. \tag{11}$$

where C_i is the cost per a unit time for an idle server.

Furthermore, we define ratio between the power and the throughput. The power vs. throughput ratios for ON-OFF (R_{pow-tp}^{on-off}) and ON-IDLE ($R_{pow-tp}^{on-idle}$) are given by

$$R_{pow-tp}^{on-off} = \frac{Cost_{on-off}}{\mathbb{E}[A]\mu}, \quad R_{pow-tp}^{on-idle} = \frac{Cost_{on-idle}}{\mathbb{E}[\widehat{A}]\mu}.$$

4.2   Power Consumption per a Served Job

In this section, we compare the performance of ON-OFF policy and ON-IDLE policy. A direct comparison of the mean power consumption does not seem fair due to the abandonment. In particular, the ON-OFF policy may consume less energy just because of the abandonment. Thus, we propose to compare the amount of energy which is needed to process one job. We fix $\mu = 1$, $\alpha = 0.1$ and $c = 50$. We further fix $K = 500$ which is large enough to ensure that there is no loss due to buffer overflow. Furthermore, we consider $C_a = C_s = 1$ and $C_i = 0.6$. This is equivalent to that the server in setup mode consumes the same energy as an in active mode and that an idle server consumes 60 % energy of an active server [4].

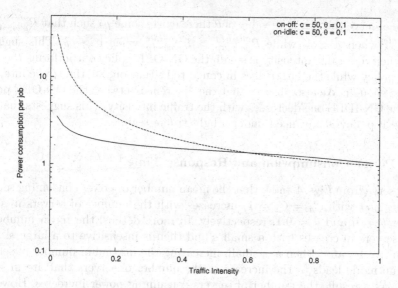

Fig. 2. Mean power consumption per a served job against ρ ($c = 50, \alpha = 0.1, K = 500$).

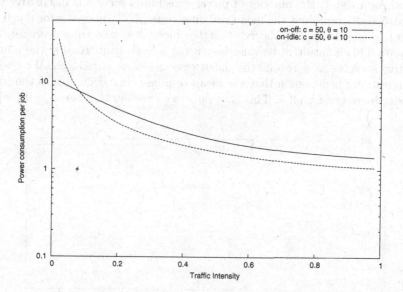

Fig. 3. Mean power consumption per a served job against ρ ($c = 50, \alpha = 0.1, K = 500$).

We investigate ratio between the mean power consumption and throughput. We observe from Figs. 2 and 3 that both R_{pow-tp}^{on-off} and $R_{pow-tp}^{on-idle}$ decrease with the traffic intensity. The reason for the ON-IDLE model is that the number of idle servers is large in the case of low traffic intensity.

We observe that for each value of θ there exists some ρ_θ such that $R_{pow-tp}^{on-off} <$ $R_{pow-tp}^{on-idle}$ when $\rho < \rho_\theta$ while $R_{pow-tp}^{on-off} > R_{pow-tp}^{on-idle}$ when $\rho > \rho_\theta$. This suggests that when the traffic intensity is small, the ON-OFF policy outperforms the ON-IDLE policy while the latter does in congested situation, i.e. the traffic intensity is large enough. We also observe that the difference between the ON-OFF policy and the ON-IDLE one decreases with the traffic intensity. This suggests that the ON-OFF policy is advanced under a light traffic regime.

4.3   Power Consumption and Response Time

We observe from Figs. 4 and 6 that the mean number of power consuming servers ($Cost_{on-off}$ with $C_a = C_s = 1$) increases with the number of servers in setup (s) for $\theta = 0$ and $\theta = 0.1$, respectively. In more details, the mean number of busy servers in creases with a small s and then is insensitive to a large s. This is intuitive because when s is small, increasing the maximum number of servers in setup mode leads to the increase in the number of servers that are in setup mode. As a result, the number of servers consuming power increases. However, when s is large enough, the necessary number of servers in setup mode is smaller than s. As a result, the number of power consuming servers is insensitive to s. Figures 5 and 7 represent the mean response time ($\mathbb{E}[T]$) against s for $\theta = 0$ and $\theta = 0.1$, respectively. We observe that the mean response time decreases as s increases. This is intuitive because increasing s leads to increasing the number of active servers. As a result, the mean response time is reduced. We observe an interesting phenomenon that the mean response time decreases as the traffic intensity increases for all s. This is because we consider the case $\alpha = 0.1$ which

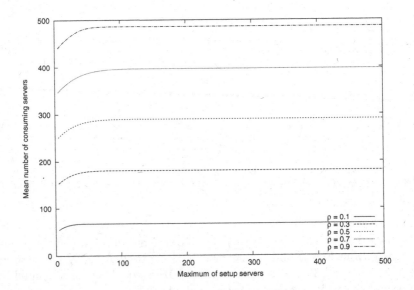

Fig. 4. Mean # of consuming servers against s ($c = 500, \theta = 0, \alpha = 0.1, K = 600$).

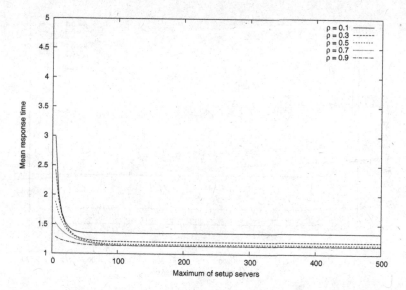

Fig. 5. Mean response time against s ($c = 500, \theta = 0, \alpha = 0.1, K = 600$).

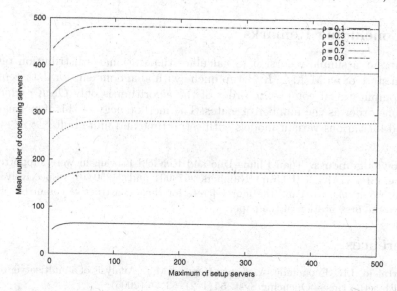

Fig. 6. Mean # of consuming servers against s ($c = 500, \theta = 0.1, \alpha = 0.1, K = 600$).

corresponds to a case of long setup time (ten times of the mean service time). In this case, a heavy load may keep a server to be in the active mode (not switched off) and thus it can process jobs. As a result, the mean response time decreases with the increase in ρ.

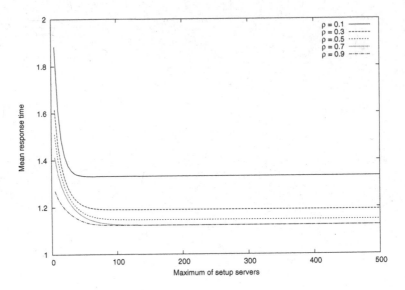

Fig. 7. Mean response time against s ($c = 500, \theta = 0.1, \alpha = 0.1, K = 600$).

5   Concluding Remarks

We present a simple recursion to calculate the stationary distribution of the system state of an M/M/c/K/Setup queue with abandonment for data centers. The computational complexity order of the algorithm is only $O(cK)$ which is the same order as the number of states. The methodology of this paper can be applied for various variant models with setup time and finite buffer.

Acknowledgements. Tuan Phung-Duc and Ken'ichi Kawanishi were supported in part by JSPS KAKENHI Grant Numbers 2673001 and 26350416, respectively. The authors would like to thank the four referees for their constructive comments which improve the presentation of the paper.

References

1. Artalejo, J.R., Economou, A., Lopez-Herrero, M.J.: Analysis of a multiserver queue with setup times. Queueing Syst. **51**(1–2), 53–76 (2005)
2. Barroso, L.A., Holzle, U.: The case for energy-proportional computing. Computer **40**(12), 33–37 (2007)
3. Gandhi, A., Harchol-Balter, M., Adan, I.: Server farms with setup costs. Perform. Eval. **67**, 1123–1138 (2010)
4. Gandhi, A., Doroudi, S., Harchol-Balter, M., Scheller-Wolf, A.: Exact analysis of the M/M/k/setup class of Markov chains via recursive renewal reward. Queueing Syst. **77**(2), 177–209 (2014)
5. Kuehn, P.J., Mashaly, M.E.: Automatic energy efficiency management of data center resources by load-dependent server activation and sleep modes. Ad Hoc Netw. **25**, 497–504 (2015)

6. Mitrani, I.: Service center trade-offs between customer impatience and power consumption. Perform. Eval. **68**, 1222–1231 (2011)
7. Mitrani, I.: Managing performance and power consumption in a server farm. Ann. Oper. Res. **202**(1), 121–134 (2013)
8. Phung-Duc, T., Masuyama, H., Kasahara, S., Takahashi, Y.: A simple algorithm for the rate matrices of level-dependent QBD processes. In: Proceedings of the 5th International Conference on Queueing Theory and Network Applications (QTNA2010), Beijing, China, pp. 46–52. ACM, New York (2010)
9. Phung-Duc, T.: Impatient customers in power-saving data centers. In: Sericola, B., Telek, M., Horváth, G. (eds.) ASMTA 2014. LNCS, vol. 8499, pp. 185–199. Springer, Heidelberg (2014)
10. Phung-Duc, T.: Server farms with batch arrival and staggered setup. In: Proceedings of the Fifth Symposium on Information and Communication Technology, pp. 240–247 (2014)
11. Phung-Duc, T.: Multiserver queues with finite capacity and setup time. In: Remke, A., Manini, D., Gribaudo, M. (eds.) ASMTA 2015. LNCS, vol. 9081, pp. 173–187. Springer, Heidelberg (2015)
12. Phung-Duc, T.: Large-scale data center with setup time and impatient customer. In: Proceedings of the 31st UK Performance Engineering Workshop (UKPEW 2015), pp. 47–60 (2015)
13. Phung-Duc, T.: Controllable setup queue for energy-aware server. In: Proceedings of the 31st UK Performance Engineering Workshop (UKPEW 2015), pp. 96–109 (2015)
14. Phung-Duc, T.: Exact solutions for M/M/c/Setup queues. Telecommunication Systems (2016). doi:10.1007/s11235-016-0177-z
15. Schwartz, C., Pries, R., Tran-Gia, P.: A queuing analysis of an energy-saving mechanism in data centers. In: Proceedings of IEEE 2012 International Conference on Information Networking (ICOIN), pp. 70–75 (2012)

Sojourn Time Analysis for Processor Sharing Loss System with Unreliable Server

Konstantin Samouylov[1], Valery Naumov[2], Eduard Sopin[1,3],
Irina Gudkova[1,3(✉)], and Sergey Shorgin[3]

[1] Peoples' Friendship University of Russia (PFUR), Moscow, Russia
{samuylov_ke,sopin_es,gudkova_ia}@pfur.ru
[2] Service Innovation Research Institute (PIKE), Helsinki, Finland
valeriy.naumov@pfu.fi
[3] Institute of Informatics Problems, Federal Research Center
"Computer Science and Control" of Russian Academy of Sciences
(IPI FRC CSC RAS), Moscow, Russia
sshorgin@ipiran.ru

Abstract. Processor sharing (PS) queuing systems and particularly their well-known class of egalitarian processor (EPS) sharing are widely investigated by research community and applied for the analysis of wire and wireless communication systems and networks. The same can be said for queuing systems in random environment, with unreliable servers, interruptions, pre-emption mechanisms. Nevertheless, only few works focus on queues with both PS discipline and unreliable servers. In the paper, compared with the previous results we analyse a finite capacity PS queuing system with unreliable server and an upper limit of the number of customers it serves simultaneously. For calculating the mean sojourn time, unlike a popular but computational complex technique of inverse Laplace transform we use an effective method based on embedded Markov chains. The paper also includes a practical numerical example of web browsing in a wireless network when the corresponding low priority traffic can be interrupted by more priority applications.

Keywords: Queuing system · Processor sharing · Egalitarian processor sharing · Unreliable server · Interruption · Probability distribution · Recursive algorithm · Mean sojourn time · Embedded Markov chain · Web browsing

1 Introduction

Processor sharing (PS) queuing systems have been widely adopted as convenient models for bandwidth sharing in computer and communication systems [1, 2]. Kleinrock [3, 4] introduced the simplest and the best known class of egalitarian processor sharing (EPS) discipline, in which a single server assigns each customer a

The reported study was funded by RFBR according to the research projects No. 16-07-00766 16-37-60103 and 16-37-00421 and by Ministry of Education and Science of the Russian Federation (No. 2987.2016.5).

S. Wittevrongel and T. Phung-Duc (Eds.): ASMTA 2016, LNCS 9845, pp. 284–297, 2016.
DOI: 10.1007/978-3-319-43904-4_20

fraction $1/n$ of the service rate when $n > 0$ customers are served. For an extensive overview of the literature on PS systems we refer to Yashkov's survey papers [5, 6]. The simplest EPS systems have Poisson arrival process and exponentially distributed service times. There are two variants of such systems: without customer waiting and with customer waiting. Both groups can be further divided into subgroups of finite capacity queues and infinite capacity queues.

Systems without waiting have no queue and if there are $n > 0$ customers in the system each gets an equal fraction $1/n$ of the service rate. Finite capacity systems have an upper limit r of the number of customers in the system. Thus, if an arriving customer finds that the limit is reached it is lost. In [7], the exact expression for the Laplace transform of the distribution of a customer's response time was obtained for infinite capacity systems. In [8], the conditional moments of the sojourn time and in [9] Laplace transform for the conditional sojourn time density was found for a finite capacity mode.

Systems with waiting have an upper limit N of the number of customers the server can serve. If an arriving customer finds that the limit is reached it waits. It also can be lost if the system has a finite capacity and there are no free waiting places. For infinite capacity models a method for calculating the moments and the distribution of the response time was presented in [10]. Nunez-Queija [11] has derived the Laplace transform of the sojourn time distribution in a system with infinite capacity and service interruptions where the on-periods and the off-periods form an alternating renewal process. General QBD model was investigated in [12], in which server unavailability and more settings regarding arrival and service time were incorporated. However, to the best of authors' knowledge, there are no works on analysis of finite buffer PS queues with waiting.

Note that most of the papers suppose to calculate characteristics via inverse Laplace transform, which is quite a complicated process. For computational reasons, a more effective procedure is the use of algorithms, e.g. see [13, 14]. In this paper, we study a finite capacity EPS system with waiting and service interruptions and apply the similar with [13] method for derivation of an algorithm for calculating the mean sojourn time.

The most challenging aspect in the analysis of sojourn time in queues with unreliable server, PS discipline and an upper threshold on the number of simultaneously served customers is how to take into account the influence of order of customers arrived during off-periods. In this paper, we propose an approach based on finite Markov chain theory to cope with that difficulty. Moreover, we apply some manual simplifications for matrix solutions based on block representation of involved matrices that significantly decrease computational complexity of the developed algorithms.

The paper is organized as follows. In Sect. 2, we propose a queuing system with unreliable server, finite buffer, PS discipline, and threshold on the number of customers as well as a method for calculating the mean sojourn time based on an embedded Markov chain. In Sect. 3, the developed model and method are applied for the analysis of web browsing in wireless networks under interruptions and some QoE influence factors. Section 4 concludes the paper.

2 Processor Sharing Loss System with Unreliable Server

2.1 Queuing Model

Consider a single-server queuing system with finite capacity r. The server is unreliable, on- and off-period durations are exponentially distributed with rates α and β respectively. Customers arrive according to a Poisson process with rate λ and service rate is exponential with parameter μ. Upon the arrival of a customer, it is placed in the queue. Customers are served according to PS discipline, but no more than N customers at once $0 < N \le r$. It means, if there are k customers in the system, only $k^* = \min\{k, N\}$ of them are served, and the rest $\max\{k - k^*, 0\}$ customers wait in the queue. If upon the arrival of a customer, the system is full already, then the customer is lost.

Behaviour of the queue with unreliable server can be described by a Markov process $\xi(t)$ with the state space $S = \{(i, k) : i = \{0, 1\}, k = 0, 1, \ldots, r\}$, where the first component indicates the on-periods ($i = 1$) and the off-periods ($i = 0$) and k is the number of customers in the system. Generator matrix of $\xi(t)$ has the following form:

$$\mathbf{Q} = \begin{bmatrix} \mathbf{Q}_{00} & \mathbf{Q}_{01} \\ \mathbf{Q}_{10} & \mathbf{Q}_{11} \end{bmatrix} \tag{1}$$

where $\mathbf{Q}_{01} = \alpha \mathbf{I}$, $\mathbf{Q}_{10} = \beta \mathbf{I}$ and blocks $\mathbf{Q}_{00}$ and $\mathbf{Q}_{11}$ are shown below:

$$\mathbf{Q}_{00} = \begin{bmatrix} -(\lambda + \alpha) & \lambda & & & \\ \mu & -(\lambda + \alpha + \mu) & \lambda & & \\ & \ddots & \ddots & \ddots & \\ & & \mu & -(\lambda + \alpha + \mu) & \lambda \\ & & & \mu & -(\alpha + \mu) \end{bmatrix}; \tag{2}$$

$$\mathbf{Q}_{11} = \begin{bmatrix} -(\lambda + \beta) & \lambda & & & \\ & -(\lambda + \beta) & \lambda & & \\ & & \ddots & \ddots & \\ & & & -(\lambda + \beta) & \lambda \\ & & & & -\beta \end{bmatrix}. \tag{3}$$

Note that the order of each block of the matrix $\mathbf{Q}$ is $(r + 1) \times (r + 1)$.

Denote $\mathbf{q}^T = (q_{00}, q_{01}, \ldots, q_{0r}, q_{10}, q_{11}, \ldots, q_{1r})$ – stationary distribution vector of $\xi(t)$ and $\mathbf{1}$ is vector of ones with appropriate size.

Proposition 1. Solution of system of equilibrium equations $\mathbf{q}^T \mathbf{Q} = \mathbf{0}$, $\mathbf{q}^T \mathbf{1} = 1$ obeys the following recurrent relations:

$$q_{01} = \frac{1}{\mu}\left((\alpha + \lambda)q_{00} - \frac{\alpha\beta}{\lambda + \beta}q_{00}\right), \tag{4}$$

$$q_{0k} = \frac{1}{\mu}\left((\alpha+\lambda+\mu)q_{0k-1} - \lambda q_{0k-2} - \frac{\alpha\beta}{\lambda+\beta}\sum_{i=0}^{k-1} q_{0i}\left(\frac{\lambda}{\lambda+\beta}\right)^{k-i}\right), \quad k = 2,\ldots,r. \quad (5)$$

$$q_{1k} = \frac{\alpha}{\lambda+\beta}\sum_{i=0}^{k} q_{0i}\left(\frac{\lambda}{\lambda+\beta}\right)^{k-i}, \quad k = 0,1,\ldots,r-1, \quad (6)$$

$$q_{1r} = \frac{\alpha}{\beta}\sum_{i=0}^{r} q_{0i}\left(\frac{\lambda}{\lambda+\beta}\right)^{r-i}. \quad (7)$$

Proof. Since matrix $\mathbf{Q}$ is irreducible, formulas (4)–(7) can be derived directly from system of equilibrium equations.

Thus, formulas (4)–(7) allow to calculate stationary probabilities up to a constant. Then, applying the normalizing condition, vector $\mathbf{q}$ is obtained.

2.2   Mean Sojourn Time

Customer's sojourn time has a phase type cumulative distribution function (CDF) $F(x)$ [16]. To find it, let us introduce a continuous-time Markov chain $\chi(t)$ that describes the behaviour of the system from arrival to departure of a particular customer. The Markov chain has an absorbing state ω, which is reached at the departure of the considered customer. To take into account the influence of the order of customers arriving during off-periods on sojourn time, we added a third component to the Markov chain $\chi(t)$. Besides the absorbing state, there are also states $(i,0,k)$, $i = 0,1$, $k = 0,1,\ldots,r$, and (i,j,k), $i = 0,1$, $j = N+1,N+2,\ldots,r$, $k = j,j+1,\ldots,r$. Here, as in the previous section, i indicates the on-periods ($i = 1$) and the off-periods ($i = 0$) and k is the number of customers in the systems. The second component $j > 0$ indicates that the customer has $(N+j)$-th spot in the line, while $j = 0$ means that the customer is in the range of first N customers in the queue.

The total number of states in the Markov chain $\chi(t)$ equal to $2(r+1) + (r-N)(r-N+1) + 1$. Let us introduce the following order on the state space:

$(0,0,1),(0,0,2),\ldots,(0,0,r),(0,1,N+1),(0,1,N+2),\ldots(0,1,r),(0,2,N+2),$
$(0,2,N+3),(0,2,r),\ldots,$
$(0,r-N-1,r-1),(0,r-N-1,r),(0,r-N,r),$
$(1,0,1),(1,0,2),\ldots,(1,0,r),(1,1,N+1),(1,1,N+2),\ldots(1,1,r),(1,2,N+2),$
$(1,2,N+3),\ldots,(1,2,r),\ldots,(1,r-N-1,r-1),(1,r-N-1,r),(1,r-N,r),\omega.$

Taking into account the order of states, generator matrix $\mathbf{A}$ of $\chi(t)$ becomes

$$\mathbf{A} = \begin{bmatrix} \mathbf{C} & \mathbf{c} \\ \mathbf{0} & 0 \end{bmatrix}, \tag{8}$$

where $\mathbf{c}$ is an exit vector to the absorbing state,

$$\mathbf{C} = \begin{bmatrix} \mathbf{C}_{00} & \mathbf{C}_{01} \\ \mathbf{C}_{10} & \mathbf{C}_{11} \end{bmatrix}, \quad \mathbf{c} = \begin{bmatrix} \mathbf{c}_0 \\ \mathbf{0} \end{bmatrix}. \tag{9}$$

Blocks $\mathbf{C}_{01}$ and $\mathbf{C}_{10}$ are diagonal matrices, $\mathbf{C}_{01} = \alpha\mathbf{I}$, $\mathbf{C}_{10} = \beta\mathbf{I}$, while $\mathbf{C}_{00}$ and $\mathbf{C}_{11}$ have the following form:

$$\mathbf{C}_{00} = \begin{bmatrix} \mathbf{A}_0 & & & \\ \mathbf{B}_1 & \mathbf{A}_1 & & \\ & \ddots & \ddots & \\ & & \mathbf{B}_{r-N} & \mathbf{A}_{r-N} \end{bmatrix}, \tag{10}$$

$$\mathbf{C}_{11} = \begin{bmatrix} \mathbf{D}_0 & & & \\ & \mathbf{D}_1 & & \\ & & \ddots & \\ & & & \mathbf{D}_{r-N} \end{bmatrix}, \tag{11}$$

where

$$\mathbf{A}_0 = \begin{bmatrix} -(\lambda+\alpha+\mu) & \lambda & & & & & & \\ \frac{1}{2}\mu & -(\lambda+\alpha+\mu) & \lambda & & & & & \\ & \ddots & \ddots & & & & & \\ & & \frac{N-2}{N-1}\mu & -(\lambda+\alpha+\mu) & \lambda & & & \\ & & & \frac{N-1}{N}\mu & -(\lambda+\alpha+\mu) & \lambda & & \\ & & & & \ddots & \ddots & \ddots & \\ & & & & & \frac{N-1}{N}\mu & -(\lambda+\alpha+\mu) & \lambda \\ & & & & & & \frac{N-1}{N}\mu & -(\alpha+\mu) \end{bmatrix}, \tag{12}$$

$$\mathbf{A}_k = \begin{bmatrix} -(\lambda+\alpha+\mu) & \lambda & & \\ & \ddots & \ddots & \\ & & -(\lambda+\alpha+\mu) & \lambda \\ & & & -(\alpha+\mu) \end{bmatrix}, \quad k = 1, 2, \ldots, r-N, \tag{13}$$

$$\mathbf{B}_k = \begin{bmatrix} \mu & 0 & & \\ & \mu & 0 & \\ & & \ddots & \ddots \\ & & & \mu & 0 \end{bmatrix}, \quad k = 1, 2, \ldots, r-N, \tag{14}$$

$$
\mathbf{D}_k = \begin{bmatrix} -(\lambda+\beta) & \lambda & & \\ & \ddots & \ddots & \\ & & -(\lambda+\beta) & \lambda \\ & & & -\beta \end{bmatrix}, \ k = 1,2,\ldots,r-N. \tag{15}
$$

The order of matrices $\mathbf{A}_0$ and $\mathbf{D}_0$ is $r \times r$, whereas the order of other matrices $\mathbf{A}_k$ and $\mathbf{D}_k$, $k = 1,2,\ldots,r-N$, is $(r+1-N-k) \times (r+1-N-k)$. The order of matrix $\mathbf{B}_1$ is $(r-N) \times r$ and $(r+1-N-k) \times (r+2-N-k)$ for matrices $\mathbf{B}_k$, $k = 2,3,\ldots,r-N$. Finally, vector $\mathbf{c}_0$ is

$$
\mathbf{c}_0 = \begin{bmatrix} \mu \\ \frac{1}{2}\mu \\ \vdots \\ \frac{1}{N-1}\mu \\ \frac{1}{N}\mu \\ \vdots \\ \frac{1}{N}\mu \end{bmatrix}, \tag{16}
$$

According to PASTA property [17], stationary distribution of $\xi(t)$ and stationary distribution of the Markov chain $\xi(t_n - 0)$, embedded at the moments t_n, $n = 1,2,\ldots$ just before the arrival, are equal. Consequently, blocking probability is given by $\pi = q_{r0} + q_{r1}$.

Proposition 2. The initial probability distribution γ of a Markov chain $\chi(t)$ has the following form:

$$
\gamma(i,j,k) = \begin{cases} q(i,k-1)/(1-\pi), & \text{if } j = 0,\ 1 \leq k \leq N; \\ q(i,k-1)/(1-\pi), & \text{if } j = k-N,\ N < k \leq r; \\ 0, & \text{otherwise.} \end{cases} \tag{17}
$$

Proof. Initial distribution γ equals to the distribution of a Markov chain $\xi(t_n + 0)$ embedded at the moments just after the arrival with the addition of appropriate j component.

Proposition 3. The mean sojourn time in the Markov chain $\xi(t)$ is

$$
m = \gamma u, \tag{18}
$$

where nonnegative vector $\mathbf{u} = -\mathbf{C}^{-1}\mathbf{1}$ is a unique solution of the system of equations $\mathbf{C}\mathbf{u} = -\mathbf{1}$.

Proof. CDF $F(x)$ of sojourn time is equal to CDF of time before absorption in the Markov chain $\chi(t)$ with the initial distribution γ. Therefore, Laplace-Stieltjes Transform (LST) of CDF $F(x)$ is $F^*(s) = \gamma(sI - \mathbf{C})^{-1}\mathbf{c}$ [16] and the average value is $m = -\gamma\,\mathbf{C}^{-1}\mathbf{1}$.

Denote $E_k = -D_k^{-1}$, put vector $\mathbf{u}$ in form $\mathbf{u} = (\mathbf{u}_{00}, \mathbf{u}_{01}, \ldots, \mathbf{u}_{0r-N}, \mathbf{u}_{10}, \mathbf{u}_{11}, \ldots,$ $\mathbf{u}_{1r-N})$ and after some simplifications we derive the following recurrent algorithm for the solution of the system of equations $\mathbf{Cu} = -\mathbf{1}$, that decreases computational complexity:

$$(\mathbf{A}_0 + \alpha\beta\mathbf{E}_0)\mathbf{u}_{00} = -(\mathbf{1} + \alpha\mathbf{E}_0\mathbf{1}), \tag{19}$$

$$(\mathbf{A}_k + \alpha\beta\mathbf{E}_k)\mathbf{u}_{0k} = -(\mathbf{1} + \alpha\mathbf{E}_k\mathbf{1} + \mathbf{B}_k\mathbf{u}_{0k-1}), \ k = 1, 2, \ldots, r - N, \tag{20}$$

$$\mathbf{u}_{1k} = \mathbf{E}_k(\mathbf{1} + \beta\mathbf{u}_{0k}), \ k = 0, 1, \ldots, r - N. \tag{21}$$

Thus, formulas (19)–(21) and (18) represent an algorithm for mean sojourn time calculation. Note that having LST of sojourn time CDF, we can obtain not only mean value, but also its variance and higher-order moments.

3   Numerical Results for Web Browsing in Unreliable Wireless Environment

3.1   QoS Characteristics for Web Browsing Under Interruptions

Nowadays, the exponential growth of mobile traffic in LTE and forthcoming wireless mobile networks leads to the need for higher bit rates. One of the promising issue regarding the problem is a flexible spectrum usage technique based on shared access to spectrum by several entities (licensed shared access, LSA). Spectrum usage under LSA operation is mainly binary by nature, i.e. at any moment, it is only used by one entity that results in user equipment's (UE) service interruptions or unavailability [18]. We consider web browsing as one of the most popular applications in mobile networks.

Let us consider an unreliable single-cell downlink channel with peak bit rate C. Times when channel is available and unavailable are exponentially distributed with averages α^{-1} and β^{-1}. We assume that UEs are stationary and perform web search in such way [19]. At the moment t_0, UE requests a webpage of average exponential size b (μ^{-1}), e.g. by clicking "Enter" after having typed the URL of the webpage in the browser's address bar or by following a link on a current webpage. Then at moment t_{Prs}, the first element of the requested page appears on the screen. Finally, at the moment t_{TPLT_2}, the page is completely rendered and displayed by the browser. After that, UE reads the page during average exponential time period Δ_R.

Two periods of time are the most important from quality of experience (QoE) point of view – these are webpage response time and webpage download time. Considering the fact that for web browsing there are constraints on the preferred and acceptable delays for a page download [19], we use the corresponding thresholds estimations for the response and download times: $T_r \geq t_{Prs} - t_0$ and $T_d \geq t_{TPLT\_2} - t_0$. Thereby, if there are N_{UEs} UEs within the cell, the overall arrival rate of UEs' requests to browse pages could be estimated as $\lambda \approx N_{UEs}(\Delta_R + T_d)^{-1}$.

To guarantee the response time requirement, a minimum bit rate guarantee during web browsing for each UE is introduced. We assume that all pages have the sizes that are identically distributed, i.e. there are no different classes, then there is the maximum allowed number of UEs $N \approx \left\lfloor \frac{b}{C(T_d - T_r)} \right\rfloor$ that could perform web search simultaneously so that to guarantee the minimum bit rate and thus response time requirements.

Threshold N on the UEs' number guarantees that a UE will download a webpage not slower than a predefined threshold on direct download delay $T_d - T_r \geq t_{TPLT\_2} - t_{Prs}$ in case there are less or equal to N UEs performing web search. If there are N UEs downloading the page (active UE) and a new UE would like to start search he will wait for some time (passive UE) up to the completion of download by any active UE. Thereby, due to the need to guarantee the requirement for the page download time we use a threshold $R \approx \left\lfloor \frac{C \cdot T_r}{b} \right\rfloor$ on the number of UEs waiting for download to start. If there are N UEs performing web search and R UEs waiting for download to start a new UE will be blocked.

We consider the proportional fairness scheduler that assume constant allocations in frequency and power domains and time division: UEs occupy portions of a time slot such as the achieved bit rates for all active UEs are equal ($r_i = r$) and $\sum\limits_{\text{all active UEs}} r_i = C$.

Web browsing via unreliable downlink channel is realized as follows. When the channel fails the webpage download is interrupted up to the channel repair. After the channel repair, the page download resumes from the interruption moment. If during the repair mode new UEs would like to perform web search, they are waiting up to the channel repair.

Regarding the quality of service (QoS), the following time characteristics are the most interesting for web browsing under interruptions:

- average download time – the average time from the moment t_0 when a UE requests to browse a webpage to the moment t_{TPLT_2} when the page is completely rendered and displayed by the browser;
- average response time (waiting time before the download starts) – the average time from the moment t_0 when a UE requests to browse a webpage to the moment t_{Prs} when the first element of the requested page appears on the screen;
- average direct download delay – the average time from the moment t_{Prs} when the first element of the requested webpage appears on the screen to the moment t_{TPLT_2} when the page is completely rendered and displayed by the browser;
- average waiting time during the download due to interruptions – the average sum of time periods within the webpage download time when the download was interrupted due to the channel failures;
- average waiting time during the webpage direct download due to interruptions – the average sum of time periods within the direct download time when the download was interrupted due to the channel failures;
- average waiting time – the sum of average waiting time before the download starts and average waiting time during the download due to interruptions.

In this paper, we propose to model the described above web browsing process in unreliable wireless environment with interruptions as a queuing system with unreliable server, finite buffer $R = r - N$, PS discipline, and threshold on the number of customers N. It was described and analysed in the previous Sect. 2 as well as the method for calculating the mean sojourn time (average webpage download time). Numerical case study concludes this Section.

3.2 Numerical Analysis of Average Webpage Download Time Under QoE Influence Factors

As we noted above, the webpage download time is a QoS characteristic, which mostly influences QoE [20]. Nevertheless, there are a lot of additional factors that also influence QoE through webpage download time changes. According to [21], these factors are grouped into three main categories – user, context, and system influence factors. User factors are UE specific, e.g. mood, weather. Context factors are as follows: UE location, interactivity of webpage, task type and task urgency. Examples of system factors are the following: channel peak bit rate, type of objects on the page (image, video, etc.), page size, caching, and so on. The purpose of the following numerical example is to analyse the influence of some system factors, namely channel pick bit rate and caching, on the average download time as well as its correspondence to the thresholds on the preferable and acceptable delays.

All necessary system parameters that are needed for the numerical analysis are figured out in Tables 1, 2, and 3. Let us note that we assume only one application, i.e. web browsing, so only a share of the overall downlink channel bit rate is available for it. According to [22], this share could be approximated as 9 %. Rare events of the channel "failures", e.g. under LSA operation, could not practically influence the webpage download time as the corresponding failure time period is sufficiently larger than the download time. Therefore, for the numerical example, we take the channel availability and unavailability time periods that describe interruptions owing to the higher priority applications.

Along with the average webpage download time, we also analyse the blocking probability, i.e. probability that a new UE would like to perform web search but the expected download time will obviously exceed the required threshold. Figures 1 and 2 depict the scenario with different downlink channel peak bit rates (see Table 2); at Figs. 3 and 4, one can see the scenario with different average webpage sizes (see Table 3).

Table 1. System parameters

Parameter description	Notation	Value
Overall downlink channel peak bit rate	C_1	See Table 2
Average time when channel is available	α^{-1}	10 s
Average time when channel is unavailable	β^{-1}	2 s
Number of UEs within the cell	N_{UEs}	20 ÷ 70
Average webpage size	b	See Table 3
Threshold on the webpage response time	T_r	2 s [23]
Threshold on the webpage download time	T_d	4 s [23]
Average time when UE reads the webpage	Δ_R	30 s [24]

For more than 30 UEs within the cell, browsing top 100 URL webpages with caching (size 493 KB), 10 MHz MIMO 4 × 4, 20 MHz MIMO 2 × 2, and 20 MHz MIMO 4 × 4 satisfy the threshold of 2 s on the preferable delay. Cases 5 MHz MIMO

Table 2. System parameters: downlink channel peak bit rate [25]

Antenna technology	Downlink channel peak bit rate		
	5 MHz	10 MHz	20 MHz
MIMO 2 × 2	43 Mbps	86 Mbps	173 Mbps
MIMO 4 × 4	82 Mbps	163 Mbps	326 Mbps

Table 3. System parameters: average webpage size [26]

Caching	URLs		
	All	Top 1000	Top 100
Without cache (total size)	2296 KB	2017 KB	1257 KB
With cache (without images)	839 KB	890 KB	493 KB

4 × 4 and 10 MHz MIMO 2 × 2 satisfy the threshold of 4 s on the acceptable delay. 5 MHz MIMO 2 × 2 is not usable for more than 30 UEs.

For more than 50 UEs within the cell with 20 MHz bandwidth and MIMO 2 × 2 (pick bit rate 173 Mbps), only top 100 URL webpages (with cache) will be downloaded within the preferable delay, and all pages (with cache) within the acceptable delay. For 30–50 UEs, the page download time satisfy the threshold on the preferable delay for all pages (with cache) and the threshold on the acceptable delay only for top 100 URL webpages (without cache).

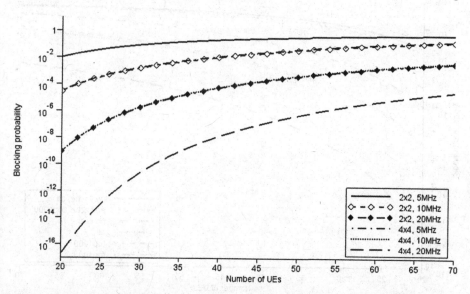

Fig. 1. Blocking probability π of a UE request vs. downlink channel peak bit rate C_1 (for top 100 URL webpage with cache)

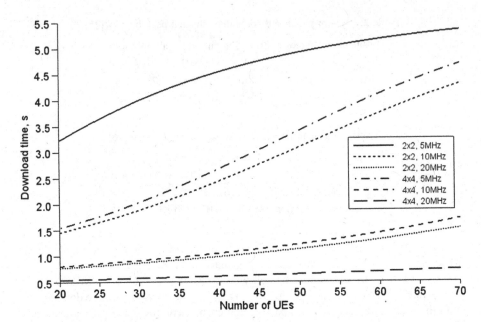

Fig. 2. Average webpage download time m (mean sojourn time) vs. downlink channel peak bit rate C_1 (for top 100 URL webpage with cache)

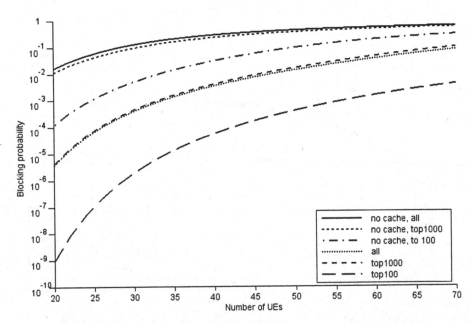

Fig. 3. Blocking probability π of a UE request vs. average webpage size b (for 20 MHz MIMO 2×2)

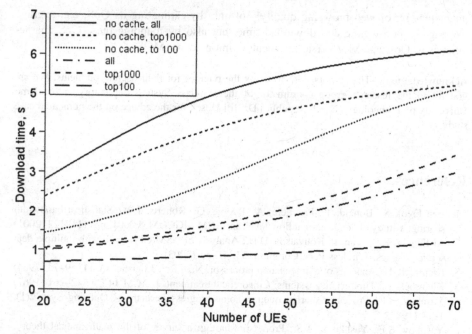

Fig. 4. Average webpage download time m (mean sojourn time) vs. average webpage size b (for 20 MHz MIMO 2 × 2)

4 Conclusion

In this paper, we propose a Markov chain based method for the analysis of mean sojourn time in the finite capacity queuing system with an unreliable server and PS discipline. We derived a recurrent algorithm for stationary probability distribution of the queue. For the analysis of mean sojourn time, the well-known absorbing Markov chain approach is used. However, we introduced a novel technique addressing some peculiarities of the queue at the end of off-periods.

The proposed queuing system with an unreliable server, a finite buffer, PS discipline, and threshold on the number of customers is applied for the analysis of web browsing in wireless networks, which radio resources are not always available. The periods of unavailability increase the webpage download time, the most important QoS characteristic. We use the developed method for calculating the mean sojourn time for the considered queuing system to analyse the page download time and the dependencies of this time on some QoE system influence factors – channel pick bit rate, webpage size, caching.

The authors see the future work via two interrelated research directions. The first one is mathematical and related to the methods for calculating other characteristics, which are listed in Sect. 3, primarily the mean waiting time. The choice of probability time characteristics for future analysis will be based on its importance for the estimation of web browsing QoS. The second research direction is applied to the understanding of web browsing quality, not only by estimating its QoS, e.g. average webpage response time and download time, but also by mapping these QoS characteristics to QoE and vice versa, i.e. mean opinion score (MOS).

Acknowledgment. The authors wish to thank the referees for their useful comments. We are grateful to the research group coordinator of the Wireless System Laboratory at the Brno University of Technology (Czech Republic) Dr. Jiri Hosek for the advice on the numerical case study.

References

1. Ben Fredj, S., Bonald, T., Proutiere, A., Regnie, G., Roberts, J.W.: Statistical bandwidth sharing: a study of congestion at flow level. In: ACM SIGCOMM 2001, pp. 111–122 (2001)
2. Al-Begain, K., Awan, I., Kouvatsos, D.D.: Analysis of GSM/GPRS cell with multiple data service classes. Wireless Pers. Commun. **25**, 41–57 (2003)
3. Kleinrock, L.: Analysis of a time-shared processor. Nav. Res. Logistics Q. **11**, 59–73 (1964)
4. Kleinrock, L.: Time-shared systems: a theoretical treatment. J. ACM **14**(2), 242–261 (1967)
5. Yashkov, S.F.: Processor-sharing queues: some progress in analysis. Queuing Syst. **2**(1), 1–17 (1987)
6. Yashkov, S.F., Yashkova, A.S.: Processor sharing: a survey of the mathematical theory. Autom. Remote Control **68**(9), 1662–1731 (2007)
7. Morrison, J.A.: Response-time distribution for a processor-sharing system. SIAM J. Appl. Math. **45**(1), 152–167 (1985)
8. Knessl, C.: On finite capacity processor-shared queues. SIAM J. Appl. Math. **50**(1), 264–287 (1990)
9. Zhen, Q., Knessl, C.: On sojourn times in the finite capacity M/M/1 queue with processor sharing. Oper. Res. Lett. **37**(6), 447–450 (2009)
10. Rege, K., Sengupta, B.: Sojourn time distribution in a multiprogrammed computer system. AT&T Tech. J. **64**(5), 1077–1090 (1985)
11. Nunez-Queija, R.: Sojourn times in a processor sharing queue with service interruptions. Queueing Syst. **34**(1), 351–386 (2000)
12. Nunez-Queija, R.: Sojourn times in non-homogeneous QBD processes with processor-sharing. Stoch. Models **17**(1), 61–92 (2001)
13. Zhen, Q., Knessl, C.: Asymptotic analysis of spectral properties of finite capacity processor shared queues. Stud. Appl. Math. **131**(2), 179–210 (2013)
14. Samouylov, K., Gudkova, I.: Recursive computation for a multi-rate model with elastic traffic and minimum rate guarantees. In: 2nd International Congress on Ultra Modern Telecommunications and Control Systems ICUMT, pp. 1065–1072 (2010)
15. Fiems, D., Bruneel, H.: Discrete-time queueing systems with Markovian preemptive vacations. Math. Comput. Model. **57**(3–4), 782–792 (2013)
16. Asmussen, S.: Applied Probability and Queues. Springer, New York (2003)
17. Wolff, R.W.: Poisson arrivals see time averages. Oper. Res. **30**(2), 223–231 (1982)
18. Gudkova, I., Samouylov, K., Ostrikova D., Mokrov, E., Ponomarenko-Timofeev, A., Andreev, S., Koucheryavy, Y.: Service failure and interruption probability analysis for licensed shared access regulatory framework. In: 7th International Congress on Ultra Modern Telecommunications and Control Systems ICUMT, pp. 123–131 (2015)

19. ITU-T G.1030 Estimating end-to-end performance in IP networks for data applications (2014)

20. Hosek, J., Ries, M., Vajsar, P., Nagy, L., Sulc, Z., Hais, P., Penizek, R.: Mobile web QoE study for smartphones. IEEE GLOBECOM **2013**, 1157–1161 (2013)

21. ITU-T G.1031 QoE factors in web-browsing (2014)

22. ITU-T M.2370 IMT traffic estimates for the years 2020 to 2030 (2015)

23. ITU-T G.1010 End-user multimedia QoS categories (2001)

24. How long do Users Stay on Web Pages? Nielsen Norman Group (2011). https://www.nngroup.com/articles/how-long-do-users-stay-on-web-pages

25. Realistic LTE Performance – From Peak Rate to Subscriber Experience. White paper, Motorola (2009). http://www.apwpt.org/downloads/realistic_lte_experience_wp_motorola_aug2009.pdf

26. Interesting Stats. HTTP Archive (2016). http://httparchive.org/interesting.php?a=All&l=Apr%2015%202016

Performance Modelling of Optimistic Fair Exchange

Yishi Zhao[1] and Nigel Thomas[2(✉)]

[1] Faculty of Information Engineering,
China University of Geosciences, Wuhan, China
zhaoyishi@cug.edu.cn
[2] School of Computing Science,
Newcastle University, Newcastle upon Tyne, UK
nigel.thomas@ncl.ac.uk

Abstract. In this paper we explore the overhead introduced by secure functions in considering a case study in non-repudiation. We present a model of an optimistic fair exchange protocol specified using the Markovian process algebra PEPA and present results derived using a fluid approximation and stochastic simulation. This system poses an interesting performance problem in that the degree of overhead of the protocol is depended on the degree of misbehaviour by the participants.

1 Introduction

Ensuring that systems are secure is a major concern for organisations and individuals, however it is essential that defensive security measures do not impose excessive overheads which might then encourage subversion of those measures in order to make the system more usable. All security measures will entail some additional work being undertaken which will impose a performance overhead. However any security measure that excessively degrades usability is obviously undesirable. It is therefore essential that any security overhead is understood, measured and minimised. In many practical situations there may be a choice of methods, such as varying protocols, algorithms or parameters, which could be employed. Changing the choice of method could have a potentially significant impact on the system performance without degrading the security. In other situations methods can be modified, e.g. by changing a key length, block size, padding or key refresh rate, which might improve performance at the cost of some level of security, thus giving a security performance trade-off [17]. However, quantifying this trade-off is not always possible, due to a lack of quantitative methods for evaluating system security. Our approach is based purely on evaluating the performance of the system and thus giving the system designer information on competing designs.

In this paper we consider a type of non-repudiation protocol known as an *optimistic non-repudiation protocol*, which utilises a trusted third party when errors occur. This leads to an interesting performance problem where the capacity of the system at a given time is determined by the degree of misbehaviour.

© Springer International Publishing Switzerland 2016
S. Wittevrongel and T. Phung-Duc (Eds.): ASMTA 2016, LNCS 9845, pp. 298–313, 2016.
DOI: 10.1007/978-3-319-43904-4_21

This paper is organised as follows. In the next section we explore performance modelling of security protocols and focus on the use of the sctochastic process algebra PEPA [8]. The main focus of the paper is then covered in Sect. 3, with a case study in secure e-commerce. Finally we make some concluding remarks and observations.

2   Performance Models of Secure Systems

A greater level of understanding of secure system performance can be gained by specifying and analysing a performance model. One objective of such analysis is to understand the trad-off that may be formed between competing require-ments for greater security and acceptable performance under variable load. The notion of the performance security trade-off has been investigated by Wolter and Reinecke [17]. In order to formalise a trade-off there needs to be measures of both security and performance whose relative values can be contrasted. Wolter and Reinecke take the view that security can be considered as a form of reliability problem; thus the system may be considered to suffer security failures and may be subsequently recovered. The performance security trade-off is then charac-terised by the performance of such a system under different levels of threat.

Researchers have used a number of modelling approaches to analyse the per-formance of secure systems. Cho *et al.* [5] used stochastic Petri nets (SPN) to investigate the potential attack in Dynamic Group Communication system (DGCs) to explore how the different rekeying methods affect both security and performance of the whole system, and optimised those methods with appropri-ate parameters. Wang *et al.* [16] formulated a queueing model for three types of attack on email systems, analysing the model for performance, dependabil-ity and information leakage. Other notable queueing models include El-Hadidi *et al.* [6,7], who evaluated the performance of the Kerberos protocol in an dis-tributed environment and Liu *et al.* [10], who studied an authentication proto-col. Recently a queueing model has been proposed by Meng *et al.* [11] which explores the relationship between encryption key refresh rate and vulnerability of the communication channel to attack.

As the correctness of security protocols is often undertaken using process algebra [1,13], it is a natural step to investigate temporal properties of protocols using stochastic process algebra (SPA). The advantage of using a formal specifi-cation for such models is that it is possible to check specific properties to ensure that the model correctly depicts behaviour which is essential to the security of the system. Thus a formal performance model and a formal security model of a given system can be shown to exhibit equivalence, giving the system designed some reassurance that the performance behaviour is valid. A process algebra allows detailed behaviour to be modelled and has the potential to be modified automatically through model transformations to facilitate alternative forms of analysis.

One of the earliest stochastic process algebra models of a secure system was that proposed by Buchholtz *et al.* [4] concerning the so-called *wide mouthed frog protocol*. The purpose of this model was to investigate the potential vulnerability of the protocol to timing attack. Thomas [14] also used stochastic process algebra, in a peformability study of a secure e-voting system. The analysis of the model did not scale well as the number of voters was increased, hence it was necessary to develop simplified models to support the analysis of larger scale systems. The issue of scale was a significant feature of three stochastic process algebra models of Internet worm attacks, proposed by Bradley *et al.* [2]. To consider scalable analysis, a fluid flow approximation based on ordinary differential equations (ODEs) [9] was employed to analyse the models. This kind of analysis approximates the original discrete state space into continuous states, and it is able to cope with models of 10^{10000} states and beyond. ODE analysis was also employed in our previous work on a *Key Distribution Centre* (key exchange protocol). This work demonstrates a rigorous approach to specifying, modifying and analysing stochastic process algebra models of security protocols by several alternative techniques [15,18,20].

3   Optimistic Fair Exchange

The case study in this section concerns a type of non-repudiation protocol known as an *optimistic non-repudiation protocol*, which utilises a trusted third party when errors occur. This leads us to model the protocol in two ways: with misbehaviour and without. We employ a modelling form in which a server has been considered as several threads, with each thread associated with a customer. Hence, the service rate of the server becomes a function of the number of threads. In the next subsection a specification of the basic version (no misbehaviour) of the e-commerce protocol is given. The subsequent section then introduces the PEPA model of this basic version of the protocol, followed by numerical results. After that, an extended version (with misbehaviour) of the e-commerce protocol is described, with the PEPA model, then some numerical results.

3.1   An E-Commerce Protocol (Basic)

This e-commerce protocol is an optimistic non-repudiation protocol, which adopts an offline TTP (*Third Tust Party*) not only to ensure fair exchange, but also to minimize the workloads from TTP server. Following the formal description in [12], the basic protocol (without misbehaviour of any principals) is illustrated below:

There is a set environment before the protocol operates, in which C (*Customer*) opens an account with B (*Bank*) and M (*Merchant*) registers with the TTP (*Trusted Third Party*). The protocol is then covered in six steps:

1. **C Selects a Product to Purchase** (*download*). The customer chooses a product, and downloads it from the Internet merchant. However, this e-product has been encrypted, and so the customer cannot acquire the product without a decryption key. This product can be used for validation later.

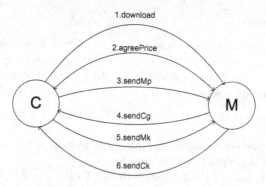

Fig. 1. The basic protocol

2. **C and M Agree Upon a Price for the Product** (*agreePrice*). Several messages may be exchanged between C and M in this step (Fig. 1).
3. **C sends PO** (*purchase order*) to M (*sendMp*). The customer sends three elements to the merchant: *(a)* the purchase order; *(b)* a digitally signed cryptographic checksum of the PO; and *(c)* the PT (*Payment Token*).
4. **M Sends Encrypted Product to C or Abort the Transaction** (*sendCg or sendCabort*). The merchant checks the purchase order which was received at the last step: if the merchant is not satisfied, then an abort message is sent to C; otherwise, the following is sent to C: *(a)* a signed cryptographic checksum of the purchase order; *(b)* encrypted product; *(c)* signed cryptographic checksum of the encrypted product; *(d)* encrypted random number; and *(e)* signed cryptographic checksum of the encrypted random number.
5. **C Sends Payment Token Decryption Key to M or Abort the Transaction** (*sendMk or sendMabort*). C checks the message from M, if it is an abort, then abort the transaction. Otherwise, C attempts to validate the product. C sends M a signed abort message if the product has failed to be validated; otherwise, sends the payment token decryption key and a signed cryptographic checksum of the encrypted product decryption key.
6. **M Sends Product Decryption Key to C or Terminates the Transaction** (*sendCk*). If M receives an abort message from C, it terminates the transaction. Otherwise, if the received PT decryption key works, M sends the following to C: *(a)* the product decryption key; *(b)* signed cryptographic checksum of the encryption product decryption key; *(c)* the multiplicative inverse of a random number; *(d)* signed cryptographic checksum of the encrypted multiplicative inverse of the random number.

To address the performance aspects of this protocol, this illustration focuses on the behaviour. Therefore the security details (which would be crucial to a security evaluation) have been eliminated from the description above. The original paper [12] gives a more detailed version.

3.2 Misbehaviour

The protocol specification above is a basic version, which operates without misbehaviour of any participants. It is necessary to investigate the performance of the TTP to observe how the protocol reacts to potential misbehaviour by participants. Following [12], several misbehaviours have been introduced as follows:

M Behaves Improperly:

- M receives the payment token decryption key in step 5, but does not send the correct product decryption key in step 6.
    1. C sends a record of the exchange to the TTP (*sendTPall*).
    2. TTP asks M to send the correct decryption key and start a timer (*notifyM1*).
    3. M send the correct key to the TTP or has no response (*sendTPk1* or *tiemout1*).
    4. if M sends the correct key, the TTP forwards the key to C; if not, the TTP sends a decryption key (which was registered before this exchange) to C and takes appropriate action against M (*sendCkbyTP1* or *sendCkbyTP2*, *takeactionM*).
- M receives the payment token decryption key in step 5, but disappears without sending the product decryption key.
    1. C's timer expires (*noresponsedelay*).
    2. C sends a record of the exchange to the TTP (*sendTPall*).
    3. TTP asks M to send the correct decryption key and starts a timer (*notifyM2*).
    4. M has no response (*timeout2*).
    5. TTP sends a decryption key (which was generated before this exchange) to C and takes appropriate action against M (*sendCkbyTP2*, *takeactionM*).
- M claims that it did not send the correct decryption key because it has not received payment.
    1. M sends the reason that he did not receive proper payment (*sendTPreason*).
    2. M still needs to send product decryption key to the TTP (*sendTPk2*).
    3. Once the TTP receives the product decryption key from M, he sends appropriate decryption key to M and C (*sendMkbyTP1*, *sendCkbyTP3*).

C Behaves Improperly: M received the payment decryption key from the TTP again after he claims the wrong key in first instance. However, he still can not decrypt the payment by the key again:

1. Notify TTP of the failure of using the payment decryption key again (*sendTPnoti*).
2. TTP gets in touch with Bank to obtain a new key (*getkfromB*).
3. Sends the new key to M (*sendMkbyTP2*).

Once again, the description above is mainly about behaviour, in order address performance and more detailed security content has been described in [12]. The terms in the brackets after each item with bold font are the action name we have used in the PEPA model below. Moreover, we would like to propose three performance questions for this extended protocol as well as our previous case studies: "how many clients can a given TTP configuration support?", "how much service capacity must we provide at a TTP to satisfy a given number of clients?" and "what is the maximum rate at which keys can be refreshed before the TTP performance begins to degrade?" These questions are answered through numerical results following the model specification.

3.3   PEPA Model

A PEPA model of the protocol incorporating misbehaviour can be specified as follows:

$$CT_0 \stackrel{def}{=} (download, r_d).CT_1$$

$$CT_1 \stackrel{def}{=} (agreePrice, r_a).CT_2$$

$$CT_2 \stackrel{def}{=} (sendMp, r_{smp}).CT_3$$

$$CT_3 \stackrel{def}{=} (sendCg, f_1).CT_4 + (sendCabort, f_2).CT_7$$

$$CT_4 \stackrel{def}{=} (sendMk, r_{smk}).CT_5 + (sendMabort, r_{sma}).CT_8$$

$$CT_5 \stackrel{def}{=} (sendCk, f_3).CT_6 + (noresponsedelay, r_n).CT_{14}$$

$$CT_6 \stackrel{def}{=} (work, r_w).CT_0 + (sendTPall, r_{stp}).CT_9$$

$$CT_7 \stackrel{def}{=} (sendMabort, r_{sma}).CT_8$$

$$CT_8 \stackrel{def}{=} (work, r_w).CT_0$$

$$CT_9 \stackrel{def}{=} (notifyM1, r_1).CT_{10}$$

$$CT_{10} \stackrel{def}{=} (sendTPk1, f_7).CT_{11} + (timeout1, r_{10}).CT_{12}$$

$$CT_{11} \stackrel{def}{=} (sendCkbyTP1, r_3).CT_8$$

$$CT_{12} \stackrel{def}{=} (sendCkbyTP2, r_4).CT_{13}$$

$$CT_{13} \stackrel{def}{=} (takeactionM, r_6).CT_8$$

$$CT_{14} \stackrel{def}{=} (sendTPall, r_{stp}).CT_{15}$$

$$CT_{15} \stackrel{def}{=} (notifyM2, r_2).CT_{16}$$

$$CT_{16} \stackrel{def}{=} (sendTPreason, f_4).CT_{17}$$
$$+(timeout2, r_{11}).CT_{12}$$

$$CT_{17} \stackrel{def}{=} (sendTPk2, f_5).CT_{18}$$

$$CT_{18} \stackrel{def}{=} (sendMkbyTP1, r_7).CT_{19}$$

$$CT_{19} \stackrel{def}{=} (sendCkbyTP3, p * r_5).CT_{20}$$
$$+(sendCkbyTP3, (1 - p) * r_5).CT_8$$

$$CT_{20} \stackrel{def}{=} (sendTPnoti, f_6).CT_{21}$$

$$CT_{21} \stackrel{def}{=} (getkfromB, r_9).CT_{22}$$

$$CT_{22} \stackrel{def}{=} (sendMkbyTP2, r_8).CT_8$$

$$TP \stackrel{def}{=} (notifyM1, r_1).TP + (notifyM2, r_2).TP$$
$$+(sendCkbyTP1, r_3).TP$$
$$+(sendCkbyTP2, r_4).TP$$
$$+(sendCkbyTP3, r_5).TP$$
$$+(takeactionM, r_6).TP$$
$$+(sendMkbyTP1, r_7).TP$$
$$+(sendMbyTP2, r_8).TP$$
$$+(getKfromB, r_9).TP$$
$$+(timeout1, r_{10}).TP + (timeout2, r_{11}).TP$$

$$System \stackrel{def}{=} TP[K] \bowtie_{\mathcal{L}} CT_0[N]$$

Where,

$$\mathcal{L} = \{notifyM1, notifyM2, sendCkbyTP1, sendCkbyTP2, sendCkbyTP3,$$
$$takeactionM, timeout1, sendMkbyTP1, sendMkTP2, getKfromB,$$
$$timeout2\}$$

$$r_1 = r_{nm1} \frac{CT_9}{\sum waitingJobs_{TP}} min \left(\sum waitingJobs_{TP}, TP \right)$$

$$r_2 = r_{nm2} \frac{CT_{15}}{\sum waitingJobs_{TP}} min \left(\sum waitingJobs_{TP}, TP \right)$$

$$r_3 = r_{scktp1} \frac{CT_{11}}{\sum waitingJobs_{TP}} min \left(\sum waitingJobs_{TP}, TP \right)$$

$$r_4 = r_{scktp2} \frac{CT_{12}}{\sum waitingJobs_{TP}} min \left(\sum waitingJobs_{TP}, TP \right)$$

$$r_5 = r_{scktp3} \frac{CT_{19}}{\sum waitingJobs_{TP}} min \left(\sum waitingJobs_{TP}, TP \right)$$

$$r_6 = r_{ta} \frac{CT_{13}}{\sum waitingJobs_{TP}} min \left(\sum waitingJobs_{TP}, TP \right)$$

$$r_7 = r_{smktp1} \frac{CT_{18}}{\sum waitingJobs_{TP}} min \left(\sum waitingJobs_{TP}, TP \right)$$

$$r_8 = r_{smktp2} \frac{CT_{22}}{\sum waitingJobs_{TP}} min \left(\sum waitingJobs_{TP}, TP \right)$$

$$r_9 = r_{kb} \frac{CT_{21}}{\sum waitingJobs_{TP}} min \left(\sum waitingJobs_{TP}, TP \right)$$

$$r_{10} = r_{t1} \frac{CT_{10}}{\sum waitingJobs_{TP}} min \left(\sum waitingJobs_{TP}, TP \right)$$

$$r_{11} = r_{t2} \frac{CT_{16}}{\sum waitingJobs_{TP}} min \left(\sum waitingJobs_{TP}, TP \right)$$

$\sum waitingJobs_{TP} = \sum_{\forall i} CT_i(t)$, $i \in \{9, 15, 11, 12, 19, 13, 18, 22, 21, 10, 16\}$.
if $N = 1$:
$f_1 = r_{scg}$, $f_2 = r_{sca}$, $f_3 = r_{sck}$, $f_4 = r_{stpr}$, $f_5 = r_{stpk}$, $f_6 = r_{stpno}$, $f_7 = r_{stpk}$,
if $N \neq 1$:
$f_1 = \frac{r_{scg}}{CT_3+1}$, $f_2 = \frac{r_{sca}}{CT_3+1}$, $f_3 = \frac{r_{sck}}{CT_5+1}$, $f_4 = \frac{r_{stpr}}{CT_{16}+1}$, $f_5 = \frac{r_{stpk}}{CT_{17}+1}$,
$f_6 = \frac{r_{stpno}}{CT_{20}+1}$, $f_7 = \frac{r_{stpk}}{CT_{10}+1}$.

Following [19], a form of functional rates has been applied to avoid over estimating the value of rates of cooperation actions, which are denoted by $r_i, i =$
$1, 2, \cdots, 11$. Each of these functions describes the actual service rate if there is
one job in the system(r_{nm1}, r_{nm2}, r_{scktp1}, r_{scktp2}, r_{scktp3}, r_{ta}, r_{smktp1}, r_{smktp2},
r_{kb}, r_{t1} and r_{t2}), or as a proportion of the number of waiting jobs (at TTP) of
each type ($CT_i / \sum waitingJobs_{TP}, i = 9, 15, 11, 12, 19, 13, 18, 22, 21, 10, 16$) and
the times of service ($min(TP, \sum waitingJobs_{TP})$), which allocates each service
with respect to its job type to eliminates the potential race.

3.4   Numerical Results

Figure 2 compares the average number of waiting customers at TTP against
initial population of customers calculated by ODEs [9] and stochastic simulation
[3]. The queue length increases when more clients are involved in the system.
However, it is not difficult to spot that the two curves seems to keep a constant
error when N is larger than 120. This phenomenon does not follow the ODE's
normal excellent accuracy when N is very large. To investigate more deeply,
we find that the population of behaviours after CT_{12} is actually very small,
due to the race that between action $sendTPk$ and $timeout1$ in component CT_{10},
and also between action $sendTPreason$ and $timeout2$. In the case where $N =$
240, it is a simple matter to calculate the functional rate of $sendTPk$ $f_7(N =$
$240) \approx 0.85397$, and the functional rate of $timeout1$ $r_{10}(N = 240) \approx 0.0020397$.
The large difference also exists between $sendTPreason$ and $timeout2$. About 400
times difference causes just a few components evolving to CT_{12} and its further
(evolving) behaviours. Thus, $N = 240$ still cannot be considered as a large scale
system with the current set of rates. That explains why the two methods do not
converge when $N = 240$. Nevertheless, the two curves will converge eventually in
some (extremely large) value of N. To take a further experiment, we set r_{t1} and
r_{t2}, the original rates of $timeout1$ and $timeout2$, to 200, and keep all other rates

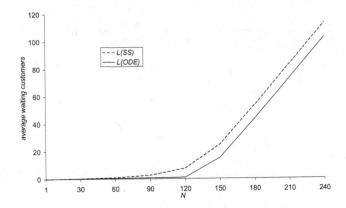

Fig. 2. Average number of waiting customers at TTP varied with population size calculated by ODEs and stochastic simulation, $p = 0.5$, $r_w = 0.01$ and all other rates are 1.

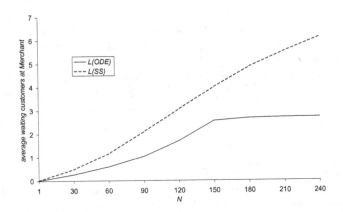

Fig. 3. Average number of waiting customers at merchant varied with population size calculated by ODEs and stochastic simulation, $p = 0.5$, $r_w = 0.01$ and all other rates are 1.

unchanged. This is in order to switch more clients to the behaviours after CT_{12}. Still in case of $N = 240$, $L(ODE) \approx 99.9595$ and $L(SS) \approx 100.0637$, illustrating the argument above.

The average number of waiting customers at the merchant is presented in Fig. 3. Generally, the more customers involved, the more customers that will be waiting at the merchant. However, the results calculated by ODEs and stochastic simulation do not converge. This is caused by the same reason as discussed above. If there is just one TTP is working on misbehaviour cases, then it becomes very busy and most of the customers are waiting for the TTP. Therefore, the scale of the queue length at the merchant remains very small. This is why results of ODE and stochastic simulation did not converge here.

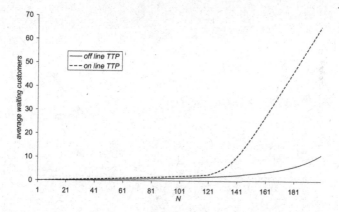

Fig. 4. Average number of waiting customers with and without TTP varied with population size calculated by ODEs, $p = 0.5$, $r_w = 0.01$ and all other rates are 1.

The total average number of waiting customers with and without misbehaviour have been compared in Fig. 4. Under the same rates for each relevant actions and the same involved number of customers, far more customers are waiting in a situation of misbehaviour, especially, when N is very large. This is an intuitive and expected result, because customers who encounter misbehaviour have recourse to the TTP for help, and then wait at the TTP for a resolution. Hence, it is clear that misbehaviours reduce the performance of the whole system, and also demonstrates that this kind (optimistic) non-repudiation protocol could perform much better than those that always employ an on-line TTP.

Figure 5 shows the average response time for the merchant at different actions. Overall, if we increase total number of clients in the system, the merchant takes longer to process each individual request. However, the response

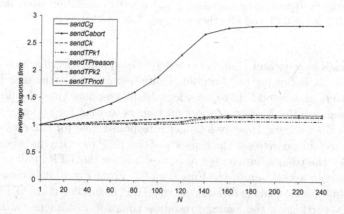

Fig. 5. Average response time at merchant varied with population size calculated by ODEs, $p = 0.5$, $r_w = 0.01$ and all other rates are 1.

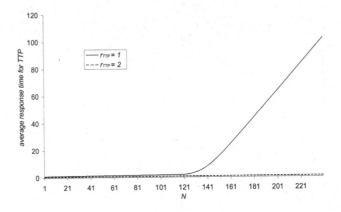

Fig. 6. Average response time for TTP varied with population size calculated by ODEs, $p = 0.5$, $r_w = 0.01$ and all other rates are 1 except for r_{TTP}.

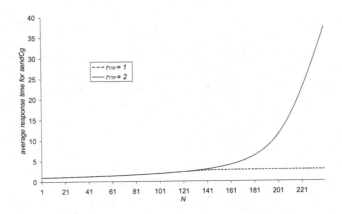

Fig. 7. Average response time for sendCg varied with population size calculated by ODEs, $p = 0.5$, $r_w = 0.01$ and all other rates are 1 except for r_{TTP}.

time increases slowly, and that is caused by the queue length which has been shown in Fig. 3. Following our functional rate definitions for the merchant (f_i), it is intuitively understood that queue length and response time should have the same increasing ratio. Moreover, more customers waiting for action *sendCg* and *sendCabort* than others, this gives longer a response time for these two actions.

We experiment to increase the capacity of the TTP to twice that shown before (2), and plot the results for average response time for the TTP in all actions and the merchant in action *sendCg* in Figs. 6 and 7. From Fig. 6, it is clear that the response time for customers waiting at the TTP is smaller if the TTP is more powerful. Nevertheless, the average response time for customers waiting at the merchant for action *sendCg* increases if we double the TTP's capacity, since the throughput from the TTP is obviously greater. A quicker response from the TTP

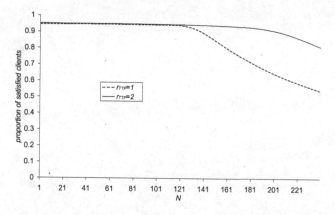

Fig. 8. Proportion of satisfied customers varied with population size calculated by ODEs, $p = 0.5$, $r_w = 0.01$ and all other rates are 1 except for r_{TTP}.

means that the number of customers waiting at misbehaviour stage decreases. Under the same total number of clients, more customers go to the normal stage without misbehaviour. Consequently, the number of customers (CT_3) waiting for action *sendCg* increases, and the average response time for these customers takes longer.

Finally, we plot the proportion of satisfied customers (been served) in Fig. 8. Generally, the proportion decreases for both case ($r_{TTP} = 1$ and $r_{TTP} = 2$) if more customers come to the system. The two curves are very close before the point, $N = 120$, and both keep a very high percentage of satisfied customers in this area. After that point, those percentages start to go down clearly. However, the proportion for $r_{TTP} = 1$ drops more quickly than the other, and it becomes 50 % when $N = 240$, while the percentage for $r_{TTP} = 2$ is still above 80 %.

3.5 Utility Function of Extended Protocol

Consider the following utility function to answer our proposed performance questions for extended protocol.

$$C = c_1 L + c_2 K r_p \ , \ c_1, c_2 \geq 0 \tag{1}$$

Here, L denotes the average waiting customers at the non-repudiation server (TTP), and K is number of servers. r_p is the response rate of the TTP. We assume the TTP server responds any type of jobs in the same rate here. c_1 and c_2 are cost rates, and they many depend on the type of system or quality of service agreement with customers.

Figure 9 shows the cost varied against the number of clients calculated by ODEs. Similar to the results of cost function in Chaps. 3 and 4, more clients results in more waiting customers with fixed service capacity. Therefore, the total cost increases along with the cost of customer waiting goes up. Furthermore, it is

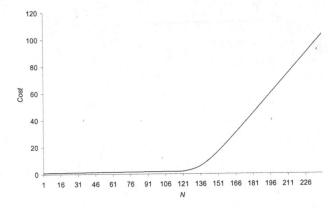

Fig. 9. Cost varied against the number of clients calculated by ODEs, $p = 0.5$, $K = 1$, $c1 = c2 = 1$, $r_w = 0.01$ and all other rates are 1.

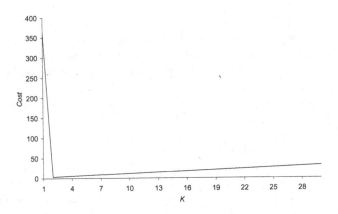

Fig. 10. Cost varied with number of TTP servers calculated by ODEs, $p = 0.5$, $N = 500$, $c1 = c2 = 1$, $r_w = 0.01$ and all other rates are 1.

a simple matter to find that the cost rises rapidly when N is around 130, and this is the maximum capacity that the TTP server can handle before performance start to significantly degrade.

Figure 10 presents the cost varied with number of TTP servers calculated by ODEs when total number of clients is 500. Again, customer waiting costs more in initial stage. Along with the system being given more servers, number of waiting clients is reduced. However, the cost of service dominate the total cost. The optimal point is around 2 in this case.

Figure 11 shows the cost varied with the rate of refresh key, r_w, calculated by ODEs. With fixed service capacity and total number of clients, more frequently refresh the session key results in more workload has been added in the system.

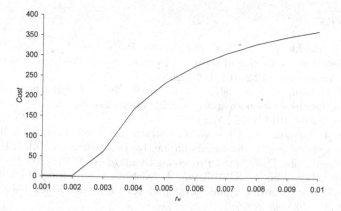

Fig. 11. Cost varied with rate of *work* calculated by ODEs, $p = 0.5$, $N = 300$, $c1 = c2 = 1$, all other rates are 1 except for r_w.

Therefore, the cost of customer waiting increases. Similarly, we can easily find that the balance point between performance and security is around $r_w = 0.002$.

4   Conclusions

This case study has investigated an optimistic fair exchange protocol in an e-commerce environment. We model the protocol when the *third trust party* is online due to misbehaviour of one or more of the participants. According to the optimistic characteristic, the protocol can work in a lighter mode when there is no misbehaviour detected, where the TTP is not engaged. However this mode is much less interesting from a performance perspective. In this work, we consider that a merchant server consists of several threads; PEPA works well in this style of modelling. The ODE solution does not always coincide with stochastic simulation when N is very large. However, in this context, this large N only gives large scale for part of the derivatives, and they are still may converge under other rates. Despite this, the ODE solutions are shown to give a good indication of expected performance and can be derived extremely efficiently for large systems.

Our analysis has focussed on identifying and employing efficient solution methods. There is considerable scope for further work to investigate the relationship between formal models of security and performance. The goal would be to create a system which could automatically produce analysable performance models from security models. However, the choice of security solution, driven by the performance security trade-off should always remain an expert task.

Acknowledgements. Dr Zhao was supported by the Fundamental Research Funds for the Central Universities, China University of Geosciences (Wuhan), CUGL150840.

References

1. Bodei, C., Buchholtz, M., Curti, M., Degano, P., Nielson, F., Nielson, H., Priami, C.: Performance evaluation of security protocols specified in LySa. Electron. Notes Theor. Comput. Sci. **112**, 167–189 (2005)
2. Bradley, J., Gilmore, S., Hillston, J.: Analysing distributed internet worm attacks using continuous state-space approximation of process algebra models. J. Comput. Syst. Sci. **74**(6), 1013–1032 (2008)
3. Bradley, J., Gilmore, S., Thomas, N.: Performance analysis of stochastic process algebra models using stochastic simulation. In: Proceedings of 20th IEEE International Parallel and Distributed Processing Symposium. IEEE (2006)
4. Buchholtz, M., Gilmore, S., Hillston, J., Nielson, F.: Securing statically-verified communications protocols against timing attacks. Electron. Notes Theor. Comput. Sci. **128**(4), 123–143 (2005)
5. Cho, J., Chen, I., Feng, P.: Performance analysis of dynamic group communication systems with intrusion detection integrated with batch rekeying in mobile Ad Hoc networks. In: Proceedings of the 22nd International Conference on Advanced Information Networking and Applications. IEEE (2008)
6. El-Hadidi, M., Hegazi, N., Aslan, H.: Performance analysis of the Kerberos protocol in a distributed environment. In: Proceedings of the 2nd IEEE Symposium on Computers and Communications. IEEE (1997)
7. El-Hadidi, M., Hegazi, N., Aslan, H.: Performance evaluation of a new hybrid encryption protocol for authentication and key distribution. In: Proceedings of the International Symposium on Computers and Communications. IEEE (1999)
8. Hillston, J.: A Compositional Approach to Performance Modelling. Cambridge University Press, Cambridge (1996)
9. Hillston, J.: Fluid flow approximation of PEPA models. In: Second International Conference on the Quantitative Evaluaiton of Systems, pp. 33–43. IEEE Computer Society (2005)
10. Liu, W., Yang, L., Li, Q., Dai, H., Hou, B.: Performance analytic model for authentication mechanism. In: Proceedings of the International Conference on Networking, Sensing and Control, pp. 1097–1102 (2008)
11. Meng, T., Wang, Q., Wolter, K.: Security and performance tradeoff analysis of mobile offloading systems under timing attacks. In: Beltran, M., Knottenbelt, W., Bradley, J. (eds.) EPEW 2015. LNCS, vol. 9272, pp. 32–46. Springer, Heidelberg (2015)
12. Ray, I., Ray, I.: An optimistic fair exchange E-commerce protocol with automated dispute resolution. In: Bauknecht, K., Madria, S.K., Pernul, G. (eds.) EC-Web 2000. LNCS, vol. 1875, pp. 84–93. Springer, Heidelberg (2000)
13. Ryan, P., Schneider, S., Goldsmith, M., Lowe, G., Roscoe, B.: Modelling and Analysis of Security Protocols. Addison Wesley, Boston (2000)
14. Thomas, N.: Performability of a secure electronic voting algorithm. Electron. Notes Theor. Comput. Sci. **128**(4), 45–58 (2005)
15. Thomas, N., Zhao, Y.: Fluid flow analysis of a model of a secure key distribution centre. In: Proceedings 24th Annual UK Performance Engineering Workshop. Imperial College (2008)
16. Wang, Y., Lin, C., Li, Q.: Performance analysis of email systems under three types of attacks. Perform. Eval. **67**(6), 485–499 (2010)
17. Wolter, K., Reinecke, P.: Performance and security tradeoff. In: Aldini, A., Bernardo, M., Di Pierro, A., Wiklicky, H. (eds.) SFM 2010. LNCS, vol. 6154, pp. 135–167. Springer, Heidelberg (2010)

18. Zhao, Y., Thomas, N.: Approximate solution of a PEPA model of a key distribution centre. In: Kounev, S., Gorton, I., Sachs, K. (eds.) SIPEW 2008. LNCS, vol. 5119, pp. 44–57. Springer, Heidelberg (2008)
19. Zhao, Y., Thomas, N.: Comparing methods for the efficient analysis of PEPA models of non-repudiation protocols. In: Proceedings of the 15th International Conference on Parallel and Distributed Systems, pp. 821–827. IEEE (2009)
20. Zhao, Y., Thomas, N.: Efficient solutions of a PEPA model of a key distribution centre. Perform. Eval. **67**(8), 740–756 (2010)

Author Index

Printed in the United States
By Bookmasters